G000271075

'IRON HARVESTS OF THE FIELD'
THE MAKING OF FARM MACHINERY IN BRITAIN SINCE 1800

Twin'd with the wreaths Parnassian laurels yield,
Or reap'd in iron harvests of the field?
Where grows? – where grows it not? If vain our toil,
We ought to blame the culture, not the soil ...

from *An Essay on Man, Epistle IV*, 'Of the nature and state of man with respect to happiness', by Alexander Pope (1688–1744)

'Iron Harvests of the Field': The making of farm machinery in Britain since 1800

Peter Dewey

Of related interest

'Factory of Dreams': A History of Meccano Ltd
by Kenneth D. Brown (Crucible Books, 2007)
ISBN: 978-1-905472-08-6 (*softback*)

Seminal Inventions: Science, Technology and Innovation in the Modern World
by John Pitts (Crucible Books, 2008)
ISBN: 978-1-905472-05-5 (*softback*)

www.cruciblebooks.com

The Millers: A story of technological endeavour and industrial success
by Glyn Jones (Carnegie Publishing, 2001)
ISBN: 978-1-85936-085-9 (*hardback*)

Our Hunting Fathers: Field sports in England since 1850
Richard Hoyle (ed.) (Carnegie Publishing, 2007)
ISBN: 978-1-85936-157-3 (*softback*)

www.carnegiepublishing.com

'Iron Harvests of the Field': The making of farm machinery in Britain since 1800

© Peter Dewey, 2008

First published in 2008 by
Carnegie Publishing Ltd,
Carnegie House,
Chatsworth Road,
Lancaster LA1 4SL
www.carnegiepublishing.com

The publisher acknowledges with thanks the support of the Isobel Thornley Bequest, University of London; the Scouloudi Historical Award from the Scouloudi Foundation, in association with the Institute of Historical Research of the University of London; the Marc Fitch Fund; and the British Agricultural History Society.

All rights reserved
Unauthorised duplication contravenes existing laws

British Library Cataloguing-in-Publication data
A catalogue record for this book is available from the British Library

ISBN: 978-1-85936-180-1 (*hardback*)
ISBN: 978-1-85936-160-3 (*softback*)

Designed, typeset in Monotype Baskerville, and originated by
Carnegie Book Production, Lancaster
www.carnegiebookproduction.com
Printed and bound by Cromwell Press, Trowbridge, Wiltshire

Contents

List of tables

Preface and acknowledgements

My interest in this subject goes back many years, but I first thought of writing about it when I realised in the late 1970s that the Rural History Centre at Reading University, which was not far from my home, was amassing the best collection of farm machinery manufacturers' records in the UK. Other projects and academic life supervened, and I did not begin work on the agricultural machinery industry until 1991. I was early assisted by a grant from the History Department, Royal Holloway, University of London. I wish to thank the Economic and Social Research Council for two grants, which enabled me to travel to county record offices and other archives the length and breadth of the UK.

I am indebted to the archivists and staff at all these offices, and especially to those at the Rural History Centre in the University of Reading, notably Dr Jonathan Brown. I am especially beholden to Ray Hooley, the then archivist at GEC-Alstom, Lincoln (successors to Ruston & Hornsby Ltd). My more personal thanks are to Geoffrey Finch, who acted as research assistant in the early stage of this work, Audrey and David Allen, Pat Gregory, Sheila Stevens and to Barry and Sheila Bransden. The libraries of Reading University and Royal Holloway were an indispensable resource, and I would like to thank the staff at both institutions. I would like to thank Richard Hoyle for his comments on the manuscript, and Alistair Hodge at Carnegie for his patience and expertise.

I am glad to acknowledge the assistance of the Isobel Thornley Bequest, University of London, in awarding a grant to defray the cost of providing illustrations.

Publication has been made possible by a Scouloudi Historical Award from the Scouloudi Foundation, in association with the Institute of Historical Research of the University of London.

I would like to express my gratitude to the Marc Fitch Fund for awarding a grant towards the cost of publication.

I am grateful to the British Agricultural History Society for making a grant in aid of publication.

I am also grateful to the holders of copyright who have given permission for the reproduction of illustrations. The largest source of these has been the Museum of English Rural Life at the University of Reading.

My personal debt is greatest to my wife Hilary, who has been a consistent source of encouragement. She even accompanied me on many of the research trips, thus making them a much more enjoyable experience.

Peter Dewey
Marlow, 2008

Introduction

TODAY only a tiny proportion of the labour force in Britain, as in other industrialised societies, is able to supply food for the rest of the population. It was not always so: at the beginning of the eighteenth century, an estimated 60 per cent of the British working population were engaged in some form of agriculture. Yet by 1850 this proportion had fallen to under a quarter, the lowest level of any country in the world. Several long-term factors came together to effect this revolution in agricultural productivity: more land under cultivation, new crops and increased yields, and increased mechanisation of the farm. It is with this last phenomenon that this book is concerned. It is an inspiring story of technological achievement, as farmers and engineers brought iron and steel to fields which had previously been the domain of locally made timber implements and horse power. Agricultural technology moved on, inexorably, from broad-cast seed and the sound of the threshing flail, via the portable steam engine and the threshing machine, right through to the modern world of giant tractors each with the power of 200 horses, combine harvesters and impressively efficient farming methods.

A huge range of influential individuals and enterprises brought this mechanical revolution to fruition. Within this story is a large element of fascinating business history; there are many instances of successful small firms growing over time to dominate particular sectors, such as ploughs, steam traction engines, lawnmowers or reapers; there are also stories of business failure, from a wide variety of causes, of periodic economic downturns and of the disruption of war. Over the period from 1800 to 2008 within the agricultural machinery industry we find the full range of business type and experience: from hundreds of small locally based workshops and craft-based enterprises, to offshoots of the giant multinational industrial firms.

The industrial revolution of the eighteenth and nineteenth centuries was born in Britain. As the first industrial nation, British technology reigned supreme; the age of Queen Victoria was the 'British century'. In no industry was this more true than in that of farm machinery. Inventors and entrepreneurs abounded, such as Robert Ransome, who revolutionised plough design; William Marshall, who built an enormous business on the back of the steam engines and threshing machines now demanded by up-to-date farmers; and James Smyth, whose small firm in the Suffolk countryside sent its impeccably built seed drills all around the world. By the First World War the iron ploughs, horse-drawn reapers and mowers, and steam-driven threshing machines made by the industry had conquered British farming, and penetrated the four corners of the world.

Many manufacturers reaped rich rewards. Great industrial dynasties arose, employing thousands of workers, such as Ransomes of Ipswich, Garretts of Leiston, Fowlers of Leeds, and Marshalls of Gainsborough. Enterprising companies such as these drove a thriving export trade, typically sending over half their output overseas, and their products were to be found around the world, from Spain to the Russian steppes, and from the veldt to the outback. British farm technology led the world. It was an amazing success story, and the industry prospered accordingly.

During the First World War the industry made high profits from military orders, producing shells, aircraft and, notably, tanks for the war effort. These revolutionary machines were conceived, designed and built in the workshops of a leading agricultural machinery manufacturer, Fosters of Lincoln. But war profits did not last, and the aftermath of the war was a deep and lingering industrial depression, as overseas markets were lost in the wake of the Russian and other European revolutions. Wedded to steam technology, many leading firms failed or were taken over. But in the 1930s there was some recovery, as tractors began to be adopted on a larger scale. The world's first modern tractor, the Fordson, flowed out from the Dagenham factory from 1932 (nearly 19,000 were produced there in 1937 alone), and Harry Ferguson began to produce his revolutionary new tractor system – with its now universally adopted three-point hydraulic tool hitch – in 1936.

After the Second World War the industry enjoyed an amazing resurgence. Continental competition had been destroyed, and Britain was by far the largest producer of tractors and other agricultural machinery outside North America. British enterprise and technology reached new heights. Boosted by the farm tractor revolution, farming and the farm machinery industry underwent a new industrial revolution. By 1963 annual tractor production and sales peaked at nearly a quarter of a million units.

From British farms the farm horse and the threshing machine were pensioned off, replaced by the tractor and the combine. The farm labourer all but disappeared. This was a golden age for the industry, which grasped the opportunities with both hands. New tractor firms were established, including those of Harry Ferguson, David Brown and Nuffield. Multinational firms from the USA and Canada, such as International Harvester and Massey-Harris, established their factories in Britain and produced advanced tractors and combine harvesters. British exports of farm machinery and tractors soared. The period also threw up the greatest-ever home-grown British industrial hero, in the person of the enigmatic Harry Ferguson, whose 'little grey Fergie' tractors still can be found in abundance at farm shows (and still working on some farms).

In the last quarter of a century, the British industry has declined, as industrial depression and sharpened international competition made inroads. The three tractor factories still remaining in Britain in 2003 were clearly vulnerable to sudden closure by their multinational owners: since that date two of them – the Ferguson factory at Banner Lane, Coventry, and the former International Harvester Doncaster factory – have been closed. At the time of publication, this leaves but one tractor factory, the Ford factory at Basildon (now part of CNH, based in the USA). The rest of the industry survives, albeit in a slimmer, probably more efficient form. Many of the great names of the nineteenth century are gone, but there is a hard core of family firms dating back a century or more, and there is still a steady influx of new enterprise.

One outstanding lesson from this history is that, be they never so competitive, firms cannot escape their economic environment. The large economic movements which follow in the train of world events will find their victims or boost their favourites. One thinks here of the effects of the First World War and the political revolutions which followed in its wake; of the resurgence of world trade in the 'Golden Age' of the third quarter of the twentieth century; finally, of the industrial devastation which followed the economic and inflationary crises of 1973–75 and 1979–82. Since then many other economies have followed that of Britain down the path of de-industrialisation. Beneficiaries have included those countries in Europe and Asia which have picked up the pieces and developed their own farm machinery industries.

Another important factor in a period of ever-quickening innovation is that technology cannot be kept at home. Bright ideas occur everywhere, and industrial espionage is rife. While the steam technology of Victoria's reign was unchallenged, British manufacturers could not compete with the USA in harvesting machinery; McCormick, and, later, International Harvester swept all before them well into the twentieth century. More recently, the tractor industry has become established in the Middle East and Asia, and the world's largest tractor producer is now India, with Japan a close second.

The industry is a salient example of the roller-coaster nature of business life. The monolithic, omnipotent façade of the great multinational company is an illusion: they can grow, but they can also atrophy and die. This has been true of the large and powerful British firms in this book – Ransomes and Fowlers – and also, more recently, of the giant multinationals – such as International Harvester – wiped out or taken over in the economic turbulence of the 1970s and 1980s.

As the British, American and European farm machinery industries decline absolutely in terms of output and even more dramatically in terms of relative size and importance, the industry in newly industrialising countries continues to expand. There is clearly still great scope for expansion of the industry in both Asia and South America; Chinese farming has barely been touched by mechanisation, and the Chinese government has recently passed a law aimed at encouraging and subsidising the manufacture of farm machinery.

What place British manufacturers will have in the future is uncertain: the British farm machine industry is unlikely ever again to be large by world standards; and it will certainly never recapture the glory years of its two periods of world domination. Yet, if we take a long view back to the very beginnings of the agricultural revolution in the UK we are also able to trace a rich vein of innovation, enterprise and technological inspiration, often taking place among the large number of relatively small-scale, craft-based workshops which were so prevalent in the early decades. Rather than mere manufacturing, therefore, perhaps it is this tradition of technical innovation and invention which actually marked out the British farm machinery industry for historical greatness, and perhaps it is this tradition which will continue to mark it out in the future.

CHAPTER ONE

The origins of an industry, 1750–1820

The agricultural revolution

The phrase 'industrial revolution' is in common currency, and refers to the wide range of technological innovations which transformed the British economy and British society in the hundred years or so after the middle of the eighteenth century. These innovations centred around the making of iron, the use of coal, the rise of the cotton industry, and the application of the steam engine to transport and manufacturing. The result was that by the later nineteenth century the British economy was the most highly industrialised in the world, and its people were the most urbanised.

Slightly less well known is the 'agricultural revolution' which accompanied the industrial. Yet they were closely bound together, and dependent upon each other. One important link between them was that the population was rising increasingly rapidly after the middle of the eighteenth century. In 1700 there were an estimated 5 million people living in England. By 1750 this had risen by 14 per cent, to 5.7 million. But in the next fifty years, the population rose much more rapidly, by over a half, and in 1801 (when the first population census was taken), the population of England and Wales was enumerated at 8.9 millions. Merely to feed the millions of extra mouths was going to require a substantial increase in food production.

Another crucial link between the two revolutions was that, over time, the process of industrialisation led to large-scale migration away from the countryside and into the towns. Many of the early textile factories in the 1750s and 1760s had been located in valleys with fast-flowing rivers, to provide the water power used by the factory machinery (as, for example at Crompton's factory at Belper, or Arkwright's at Cromford – both in rural Derbyshire). But by the end of the eighteenth century, the coming of steam engines to provide factory power meant that industry would increasingly locate in the towns. Thus urban jobs grew at the expense of rural jobs, and the national proportion of workers in farming fell. In order to provide food for the increasing town populations, and to maintain the momentum of industrialisation, farming had to become more productive. Failure to do so would bring nearer the spectre raised in 1798 by Thomas Malthus in his *Essay on the Principle of Population*, of population growth outstripping the increase in food production, with consequent food shortages, or even famine, disease and death.

There was thus great pressure on farming to become more efficient. The British agricultural system was by no means backward by contemporary European standards – the farming of England was probably second only to that of the Low Countries in efficiency – but it did have a lot of slack which could be taken up. Most of this lay in the persistence of a large amount of 'open-field' farming. This essentially medieval, feudal pattern of land distribution involved organising the arable land round a village into two or three large fields; these were divided by community decision into strips, farmed by each villager. A farmer would cultivate a mixture of scattered strips, located in each of the fields. Usually, one of the fields would be left fallow; the other two would grow crops. The result was a two- or three-course rotation; fallow/corn, or fallow/corn/corn. Livestock were accommodated on common pastures or other common land.

The advantage of the open-field system was its equitable distribution of good and poor land. But there were disadvantages. The first was that in any particular year a high proportion of the arable land was not growing a crop, but was merely fallow. In that fallow year, during which the land was allowed to rest and its natural fertility restored, the land still had to be ploughed, or at least cultivated in some fashion, in order to keep down the weeds, but it was not sown with a crop. The other main disadvantage of the open-field system was that the admixture of land (and livestock) meant that farming practice had to be that of the slowest and most backward farmer. The contiguity of strips meant that one man's weeds infested his neighbour's strip; on the commons and common pastures, each man's cattle or sheep intermingled, and thus selective breeding was not possible.

The open-field system had never been universal, being more suited to the lowland region of central and southern England. In the rest of England, the hill and pastoral systems of farming seem to have been best served by an enclosed agriculture where the scattered strips had been consolidated into larger, discrete holdings, often accompanied by the right to run animals on the still-open commons on the higher ground and mountains. Further, there had been an enclosure movement of sorts since the medieval period, especially in the fifteenth and sixteenth centuries, chiefly to create new pasture, to increase the number of sheep kept at a time when English wool was in high demand at home and overseas. The end result was that by 1700 about half of the agricultural land of England was already enclosed. The rest was still open-field, and ripe for more efficient exploitation.

By the late eighteenth century, the twin pressures of population growth and, after 1793, war, were coming together to exert pressure on the supply of food in Britain, and thus on the price of the staple bread cereal, which by that time was wheat. The rising population came up against a farming system which was slow to respond, as long as a large part of the arable land was still cultivated under the open-field system. In addition, the long periods of war, especially the French Revolutionary and Napoleonic wars, which ran from 1793 to 1815, with only a brief truce in 1802, caused further strain. Before 1760, there had been an export of wheat from England. After this, the export surplus soon faded away, to be replaced by a small volume of imported wheat – perhaps about 5 per cent of total supply. This import, although small, was critical in the tightening supply situation of the 1790s. Pressure on prices was enhanced by a series of particularly wet summers in the 1790s and in 1810–12. The upshot was that cereal prices, especially wheat, rose rapidly. The average annual price per quarter* of British wheat rose from 47.9 shillings† in the 1770s, to 63.5 shillings in the 1790s, and to 83.9 shillings in the first decade of the nineteenth century. Decadal peaks were much higher than this; the two peaks of the war period were of 119.5 shillings in 1801, and 126.5 shillings in 1812.[1]

The market was thus pushing farmers towards cereal cultivation. To profit fully from this market situation, landowners turned increasingly to the enclosure of the open fields. This new, great 'enclosure movement' began before the 1770s, and continued until the 1850s, but its peak was in the years from 1790 to 1815. By the end of the movement in around 1850, some 8.4 million acres of land had been enclosed by Act of Parliament in England and Wales, amounting to nearly 24 per cent of the total acreage of agricultural land.[2] This enormous movement, in which the scattered strips of the open-field farmers were combined into separate holdings, affected profoundly the way in which farming operated. Farmers were now free to pursue their own cropping and stocking policies; to drain their land; to use newly introduced animal fodder crops such as clover and turnips; and to use the new implements which were now a more worthwhile investment.

Before the agricultural revolution, a large part of the tractive power employed in the field had been supplied by oxen. These had various advantages: they were very powerful, docile, fed on grass rather than oats, and could be sold for meat when too old for work. But compared with horses they worked too slowly, and so between 1500 and 1800 the horse slowly replaced the ox. By the early nineteenth century, there were some 800,000 horses working in British farming, and their numbers were increasing, with the development of new forms of horse-drawn machinery and implements. As new machines came into use, so the horse power

* A quarter was 28 pounds or 2 stones in weight.

† Before decimalisation of sterling in the early 1970s, there were 12 pence (12*d.*) in 1 shilling (1*s.*), and 20 shillings (20*s.*) in £1. Throughout this book old-style money is given in the form £1 16*s.* 8*d.*

available to use them also rose; by the third quarter of the nineteenth century, there were some 940,000 horses working on British farms.[3]

Farming implements before 1800

The main field tasks of sowing, harvesting, and crop processing had of course developed over time since the neolithic period, when mankind

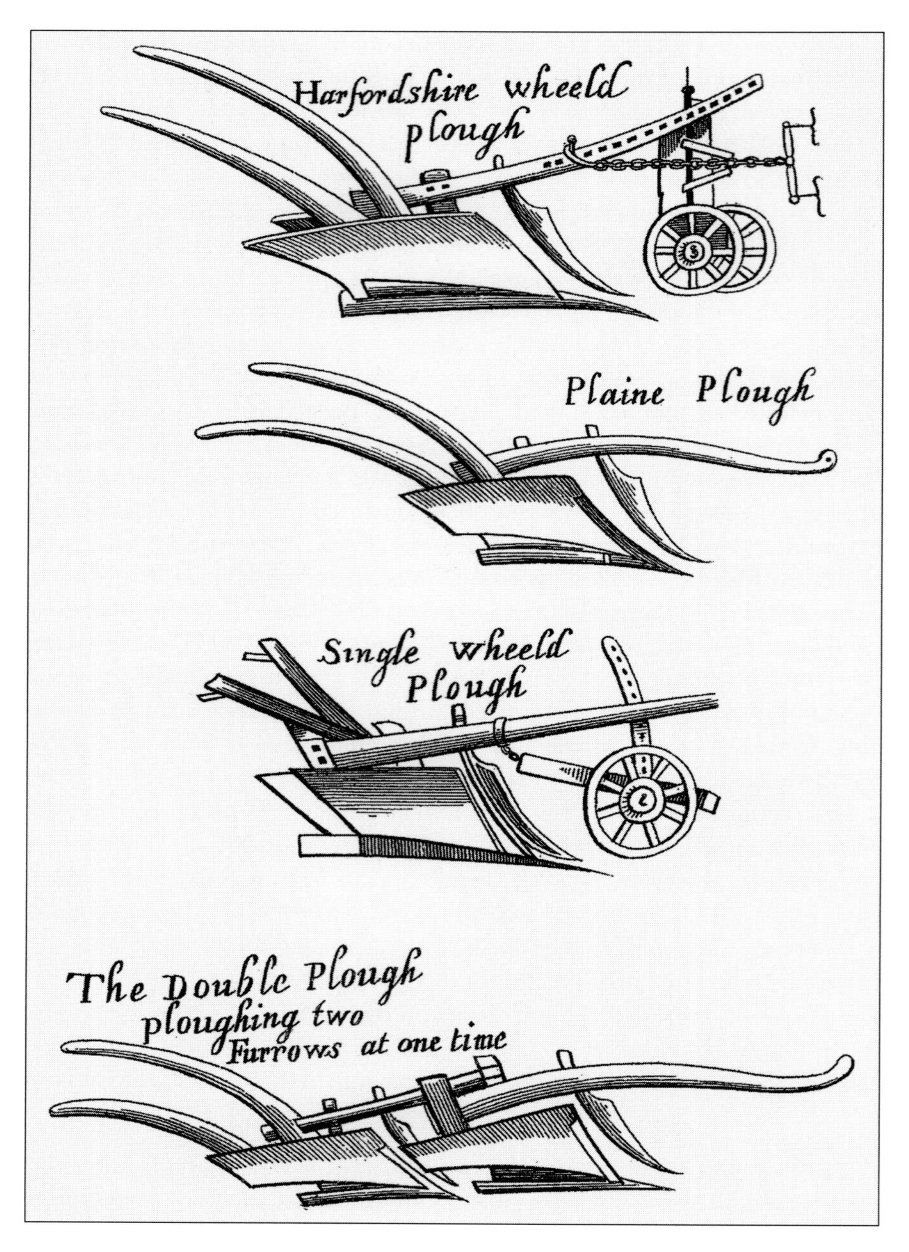

Seventeenth-century ploughs, from Walter Blith, *The English Improver Improved* (3rd edn, 1653).

had first begun to abandon hunting and gathering in favour of settled agriculture. Instruments such as the plough, the cultivator (a crude plough, used for breaking up land prior to ploughing), the harrow, and harvesting hand tools such as the sickle and scythe had all developed in various ways. Yet these tools and instruments were still the product of village craftsmen, and almost identical to those which had been used several hundred years earlier; and they were still constructed largely of wood. The relatively slow rate of development can be seen from the design of that most fundamental of tools, the plough: a comparison of the ploughs illustrated in the book of the seventeenth-century agricultural writer Walter Blith, *The English Improver Improved* (3rd edn, 1653) with the contemporary ploughs in Lord Somerville's book of 1809, *Facts and Observations relative to Sheep, Wool, &c.* shows how little the common run of plough had changed in the intervening 150 years.[4]

It was not until well into the eighteenth century that serious experiments were made with all-metal ploughs. Until then, the parts of the plough which came into contact with the soil (the coulter, share, mouldboard),

Lord Somerville's illustration of contemporary double-share ploughs (1809).

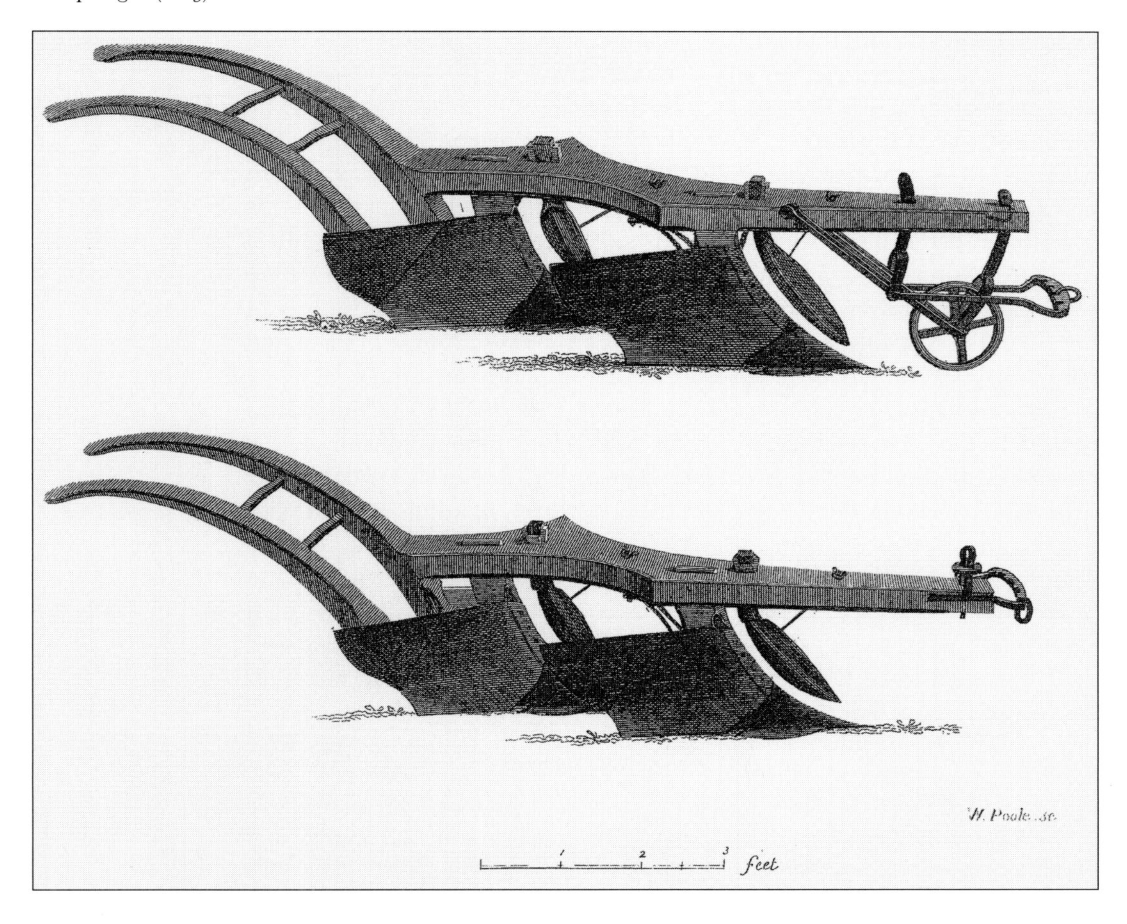

had been made of wood, with their leading edges tipped or sheathed with iron. This feature is evident in the illustrations from Blith and Somerville. The earliest reference to a plough with an all-iron mouldboard occurs in a book published in 1707 by John Mortimer, who had purchased an estate in Essex. At Colchester he found a light plough of East Anglian type, which was 'very peculiar for its Earth-Board being made of Iron, by which means they make it rounding; which helps to turn the Earth or Turf, much better than any other sort of Plough'.[5]

Some years later, in 1730, a type of plough referred to as the Rotherham plough was patented. This had a coulter and share made of iron, and a mouldboard covered with an iron plate. It was a light, general-purpose swing plough (i.e. without wheels) which was easy to make, cheap to produce, and yet stronger than other contemporary ploughs. Fewer horses were needed to pull it (the conventional plough was rather massive, employing a lot of timber in its construction), and there was consequently less need for a second man or boy to tend the horses. It appears that this plough was made from standard patterns and that the parts were interchangeable. Advertisements for it claimed variously that it could reduce ploughing times by a third, or save one-third of the horsepower. It enjoyed wide popularity, and attempts were made to improve on it. In the 1770s the Society of Arts was holding experiments into the performance of ploughs, and tested a novelty for the time, an all-iron plough made by Mr Brand, of Manningtree, Essex, which seems to date from 1773. But when in 1782 the Society published a book illustrating the types of plough which had its approval, they all still had bulky wooden beams.[6]

In spite of the increasing popularity of ploughs made partly or all of iron, there is no doubt that the great majority of ploughs at work in *c.* 1800 were still of mainly wooden construction. The contemporary reports of the Board of Agriculture* give full descriptions of a multitude of local and regional types, in which the all-iron plough is clearly an exception.

Rather more technical progress had been made by 1800 in the matter of sowing the seed onto the ploughed land. From time immemorial, the dominant method had been the broadcast sowing of seed by hand: the sower walked along the furrows, distributing large handfuls of seed with a sweeping motion of his arm. There seems little doubt that in 1800 this was still the prevalent method, but for several decades improving farmers had been experimenting with mechanical sowing. Although such experiment went back for centuries, the first recognisably modern seed drill, designed to be horse-drawn, was certainly the work of Jethro Tull, who published a description of it, with illustrations, in his 1733 book, *Horse Hoeing Husbandry.* Although this made an impression at the time,

* The Board of Agriculture was a private body, whose aim was agricultural improvement. Its series of reports did not cover all English counties, but they supply an invaluable guide to contemporary farming practice

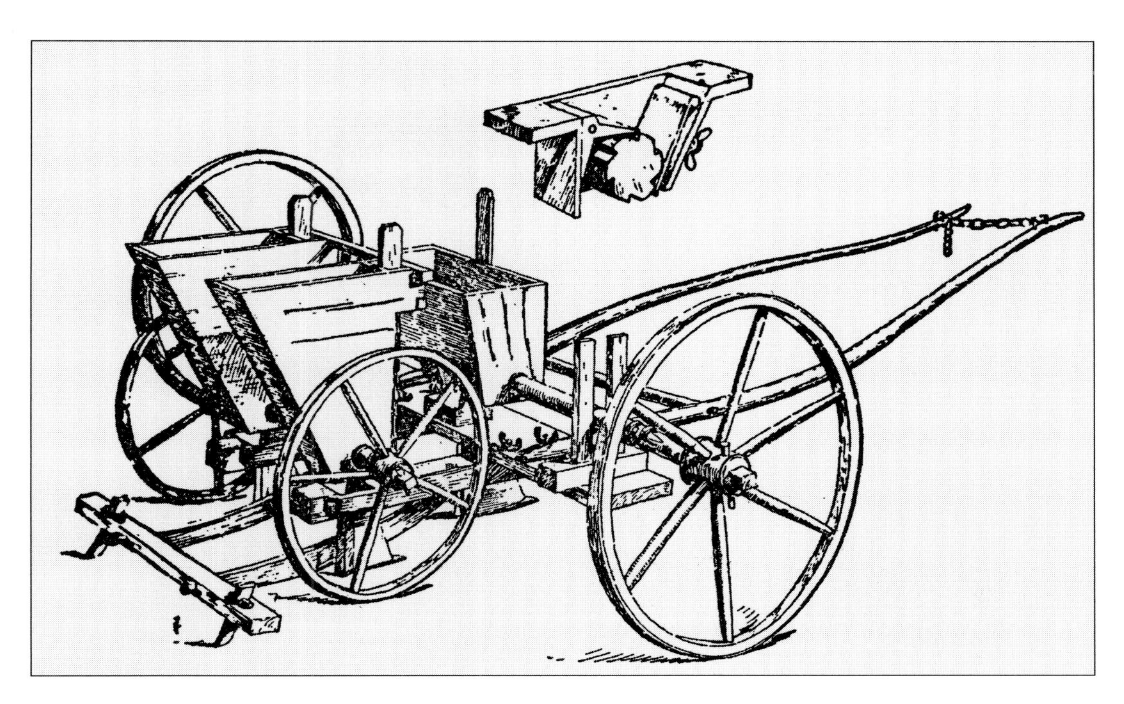

Jethro Tull's seed drill, drawn from his sketch and description in *Horse-Hoing Husbandry* (1733), by R. H. Anderson, 'Grain Drills through Thirty-Nine Centuries', *Agricultural History* (Oct. 1936).

interest in the subject died out for several decades thereafter. It revived in the 1770s, and various inventors produced their own models, some of which were recognisably what was to become the common pattern by the early decades of the nineteenth century. The most notable was that of the Rev. James Cooke, who patented his machine in 1782, and whose principles were to be acknowledged by J. Allen Ransome, in his magisterial book, *The Implements of Agriculture* (1843): 'to have been adopted in the construction of some of the most approved of the present day.' But the use of seed drills was definitely the exception rather than the rule in 1800, and, despite making good use of iron components, the machines were still mainly of wood.[7]

The idea of mechanising the arduous process of grain harvesting dates to antiquity, but the first English patent for a reaping machine did not appear until right at the end of the eighteenth century, in 1799. This machine, by Joseph Boyce of London, seems to have been intended to be pushed manually into the standing crop. It was only partly successful, and represented for the time being a technological dead end, although its principles were to be resurrected in the Australian 'stripper-harvester' of the early twentieth century. In 1800 the vast majority of the crops harvested in Britain were got in by use of the scythe, sickle or reaping-hook in the case of grain crops or hay, or by digging with a spade in the case of root crops.[8]

The final process in getting the grain was threshing, to separate the ears of grain from the stems of straw. A variety of methods had been

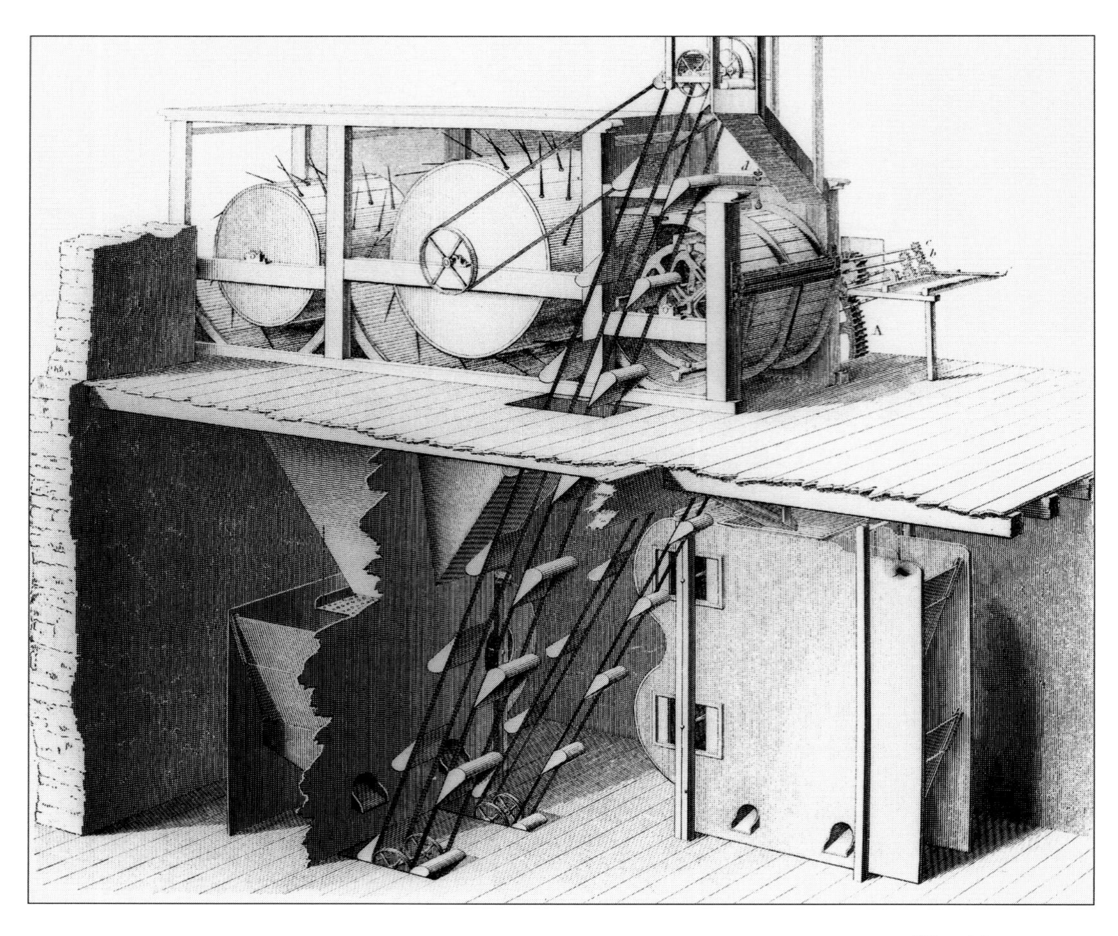

Threshing machine on Meikle's principles. (*Museum of English Rural Life*)

tried: treading of the ears by farm animals walking over the cut crop on the threshing floor; the pulling of a sledge or wooden framework across the cut crop; beating the grain out of the ears by men using flails.* The first two methods were more suited to dry climates. In Britain by 1800 the almost universal method was the flail. The flail had the advantage also that, although the work was arduous and badly paid, it could be done in the barn during the winter, thus providing work for farm labourers who otherwise would have been unemployed. The use of the 'poverty stick' had obvious social advantages for the landowning gentry and larger farmers, who feared popular insurrection in the hungry years of the 1790s. Some attempts to mechanise threshing were made, principally in Scotland around 1735, where experiments were carried out using a water-wheel to drive sets of flails. Later machines were manually operated, and modest in size, but worked. They were becoming general in Northumberland by 1794. The principle of the modern threshing machine was hit upon

* Flails in this sense are heavy, hinged sticks.

by Andrew Meikle, who used two rollers between which the sheaves passed. His first machine was made in 1786. Failing to patent it, Meikle saw it widely imitated. It could be operated by water or by horse power. By 1800 threshers (and winnowers, which were a derivation of the thresher) were probably in fairly widespread use in lowland Scotland and Northumberland, but had hardly penetrated lowland England. They were either small and manually operated, or larger and worked by water or horse power, when they were more usually fixed permanently in place in the barn or outbuilding. But the addition of wheels to make the thresher portable, which seems to have occurred in Norfolk about 1803, was destined to make it of much greater utility for the English farmer.[9]

The key arable processes of ploughing, seeding and harvesting were thus very far from modernised by 1800. The same could be said of any of the other field, barn, dairy, or livestock operations which formed the routine of farm work. Human ingenuity was clearly not lacking; nor was the commercial incentive. What was necessary to improve the tools of agriculture to the point where they could take their place in the coming industrial age was the technological revolution brought about in the iron industry in the age of the industrial revolution.

The technological revolution

Like steel, iron already had a long history by the beginning of the eighteenth century, but it was about to undergo a technological revolution which would simultaneously reduce its price and raise its quality. In effect, it would become a new product, and gain enormous new markets, among which was to be the manufacture of farm machinery.

There are two basic processes involved in producing iron. The first is to melt the iron ore, in conjunction with a suitable fuel. Before the industrial revolution, the most usual fuel was charcoal. The result of this process is a crude product, known as cast, or pig iron. Although cast iron has its uses, being able to withstand compression (and thus often selected for uses such as supporting columns in buildings), it performs poorly at resisting sideways compression, or sudden shocks. For this, the better product was wrought iron (or 'bar iron' as it was commonly known), which was a type of refined pig iron. This had fewer impurities than cast iron, and in particular a lower carbon content, and was more flexible and able to take more bending stresses. It thus had many more applications than cast iron, from horseshoes to hinges and hand tools. The method of converting the pig iron into wrought/bar form was by a combination of heating and beating, as practised by blacksmiths for centuries.

The output of the charcoal iron industry was insufficient for the growing demands made on it; these came a wide variety of quarters, from merchant shipping and the Royal Navy to canal construction, weapons, cutlery, tools, building hardware, and domestic utensils. But the home iron industry was running out of fuel; the supply of timber suitable for making charcoal was limited and awkwardly located. The largest section of the iron industry in the early eighteenth century had been in the Kentish Weald, by very reason of the adequacy of the timber supply, but it was far from British sources of iron ore. Much of the increase in demand was thus met by imports, chiefly of Swedish and Russian iron.

But the technical changes of the eighteenth century would transform the industry. The first of these was the replacing of the dwindling supplies of charcoal with coal. This was pioneered by several industrialists, the best known being Abraham Darby of Coalbrookdale, in Shropshire. He first converted the coal into coke, thus reducing its impurities (chiefly sulphur). He patented his method in 1707, but, partly due to the close-knit and secretive nature of his Quaker family business, the process was slow to diffuse, and only became general in the 1750s. The result was a reduction in costs, as the technology improved through experiment, and as fresh coalfields were opened. The second innovation was the process patented by Thomas Cort in 1784, which permitted the use of coke in the refining process, thus reducing costs still further. This, the 'puddling and rolling' process, involved stirring (with an iron bar) the mass of semi-molten iron to bring it into contact with enough air to burn out impurities, and then rolling the pasty iron through grooved rollers. This squeezed out the slag, and allowed the iron to emerge as bar iron. From then on until the middle of the next century, most iron produced in Britain was to be wrought/bar iron.[10]

These revolutionary developments in effect created a new product, at a lower price and with higher and more consistent quality. Riden estimated that in 1750–54 the average output of pig-iron in Britain was about 28,000 tons a year. Most of this was charcoal iron (26,000 tons), the rest being coke iron. By 1790–94, the total output had risen to 80,000 tons a year, of which 70,000 tons was coke iron. The Napoleonic wars stimulated the industry to new heights (400,000 tons in 1814), and although the output fell after the conclusion of peace, it had risen again, to 690,000 tons, in 1829. In the mid 1830s, the million-ton mark was passed. By this time, imports were a negligible part of the British market. Truly, the nineteenth century was to be the century of iron.[11]

The consequence of the shift away from charcoal, which had been in increasingly short supply, was that the price of iron fell. As early as the 1760s, coke pig was perhaps as much as £3 a ton cheaper than charcoal pig.[12] While the price of pig-iron remained stable during the period 1790–1810, this was at a time of high inflation, during which the general price level approximately doubled, so that in real terms the price of iron fell. The cost reduction continued after the end of the wars in 1815, as fuel economy continually improved. In the 1830s pig iron prices were about £5–£7 a ton, and they fell to between £2 and £3 by the 1850s.[13]

The entrepreneurs and their businesses

Prior to the last quarter of the eighteenth century, the business of making agricultural implements and machinery hardly existed beyond the local village workshop. Foundries and their forges existed both in towns and in the countryside, but mainly to provide the raw materials for the machinist or millwright. While local millwrights and carpenters were engaged on the maintenance of machinery, and the erection of large items such as wind and water mills, they were seldom involved in the business of making metal farm implements. Insofar as metalwork was involved in the making of farm implements and machinery, this was usually the province of the village blacksmith. The larger farm or home farm of an agricultural estate would have its maintenance men, and could call on the services of the local smith. There were also various ingenious inventors and farming innovators who got the local smith to make up an implement to their own design, but manufacturing in the sense of producing a large number of identical machines or implements did not exist. Up to about 1780 the business was with few exceptions largely the province of the local smith, who would serve at most an area of about five miles around him.[14]

The blacksmithing origins of the farm machinery industry can be traced in the history of what was for over 200 hundred years the oldest firm in the business: Hedges, of Bucklebury, Berkshire. Founded as a blacksmith/foundry in 1739, it remained essentially a foundry; their implements probably had only a local market, even though the business was was still in operation in 1970. As Blake observed, such firms largely, '... operated at the local level, repairing and servicing implements made by other manufacturers and producing occasional pieces of the less complex machinery themselves.'[15]

The same story of early start and restricted growth, in spite of a long history, can be seen in the rather larger firm, almost as old as Hedges, of L. R. Knapp & Co., of Clanfield, Oxon. Founded in 1745 as wheelwrights and smiths, the Knapps began in business at Faringdon, Berks., where they made wagons and 'simple implements for local farmers'. It remained a small firm, employing only about 40 people as late as 1962. For most of this period, it did not have its own foundry, except when it acquired and operated a small one from 1890 to 1960. At some time, perhaps towards the end of the eighteenth century, it moved into corn drills, and later steam threshing machines. Although it made other products at various times, corn drills remained its specialism into the post-Second World period.[16]

In spite of these precursors of the specialist farm machinery industry, the first period of rapid growth in specialist firms only took place after *c.*1780. In this first phase was established the first generation of proper agricultural engineering firms, a high proportion of which formed the backbone of the industry for almost two centuries. It is worth listing them:

Table 1.1 *Agricultural engineering firms founded between 1770 and 1815*

Firm	*Founded*	*Location*	*County*
Hedges	1739	Bucklebury	Oxfordshire
Knapp	1745	Faringdon	Berkshire
Hillson	*c.*1750	Stratford-on-Avon	Warwickshire
Burrell	1770	Thetford	Norfolk
Garrett	1778	Leiston	Suffolk
Corcoron	1780	London	London
Small	1783	Dalkeith	East Lothian
Ransome	1785	Norwich, later Ipswich	Norfolk Suffolk
Ball	1798	Rothwell	Northamptonshire
Hanford	1798	Hathern	Leicestershire
Rumsby	1799	Bungay	Suffolk
Cooch	1800	Harleston	Northamptonshire
Holyoak	1800	Lutterworth	Leicestershire
Plenty	1800	Newbury	Berkshire
Smyth	1801	Peasenhall	Suffolk
Bentall	1803	Maldon	Essex
Wedlake	1803	Romford	Essex
Bamford	1805	Uttoxeter	Staffordshire
Hunt	1808	Earls Colne	Essex
Howard	1813	Bedford	Bedfordshire
Wood	1813	Bury St Edmunds	Suffolk
Tasker	1813	Andover	Hampshire
Gibbons	1814	Hungerford	Berkshire
Hornsby	1815	Lincoln	Lincolnshire
Tett	1815	Warminster	Wiltshire

Source: Blake, S., 'An Historical Geography of the British Agricultural Engineering Industry, 1780 to 1914' (unpublished Ph.D. thesis, University of Cambridge, 1974), Table 4, pp. 79–80.

The interesting points about this listing of 25 firms are the increasing rate of firm formation as time went on, especially after 1790; the heavy concentration of firms in East Anglia (and to a lesser extent southern England), and the fact that a large number (9 of the 25) became major concerns in the nineteenth century – Ransomes, Smyth, Howard, Hunt, Garrett, Bamford, Bentall, Hornsby, Tasker. There was a high survival rate, the majority of the 25 firms surviving until after 1945. Only two ceased business before 1914 (Plenty in 1874; Gibbons in the 1890s). Twelve survived into the 1960s, and some beyond. It was a remarkable record of commercial success.[17]

The blacksmithing origins of these founding firms can be seen in many cases. Leaving aside Hillson (later Troth & Hillson), who seem to have confined themselves to making edge tools,[18] the next one on the list chronologically is Burrells, whose date of foundation is given as 1770. However, this would seem to be misleading. The Burrells entered the blacksmithing business in the 1750s, probably combining it with whitesmithing and locksmithing. But the head of the business, Joseph, probably did not begin the development and manufacture of agricultural machinery until after the death of his father in 1793. The first notable impact he made on the machinery scene was the winning of a silver cup at the Holkham Sheep Shearing and Agricultural Meeting of 1803 for the excellence of his seed drill. Following this success the firm advertised itself as, 'Manufacturers of Chaff engines, Dressing machines, Drill rolls and Drill machines, Kitchen ranges, Stoves, Grates, Iron fencing and Iron work of every description and dealers in Gutta Percha*.'[19]

Garretts also had clear smithing origins, there having been three previous generations in the family craft of bladesmith (making sickles and hoes) and gunsmith before Richard Garrett set up in business in Leiston in 1778. The forge which he took over was already a going concern. The works, which were not large, employed at the most eight or ten men, and used a horse to drive a grindstone.[20]

James Small, who achieved fame as the first ploughwright to publish a book on the scientific principles of plough design – *Treatise of Ploughs and Wheel Carriages* (1784) was an 'operative mechanic' in the factory which made the Rotherham plough, and the family remained in the plough business until 1934.[21]

The technical origins of the Ransome business were more sophisticated than those of the local blacksmith. Here, the Quaker connection was important. Richard Ransome, a Quaker schoolmaster of Wells, Norfolk, apprenticed his son Robert (b. 1753) to a Norwich ironmonger. This, it is suggested, may have been inspired by the growing national reputation in the iron business of the Darby family of Coalbrookdale, who were also Quakers. It was a time of intense experimentation with the design and performance of ploughs. Ploughing contests were in vogue, and premiums were being offered for new designs. This may have been a consideration in Robert Ransome setting up on his own in Norwich in the early 1780s, as one of the first iron and brass founders in East Anglia. An early result of his experiments was his 1783 patent for a cast-iron roofing tile, 'such plates being superior to tiles, slates and lead'. More importantly, in 1785 he took out a patent for a tempered cast-iron plough share. An advertisement from the *Norwich Mercury* of Saturday 23 April 1785 describes the advantages of the product, and informs the readers where it was to be sold. (See page 14.)

This was the beginning of the revolution in the technology of farm implements manufacture. Until this time, cast iron had not been used for farm implements because it was too brittle and unable to withstand the

---

* Gutta percha was an early natural plastic, a precursor of rubber.

constant wearing action of the soil. Thus the metal parts of a plough were made of wrought iron, and any repairs, most often to the plough share, necessitated a time-consuming trip to the blacksmith. It may be noted that Ransome's advertisement calls attention to his share's superiority over 'hammered iron' (i.e. wrought iron), and claims that it lasted six times as long. It is also evident that the distribution was much wider than might have been the case for a simple local blacksmith enterprise, Ransome having established a network of agencies in Norfolk and Suffolk.

Ransome moved his business to Ipswich in 1789. Better transport (to the sea, via the River Orwell) for products and raw materials, and for marketing the finished products, may have been the reason. In addition, it made it easier to get coal from the coasting colliers which came down from Newcastle, there being no local sources of coal.

The 1785 patent was followed in 1803 by another, which covered the process of chilling the cast-iron plough share, so that the soft, upper side of the share was worn away at a quicker rate than its harder under side, so automatically maintaining the sharpness of the cutting edge. In effect, the share was self-sharpening. With these two patents, for the first time the cast-iron ploughshare was a practical proposition.[22]

By the early nineteenth century, it may not be fanciful to detect a change in the origins of the firms, and their mode of evolution. It was now less often a case of an existing blacksmith (or, more rarely, foundry) moving from its original products into the developing areas of farm implements and machinery. It was now more the case that firms were being founded from the start as makers of farm equipment. This is so in the case of both the firms that began in a small way and did not grow much, and the ones that became national names. Thus William Plenty of Newbury, Berkshire, started in business in 1790 with a workshop/foundry which by 1800 was already noted for its ploughs, a patent for one of which was taken out in that year. The firm moved into other products as early as 1816, when they began making lifeboats, and later marine and other steam engines.[23]

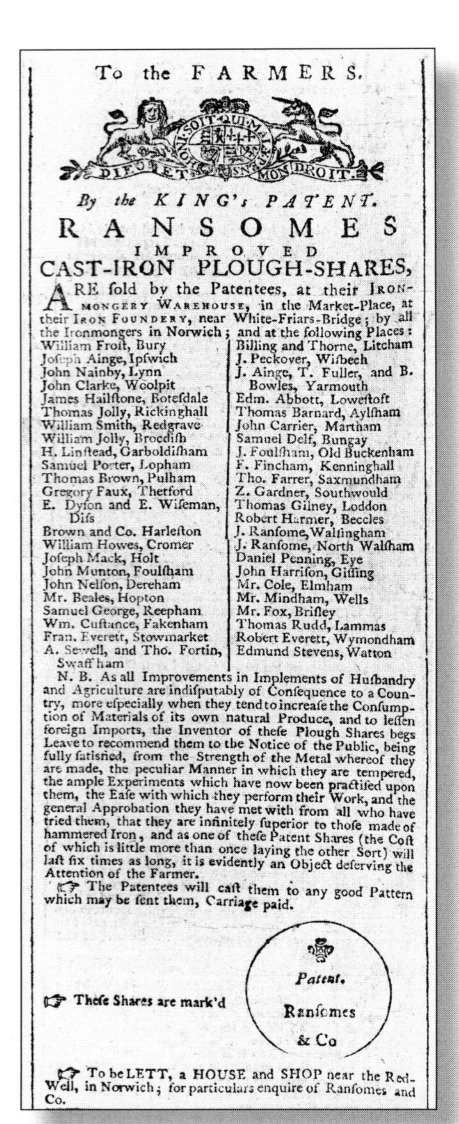

To the FARMERS.

DIEU ET MON DROIT. HONI SOIT QUI MAL Y PENSE

By the KING's PATENT.

RANSOMES

IMPROVED

CAST-IRON PLOUGH-SHARES,

ARE ſold by the Patentees, at their IRONMONGERY WAREHOUSE, in the Market-Place, at their IRON FOUNDERY, near White-Friars-Bridge; by all the Ironmongers in Norwich; and at the following Places:

William Froſt, Bury	Billing and Thorne, Litcham
Joſeph Ainge, Ipſwich	J. Peckover, Wiſbech
John Nainby, Lynn	J. Ainge, T. Fuller, and B. Bowles, Yarmouth
John Clarke, Woolpit	Edm. Abbott, Loweſtoft
James Hailſtone, Boteſdale	Thomas Barnard, Aylſham
Thomas Jolly, Rickinghall	John Carrier, Martham
William Smith, Redgrave	Samuel Delf, Bungay
William Jolly, Brocdiſh	J. Foulſham, Old Buckenham
H. Linſtead, Garboldiſham	F. Fincham, Kenninghall
Samuel Porter, Lopham	Tho. Farrer, Saxmundham
Thomas Brown, Pulham	Z. Gardner, Southwould
Gregory Faux, Thetford	Thomas Gilney, Loddon
E. Dyſon and E. Wiſeman, Diſs	Robert Harmer, Beccles
Brown and Co. Harleſton	J. Ranſome, Walſingham
William Howes, Cromer	J. Ranſome, North Walſham
Joſeph Mack, Holt	Daniel Penning, Eye
John Munton, Foulſham	John Harriſon, Giſſing
John Nelſon, Dereham	Mr. Cole, Elmham
Mr. Beales, Hopton	Mr. Mindham, Wells
Samuel George, Reepham	Mr. Fox, Briſley
Wm. Cuſtance, Fakenham	Thomas Rudd, Lammas
Fran. Everett, Stowmarket	Robert Everett, Wymondham
A. Sewell, and Tho. Fortin, Swaffham	Edmund Stevens, Watton

N. B. As all Improvements in Implements of Huſbandry and Agriculture are indiſputably of Conſequence to a Country, more eſpecially when they tend to increaſe the Conſumption of Materials of its own natural Produce, and to leſſen foreign Imports, the Inventor of theſe Plough Shares begs Leave to recommend them to the Notice of the Public, being fully ſatisfied, from the Strength of the Metal whereof they are made, the peculiar Manner in which they are tempered, the ample Experiments which have now been practiſed upon them, the Eaſe with which they perform their Work, and the general Approbation they have met with from all who have tried them, that they are infinitely ſuperior to thoſe made of hammered Iron, and as one of theſe Patent Shares (the Coſt of which is little more than once laying the other Sort) will laſt ſix times as long, it is evidently an Object deſerving the Attention of the Farmer.

☞ The Patentees will caſt them to any good Pattern which may be ſent them, Carriage paid.

☞ Theſe Shares are mark'd

Pattnt. Ranſomes & Co

☞ To be LETT, a HOUSE and SHOP near the Red-Well, in Norwich; for particulars enquire of Ranſomes and Co.

Robert Ransome's advertisement for his patent cast-iron ploughs, in the *Norwich Mercury*, 23 April 1785.

One consistent feature in the industry is the influence of farmers turned manufacturer. The earliest ones from Table 1.1 include John Cooch of Harleston, Northants, who farmed some 500 acres and made and repaired ploughs and other implements. In 1800 he patented a corn and seed dressing and winnowing machine, and from then on was a combined farmer/manufacturer, the family specialising in manufacturing only as late as the 1880s. A similar road was trodden by William Bentall, who farmed at Goldhanger, Essex. In *c.*1790 he made an improved plough for himself, and was induced by the local demand for it to set up a foundry-cum-smithy to make it for his neighbours. In 1797 he decided to specialise in plough manufacture, and ceased farming. In both these businesses, the 'take-off' was sudden, and the firm did not pass through a craft/blacksmithing phase.[24]

That the blacksmithing phase was not yet quite over is evident in the origins of one of the last firms on the list, which was to become very large indeed. This was Hornsby's. It began as the blacksmith's shop in Grantham, Lincs., of Richard Seaman, who probably made metalwork for farm carts, and simple implements. In 1810 he took on the young Richard Hornsby, who had just completed his apprenticeship as a wheelwright. In 1815 they moved premises, setting up as blacksmiths and makers of horse-drawn thrashing machines, winnowing machines and drills. Richard Hornsby showed both technical ingenuity and business flair. It was not until about 1826, when the works had expanded further, and by now included a brass foundry, that the manufacture of agricultural implements formed the greater part of their firm's output.[25]

The primacy of East Anglia

The increasing rate of formation of firms after the 1780s, and the concentration of new firms in East Anglia, deserve some explanation. The former was a product of two factors: the modernisation of agriculture which occurred in the agricultural revolution, and the increasing profitability of farming, especially of corn. The latter was a function of nearness and accessibility to markets, and of improvements in transport so as to ensure easier and cheaper supplies of fuel and raw materials, and marketing of the products.

The agricultural revolution was particularly evident in East Anglia. The rapid enclosure of the formerly open fields was followed by the implementation of the new forms of agricultural improvement. The former commons and open fields were first drained and marled, then ploughed. Then the new grass fodder crops such as sainfoin, lucerne, and clover would be planted. These enabled a higher proportion of the land to be devoted to keeping animals, and less to fallow. The increasing use of turnips and other roots provided winter keep for animals, and the tilling and singling (by the use of much manual labour) of the root crop led to the suppression of weeds. The raising of the livestocking ratio per acre produced more dung, and thus soil fertility was raised and soil structure was improved. The upshot was greater efficiency and crop yields per acre, and cereal farming became more profitable.

At the same time the twin pressures of rising population and the restriction of imports of grain during the French wars caused a rapid rise in cereal prices, especially wheat. The incentive for farmers and landowners to move into cereals was therefore great. While satisfactory national or even regional statistics are lacking, there is agreement that enclosure led to substantial increases in farm profits, and thus also in rent and land prices, so that the period *c.*1780–1815 was a very good time for farmers and landowners, William Marshall, the late eighteenth-century agricultural writer, thought that land intermixed in open fields or commons was worth one-third less than equivalent enclosed land. The general consensus is that rents doubled with enclosure. F. M. L. Thompson considered that Norfolk rents doubled between 1778 and 1815, the increase being rather smaller in Northumberland, about the same in Yorkshire, and rather greater in Durham and Shropshire. However, recent research suggests that this might be an exaggeration, and the overall rise may have been only about 30 per cent. However, there is little doubt that, in spite of rising rents, farming profits increased greatly during the war period, and farmers who formerly thought of themselves as manual workers now felt tempted to ape the gentry. Cruel fun was had by the cartoonist Gillray at the expense of farmers' conspicuous consumption, such as the learning of the piano by their daughters.[26]

James Gillray's cartoon, 'Farmer Giles and his Wife showing off their daughter Betty to their Neighbours on her return from School' (1809).

East Anglian blacksmiths and engineers could take advantage of the increased demand for the new agricultural implements within their region. By the 1820s it was noted that ploughs made entirely of cast iron were in widespread use throughout East Anglia. Threshing machines were making their appearance in small numbers, and progressive farmers were using seed drills.[27] East Anglia was, if not the cradle of the agricultural revolution as was thought at one time, certainly a major focus of agricultural innovation and engineering inventiveness. Leading aristocrats held large annual gatherings, at which the new ideas were discussed, and prizes offered for new machines, improved breeds of livestock, and the performance of agricultural tasks. Foremost of these gatherings were the Woburn (Bedfordshire) and Holkham (Norfolk) sheep-shearings, held by the Duke of Bedford and Thomas Coke respectively.[28] It was also a prime cereal-growing region. By the mid-eighteenth century, it had a large export trade in wheat to continental Europe. Its proximity to London, which was by far the largest market for agricultural produce in Britain, provided a large and rapidly growing demand for foodstuffs, especially wheat. In the two decades between 1801 and 1821 alone the population of Greater London rose from 1,117,000 to 1,600,000, or 43 per cent.[29]

During the early industrial revolution the national road system was inadequate for the movement of bulky commodities, such as coal, so most industry came to be concentrated in the coal regions – roughly north of a line from the Wash to the Bristol Channel; for the movement of manufactured goods industry relied heavily upon the canal and river system, and upon coastal shipping. But agricultural machinery was different, because its markets were in the arable regions of Britain, chiefly in the southern and eastern counties. The water transport systems of East Anglia made this region a viable site for the new agricultural machine industry. The southern and central East Anglian coast is indented by navigable river systems, giving access to large towns. From south to north they are the rivers Crouch and Blackwater (for Maldon and Chelmsford), Colne (for Colchester), Stour and Orwell (for Ipswich), Deben (for Woodbridge), Waveney (Beccles, Bungay), and Yare (for Norwich). These were all navigable by substantial ships (up to about 30 tons) by the late eighteenth century. The removal of Robert Ransome's firm from the inland town of Norwich to a new waterside location (the 'Orwell Works') at Ipswich in 1789 was clearly dictated by a desire for readier access to markets and raw materials. William Bentall, having begun to make his ploughs at Goldhanger, looked to find a site with better transport for his materials (coal or coke, pig-iron and timber), which were all brought in by sea:

> ... with the expanding business requiring ever larger quantities, the unloading and transport of his raw materials from ship to lighter, from lighter to quay, and from quay to Foundry Field became too great a strain on the rudimentary facilities that existed in the little

waterside village of Goldhanger ... He found the position he wanted three miles away at Heybridge. Land was available alongside the recently opened Chelmer and Blackwater Canal, and by the end of 1804 he had completed arrangements for its purchase.

The new site gave easier access to the Blackwater estuary, and was nearer to the local market town, Maldon. Bentall seems to have imported his coke direct, rather than process imported coal. He probably also chose the new site with exports in mind. Castings could be loaded into barges at the works, transhipped into coastal freighters at Heybridge Basin, and delivered around England or to the Continent.[30]

The importance of water transport is also evident in the history of the iron foundries of Colchester, Essex. The Hythe dock area of the town's riverside was the most flourishing port in Essex, and permitted the import of coal and iron. As early as 1775, coke was being produced in purpose-built ovens there, and the first documented Colchester foundry, established in 1792 by Joseph Wallis, was probably the reason for the advertisement in 1795, by one Samuel Cooke, merchant, and George Round, banker, that their established 'coal, coke, bottle and stone trade' now included the bulk importation of iron. The import of pig iron was arranged by merchants and importers, who also distributed bar iron to the forges. In 1817 William Bentall got pig iron (at 24 shillings per hundredweight)* from Robert Greenwood, a Maldon iron merchant. Later, he imported his pig iron direct. A small addition to supply was had by taking old iron back from customers, and collecting scrap from agents. In Maldon, the import of coal dated back to the late sixteenth century, and by 1793 there were five coal-merchants in business.[31]

Conclusion

In the period from *c.*1770 to 1815, the agricultural machinery industry began to emerge as a distinct business. Hitherto the blacksmith and wheelwright provided implements and a few machines, serving local markets. But the coming of a mass iron market, and the technical improvements in methods of casting and working iron, changed the situation. By 1815 there had emerged a large number of businesses which were brought into being by, among other things, the demand for the new agricultural implements. Many of these firms were located in East Anglia, for reasons of access to markets. While most of these businesses had to rely on other products as well as the agricultural ones, and there was only one (Ransomes) which operated on a major industrial scale, the industry was moving rapidly away from its blacksmithing origins, and was making its own contribution to the wider industrial revolution of Britain.

* One hundredweight (cwt) = 112 lbs. There were 20 cwt in one imperial ton.

CHAPTER TWO

Towards a national market, 1820–1850

The economic background

Until the late 1930s, the economic experience of most nations was dominated by long periods of either inflation of deflation. The end of the second decade of the nineteenth century brought an end to a long period of rising inflation, which had reached frenetic rates during the French and Napoleonic Wars. This may be seen in the index of English consumer goods prices (mainly food, fuel and cloth) compiled by Schumpeter and Gilboy:

Table 2.1 *English consumer goods price index, 1750–1820 (1701 = 100)*

1750	1760	1770	1780	1790	1800	1810	1820
95	98	100	110	124	212	207	162

Source: Mitchell, B. R., *British Historical Statistics* (Cambridge, Cambridge University Press, 1988), p. 720.

Even this does not convey fully the extreme fluctuations in the course of this period, with wartime peaks in 1812 and 1813 (index values of 237 and 243). Prices began to come down after the peak of 1812, despite increases in the post-war boom of 1817–19 and again during the trade booms of the late 1820s, 1830s, and 1840s. By 1822 the price index had fallen to 125, and by 1826 prices had fallen to a level not seen since the early 1790s; by 1850 the price level had fallen by another one-quarter below the level of 1826.

The economic background was thus not necessarily propitious for industrial enterprise. A time of deflation is usually one in which the prices of goods sold fall faster than manufacturers' costs, thus squeezing profits, and making it harder for entrepreneurs to find capital for expansion. On the other hand, if the fall in the sale price of their products had itself been produced by the use of their own cost-reducing innovations, then manufacturers would be in a commanding position. They would be leading the deflation on, keeping up their rate of profit, and finding the cash for further expansion, since the falling sales prices would be widening and deepening their markets. Clearly much would depend on what sort of industry it was, and whether it was producing goods with a growing market, with growing efficiency and at a lower price. The evidence is overwhelmingly that the manufacture of farm implements and machinery fell into this category.

The agricultural background

Wheat was an important component of farmers' incomes and of consumers' spending, and its price therefore had a strong influence on the general level of prices. The price of wheat had been high during the war period, but on the return of peace a number of factors contributed to lower them: the area under cultivation had risen during the war; and imports were resumed, albeit constrained by the Corn Laws which were designed to protect British producers. The good times were over for the wheat producer and his landlord. The average price of British wheat in the 1780s had been 46.1*s.* per quarter. In the 1830s it was 56.7*s.*[1] Given the rise in the cost of living, and the substantial investments made in their land by landowners and farmers since 1790, it is doubtful if the wheat price was any higher in real terms than forty years previously. Deflation and retrenchment were now the rule. Farmers would have to try to reduce their costs; wages would be lowered. The incentive to cut costs and lower wages was enhanced because, in spite of the general deflation, there is little evidence of any substantial fall in rents. F. M. L. Thompson has suggested that rents per acre fell between 1816 and 1833 by about 10 per cent, but that would not have brought them down to their pre-war levels. An earlier estimate by R. J. Thompson showed rents on some 60,000 acres of land in Lincoln, Essex, Hereford and North Wales as rising from about 11*s.* an acre in the first decade of the nineteenth century to 15*s.*–16*s.* by the 1820s, and holding that level for the next two decades. Another set of estates, covering 119,000 acres, chiefly in the eastern counties, saw rents per acre at about 18*s.* by 1816, and slowly rising thereafter until the 1870s.[2]

The result of the price and rent trends must have been a financial squeeze on farmers, and to a lesser extent, landowners. In such a situation, it might have been presumed that the rate of adoption of new machines and implements would be slowed down, as farmers cut their financial coat according to their cloth. But the evidence is that the new machines continued to make headway. Few quantitative studies are available. An early one is by Walton (1976), who used the evidence of the implements and machines listed on farm sale advertisements in Oxfordshire to infer the rate at which the new technologies spread. Steady growth was seen between 1820 and 1850 in the case of barn machinery (turnip cutters, winnowing machines, chaff machines), in which the proportions reported in sale advertisements by 1850 were about 30%, 60% and 20% respectively. Threshing machines reached about 20% in 1820, but then declined to very few in the 1830s, before rising again to about 20% by 1850. Of field implements, the only ones of any significance by 1850 were drills, which were reported in about 30% of the sale notices by 1850. It was not until after 1850 that horse hoes began to spread. That apart, the only change of significance in field work was the continued diffusion of the improved iron ploughs, and a change in corn harvesting hand-tools in which the bagging hook, and later the scythe, replaced the sickle. Reaping machines did not appear until the 1850s.[3]

Horse hoe, 1850, by Garrett of Leiston (Suffolk)

The principle of the horse-drawn hoe had been first described in the 1730s by Jethro Tull in his book *The Horse Hoeing Husbandry.* Tull's book described and illustrated his two inventions: a seed drill and a hoe, both drawn by a horse. While the principles of both were sound, their common implementation had to wait until the advent of modern iron-founding and working techniques. The horse hoe shown here is that made by Garretts of Leiston (Suffolk) in 1850, a common type in use at the time. The hoeing was done by the angled iron blades at the rear of the machine, which cut the weeds as the machine passed along. The blades could be raised or lowered individually, or all at once by means of the large lever above the machine. The blades were held down by weights. This model could hoe four rows of turnips, or six of beans, or nine of wheat. It was estimated that it could easily work ten acres a day, using two horses (in relays), and a man and a boy, for a total cost of 10s., which was about half the cost of hand labour alone.

The horse hoe could be accurately steered between the rows of the growing crop so as to cut out the weeds growing there, without damage to the crop itself. In the case of root crops, which require space all around each plant, and thus had to be reduced in numbers by hoeing, two hoeings could be given; one between the rows, and one across the rows. The first hoeing cut out the weeds; the second hoeing cut out plants, leaving room for the remainder to grow unchecked by their neighbours.

Garrett's horse hoe, from the firm's 1862 catalogue.

The threshing machine was a special case. Its adoption had been stimulated by wartime labour shortage. By around 1810 it was probably known in most counties of England, although more frequently seen in the north-east and the south-west. After the peace of 1815, when the labour shortage abated, its adoption was slowed. It was slowed further by the labour unrest which attended its use in the 1820s. The surplus of labour was greatest in the corn-growing counties of southern England, and so the pressure on living standards was also greatest there. By this date it was clear to many farmers and landowners that while adopting the threshing machine could save them money, it also risked provoking serious and possibly violent discontent among the labourers, who saw it as a threat to their livelihoods. A memorandum from the Middleton estate office in Suffolk in 1830 remarked:

> But admitting that by machinery the farmer saves 10 per cent surely, with a population much on the increase and generally speaking willing to work and obey orders if fairly treated, it is if not incumbent highly necessary that both Landlord and Tenant should go hand in hand in reverting to Manual Labour in every case compatible therewith.*

The discontent of the labourers reached crisis levels in the 'Captain Swing' riots of 1830, when there were outbreaks of rioting, arson, animal maiming and machine breaking. Threshing machines in particular attracted the odium of the rioters, since they were seen as reducing the opportunities for them to do winter work, using hand labour and flails to thresh the grain. It is thus not surprising that the use of the threshing machine seems to have declined sharply in the 1820s and 1830s. Farmers preferred social and industrial peace to lower labour costs. It was not until the 1840s that the use of the threshing machine revived, and by then the fixed machine was being replaced by the larger, usually steam-driven and mobile sort.[4]

The financial position of farmers was improving by the 1840s. The rapidly rising population had by then probably absorbed the surplus corn production which had existed immediately after the wars. Another factor which was working in favour of the machinery industry was the reduction in the price of iron, evident from the late 1820s. In 1825 the price of English merchant bar iron at Liverpool was £14. By 1831 it was down to £6 5*s*. Some of this reduction was due to the depression in trade at that time, and there was some revival later. But the long-term trend was definitely downward. This, and the use of better machine tools, made the implements and machinery both better quality and cheaper.[5]

---

* Quoted in Wade Martins, S., and Williamson, T., *Roots of Change: Farming and the Landscape in East Anglia, c.1700–1870* (The Agricultural History Review, Supplement Series 2, University of Exeter, 1999), p. 118.

Growth of the industry

These factors contributed to make this period one of continuing growth. Existing firms grew, and new ones were founded. A list of the new firms is given by Dr Blake:[6]

Table 2.2 *Agricultural engineers founded c.1815–1835*

Firm	*Founded*	*Location*	*County*
Carson & Toone	1816	Warminster	Wiltshire
Wilder	1818	Wallingford	Berkshire
Batting	1820	Maidenhead	Berkshire
Headley & Edwards	1824	Cambridge	Cambridgeshire
Maynard	1824	Whittlesford	Cambridgeshire
Nicholson	1825	Newark	Nottinghamshire
Kearsley	1826	Ripon	Yorkshire
Tuxford	1826	Boston	Lincolnshire
Holmes	1827	Norwich	Norfolk
Twose	1830	Tiverton	Devon
Brown	1830	Leighton Buzzard	Bedfordshire
Reeves	1830	Bratton	Wiltshire
Maldon Ironworks	1833	Maldon	Essex
Holman	1834	Penzance	Cornwall
Jack	1835	Maybole	Ayrshire
Bissett	1835	Blairgowrie	Perthshire
Blackstone	1837	Stamford	Lincolnshire

This list also displays a concentration in the eastern counties, but it is noticeable how few of these firms were to achieve the first rank in the industry. By the First World War, the greatest success stories were to be those of Holman and Blackstone, but by then Holman had entirely moved out of the farm machinery business, and Blackstone had developed a large trade in oil engines, quite apart from their use in farming. Most of the others either remained in the farm machinery business and failed to grow significantly, moved out of farm work, or failed. Those who succeeded outside the farm machinery business usually developed into specialist steam engine makers or general engineers. National reputations in farm machinery were achieved only by Nicholson, and perhaps also Reeves, Wilder and Bissett. However, opinions differ on what constitutes a national reputation: Dr Blake's list of those with a national reputation includes Tuxford, Reeves, Nicholson and Maynard.

Further names could be added to the list. In Norfolk there were the firms of Soame (Marsham, Cambs.), founded in 1782; Randell (North Walsham) from 1828; Smithdale (Norwich) from at least 1853, but probably much earlier, and Barnard (Norwich) from some time before 1844. None of these was to grow significantly, and none could reasonably be described as specialists in agricultural engineering. Probably the best description of the type of business is that adopted by Randells – 'Ironmongers and agricultural and general engineers'.[7]

The failure of most of these firms to grow rapidly, or at least not to grow until much later in the century is striking. Nicholsons of Newark (Notts.), although listed by Blake as having been founded in 1825, seems to have been in a small way of business until after 1850. One source describes it as having been founded in 1842 following the dissolution of a partnership of 'ironmongers and bar [iron] merchants'. At first it seems to have concentrated on cast-iron grates for cottages and houses, as well as some small agricultural implements. At the population census of 1851 it had only four employees. It was not until 1858 that a larger works was established, and growth became rapid, turnover rising to £14,463 by 1862.[8]

A similar tale of deferred growth comes from the firm of Reeves, of Bratton, in north Wiltshire. Although Blake gives the founding date as 1830, there had been a Reeves in the blacksmith business in Bratton since 1774. Probably about 1819, the business branched out into making ploughs, and around 1829 into corn-drills. Growth was then steady. By 1849 a valuation of the works shows a sophisticated foundry with steam engine and boiler, valued at £110, as well as a smithy with five bellows (and thus presumably up to five smiths). By then the product range was wider, including ploughs, rollers, harrows, wagons, threshing machines and winnowing machinery. However, in 1845 the firm still only produced 30 ploughs, only two of which were made of iron. By 1857 the annual total of ploughs produced had risen to 113.[9]

Meanwhile, some firms which had started in business earlier were progressing steadily. Thus Smyth of Peasenhall (Suffolk), which had begun in 1797/1801, and had specialised in seed drills from the start, had by 30 April 1836 – at which date its earliest sales ledger survives – reached machine number 1,658. Whether this meant that 1,657 machines had been sold to that date cannot be ascertained, since it is not known when the practice began among manufacturers of amending machine numbers to either impress their customers or deceive competing firms. But from then on Smyths were selling 200–300 drills a year.[10]

Of the firms founded in the first period, before 1820, the one which had risen to a commanding lead was Ransomes. Their outstanding position had been achieved by 1815. Taking advantage of the wartime boom in farming, the firm concentrated on making metal ploughs. Ransomes' third patent, in 1808, was for the manufacture of ploughs having interchangeable parts. Thus a large manufacturer could now produce a wide variety of ploughs suitable for local conditions, and corresponding to particular local designs, by the use of detachable parts. Thus the barrier to large-scale production was broken. The earliest surviving ledgers show the firm's receipts almost doubling between 1809 and 1815. However, the post-1815 depression in agricultural incomes was a serious threat to the

Seed drill, 1850, by Smyth of Peasenhall, Suffolk

By 1850 seed drills had largely replaced the ancient method of sowing seed, scattering broadcast from the hand. Hand sowing did still survive for small patches of land, and for fine, small seeds such as clover, and it was not uncommon for a manual, one-wheeled seed barrow to be used here. But for larger fields and for corn crops, the horse-drawn drill was now paramount. There were two main sorts: the cup feed drill and the force feed drill. The Suffolk drill shown here was popularised by Smyth of Peasenhall. Its superstructure consisted of two wooden boxes; the upper one contained the seed, which fell into chambers in the lower box. Through this box ran a spindle, bearing in each chamber a vertical disc, from the periphery of which a series of small cups protruded. The spindle rotated, picking up in each cup little loads of seeds, and emptying them into the funnels leading down to the hollow coulters, through which the seed fell into the ground. The rate of seed delivery could be varied by changing the cog wheels, thus varying the rate of rotation of the spindle. The force feed drill usually had a toothed wheel running in the bottom of the seed-box, forcing seed out between its teeth. In both types, the seed fell down the tubes to the coulters, which cut tracks in the ground for it. The force feed drill had advantages on hilly ground, where cup loads were subject to great variation, although force feeds could break and damage the seed. The coulters, suspended from the chains at the rear of the drill, could be raised or lowered by varying the pressure exerted on them, either by weights, springs, or by a bar running transversely across them. Cup feed drills were usually fitted with weights and a pressure bar. Disc coulters could be fitted, and were useful for sowing seed in across plough furrows, but offered a temptation to the farmer to drill before the land was properly worked into a seedbed. Later, combine drills were developed, which could sow artificial fertiliser at the same time as the seed, either via the same coulters, or via a different set.

A four-row seed drill made in 1928 by Smyth of Peasenhall. (*Museum of English Rural Life*)

firm. It seems to have survived the period 1815–30 by diversifying into civil engineering. In 1812 William Cubitt, the famous civil engineer, had entered into a contract as the firm's engineer, and under his direction the scope of the firm was expanded to take in bridge-building and millwrighting. A surviving account states that in his first four years at Ipswich the firm gained work valued at almost £5,000, which, it was claimed, would not have occurred without him. In 1819, under his direction, the firm began work on the Ipswich town gas supply, a contract for which in that year accounted for almost 20 per cent of the firm's income. The civil engineering side underpinned the firm's finances while agricultural work was at a low ebb, so that the firm carried on, with employment for 50–60 men and boys in the early 1820s. There was, as far as is known, no other firm in the industry approaching this size at the time.[11]

The difficult conditions of the post-1815 period must have spurred the search for new products to add to Ransomes' list. One such was the lawnmower. Invented by William Budding at Thrupp, near Dursley, Gloucestershire, in 1830, it was the subject of a partnership agreement between himself and one John Ferrabee in 1831, by which the two agreed to manufacture the new machine. As well as producing their own machines, in 1832 Ferrabee issued a licence to Ransomes to manufacture the Budding machine. However, Ransomes sales were slow in the early days. By the time of Budding's death in 1846, Ransomes had sold only 1,200, which is an average of only around 80 a year. But the long-term future of the lawnmower was to be much more buoyant.[12]

That the difficult economic conditions of the 1820s forced even a highly specialist firm like Ransomes to diversify makes it less surprising that the more generalist firms continued to do a variety of work. Taskers of Anna Valley (near Andover, Hants.) seem to have started their foundry work in 1813, and a billhead describes their products in that year as:

> Tasker's Improved Ploughs; articles of every description to fit all ploughs in general use in arms and boxes; tire-iron for wheels; iron nine shares and Drag ploughs: Iron Gates, Palisades and Park Railings; Field, Garden and Park Rollers; Drills and Broad Cast Machines on the most improved principles, etc., etc.*

However, the firm's historian, L. T. C. Rolt, considered that, 'iron castings for general structural use probably absorbed most of the foundry's limited capacity', and cites the case of the cast-iron columns supporting the floor of the Guildhall at Andover, which were cast by the firm in 1825.[13] As late as 1842, the firm was responsible for a large cast-iron bridge at Clatford. The older firm of Burrell changed its name to 'Charles Burrell, Engineers and Agricultural Machinists' only some time after 1837, when Charles

* Rolt, L. T. C., *Waterloo Ironworks: A History of Taskers of Andover, 1809–1968* (Newton Abbot, David & Charles, 1969), p. 35.

Burrell succeeded his father, but was still involved in non-agricultural work, building the main (cast-iron) bridge in Thetford, probably in the late 1820s, and was involved in the establishment of the town gas works.

The growth of the industry since 1800 had been spectacular, as had been the growth of manufacturing in general, but even in 1851, at the time of the Great Exhibition, it could not be said that it was a large-scale industry, despite the fact that the term 'agricultural engineer' can be applied to a wide range of firms, from the specialist or semi-specialist producer, down to the general engineering works and foundries which might have an interest in making farm machinery, to the establishments which were essentially acting as manufacturers' agents, and finally the blacksmith/repairer, who by now had ceased to be involved in manufacturing to any degree. In 1851 there was really only one dominant firm, outstanding by reason of its size and technical excellence – Ransomes – which was into its third generation of family managers, and employed 900, more than twice as many as its nearest rivals. The other leading agricultural engineers were Hornsby of Grantham, Clayton and Shuttleworth (Lincoln), and Howard of Bedford, each employing about 400 in 1851. The only other East Anglian producer of a similar size was Garretts (Leiston, near Ipswich), employing 300. Ransomes other East Anglian rivals were smaller: Burrells (Thetford) with 160; Hunt of Earls Colne (Essex), with 100. These apart, the firm employing more than 50 men was probably a rarity.[14]

The first surviving estimate for the number of firms engaged in the industry dates from 1846, and was drawn up by Jabez Hare. His *List of Engineers and Agricultural Implement Manufacturers* was published in poster format, as an advertising device. He was an 'Engineering and Perspective Draughtsman, and Engraver on wood', who was employed by several of the larger agricultural engineering companies in the preparation of their catalogues, and presumably hit on the idea of listing the companies in the industry as, at least in part, a way of publicising his services. The list has 288 names for England, to which Dr Blake added 21 further names known, but omitted in Hare's list, making 309 in all. Hare's distribution of the 288 was mapped by Dr Blake: 98 in the Midlands, 59 in the North, 49 in East Anglia, 48 in the South, and 34 in the South-West. Hare's list has a particular concentration in the Midlands. The inclusion of the known manufacturers which he omitted would not significantly alter this distribution, since of the 21 omissions, 14 are from East Anglia, three from the North, three from the Midlands, and one from the South. In all probability his bias to the Midlands reflected the bias of his commercial contacts rather than being a true reflection of the national distribution of manufacturers. Since Hare had proceeded by asking firms to write to him and tell him of their existence, it is uncertain as to how far he covered the whole industry, or to what extent he included firms which had their interests more in general engineering or foundry work, or even just repairing farm machinery. There is the further problem that many non-specialist firms were sub-contractors to the agricultural engineering industry, but may still have felt that they were entitled to be included, if only for the purposes of their own advertisement. On the other hand, Hare's grasp is unlikely to have extended to the very small one-man or small family firm not employing non-family labour. All that can safely be concluded is that there was a substantial number of firms in the industry, and enough for several hundred to wish to be classed as manufacturers of agricultural machinery – in itself an intriguing indicator of industrial maturity.[15]

Products and technical change

The momentum established in the development of agricultural machinery in wartime carried the industry forward even through the difficult times of the 1820s and early 1830s. New firms were founded (even if their growth may have been inhibited for the time being), and new products brought on to the market. The contrast between what was available to farmers by the mid-1830s, compared with the range of implements and machines available at the end of the French wars was considerable. This may be illustrated by the front page of the first catalogue (1834) issued by Richard Hornsby of Grantham, who now described himself as: 'Agricultural Machine Manufacturer, & Iron & Brass Founder'. The machines illustrated are: a horse-geared threshing machine; a ridging drill; a deep drilling machine; a drilling machine mounted on rollers; a cultivator; two further drilling machines; a winnower; a chaff cutter; a rape and linseed cake crusher; a land press; and a barley huller.

A similar story of increasing specialisation, and a wider agricultural product range, is evident in the poster produced by Garretts in *c.*1828, which described the firm as being a 'General Iron Works and Agricultural Implement Manufactory'. It illustrated a large pair of wrought-iron gates, two different types of cast-iron grates, a horse-gear threshing machine, four types of plough, a chaff cutter, a rape and linseed cake crusher, two winnowing machines, a turnip cutter, a haymaker, and a large seed drill. The firm was still not particularly large; even in 1837 there were only 60 employees.[16]

The diffusion of knowledge of the new products to the farming community occurred steadily, stimulated in part by the formation of the Royal Agricultural Society of England in 1838. This was the first national (English) body to be devoted to furthering the progress of agriculture since the demise of the Board of Agriculture in 1822, although in Scotland there was the example of the Highland and Agricultural Society, which had been established in 1783, to draw upon. The motto of the Royal Agricultural Society (RASE) – 'Progress with Science' – sums up its aims neatly. From the first, it was keenly interested in technical advance, and through the medium of its annual, peripatetic shows provided a powerful means of publicising and testing the latest machinery and implements. The membership was an elite one, seemingly composed of landlords, the larger farmers, land agents and aristocrats, and was never more than about 7,000 until the 1880s, but it was an influential body, and the attendance at its shows was much larger. In 1852, the first year in which figures were kept, the attendance at its Gloucester show was 36,245.

There can be little doubt that the Society was a powerful force for change. Dan Pidgeon, the later nineteenth-century farming expert and writer held that the opportunity it provided for viewing, testing and evaluating agricultural machinery was a radical change of environment for the implement makers, and that the shows provided a focus for what he termed 'storm centres', around which 'successive hurricanes of interest' in types of machinery 'gyrated'. In the 1840s the emphasis was on cultivating equipment and tile-making machines for drainage. In the later 1840s and 1850s it was stationary steam engines for the farm. In the 1850s and 1860s steam tillage received much attention.[17]

Whatever the influence of RASE, the 1840s was a watershed for the industry. It was the decade in which many new firms entered the business, older firms suddenly grew enormously, and new products which were to be of central importance to its fortunes up until 1914 were added to the industry. This may be seen by some examples. The firm of Bentall, although supported by its original innovative 'Goldhanger' plough, was still in 1840 essentially a local firm. Its new head, the young Edward H. Bentall, proved to be an engineering genius, with a practical and enquiring mind. In 1842 he patented an improved 'Goldhanger' plough, to protect the firm against imitators. In 1842 he made a technical breakthrough, which led to the patent in 1843 for his 'broad-share and sub-soil plough'. This was an immediate success, and proved to be the turning-point of the firm. In the next eight years over 14,000 were sold, nationally and overseas. The success was underpinned by a change of name; the firm became 'E. H. Bentall & Co.' This was designed to instil customer confidence, and gave the firm a spurious prestige – fictitiously, since Bentall was still the sole proprietor, and at that time an Act of Parliament would have been necessary to acquire corporate form. (The only companies then in existence were the great overseas trading corporations such as the East India Company, and, more recently, the canal and railway companies; it was not until the Companies Acts of 1856 and after that it became easy and cheap to incorporate a firm.) It would seem that Bentall was trying to have some of the prestige attaching to a large-scale organisation rub off on his still relatively small family firm.[18]

While the success of Bentall might have been held to be the result of one individual's efforts, other firms also grew rapidly in this period. Ransomes, buoyed by the railway boom of the 1840s, transferred its railway engineering business to a new, larger works (the Orwell Works) in

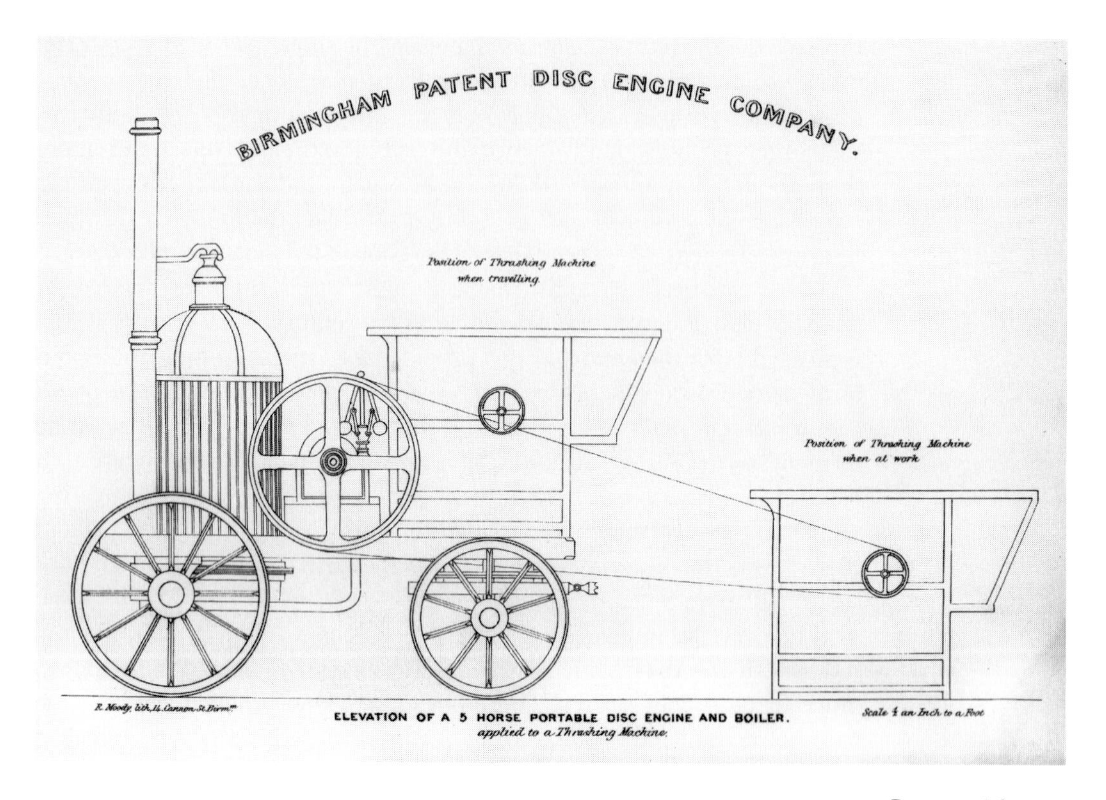

Ransomes' first portable steam engine, and its portable threshing machine (1842). (*Museum of English Rural Life*)

1849. At the banquet celebrating this event it was noted that there were 1,130 men on the payroll, making it by far the largest in the industry. The range of agricultural implements by this time (from the 1844 catalogue) was centred around ploughs, but included also Biddell's scarifier (made under licence), horse rakes, rollers, turnip cutters, grain crushers, thrashing machines and horse-works.*

The new age of steam was coming to farming. In 1841 Ransomes had been the first manufacturer to show a 'portable' steam engine (i.e. not fixed, but capable of being moved around the farm on its own wheels), and its associated thrashing machine, at the Royal Show. In 1842 the firm entered a portable engine again, but this time with a chain drive to make the engine self-moving.[19]

Steam engine technology had been developing for many years, and by the 1830s the problems of how to generate and use high-pressure steam safely had been solved. Early experiments with steam engines for farming were made by Cambridge of Market Lavington (Wiltshire) and Nathan Gough of Salford in 1830. Another early maker was Howden of Boston (Lincolnshire) who 'built the first really practicable portable engine' (Ronald Clark) in 1839. This was only incidentally for agricultural

---

* Gearing driven by one or two horses, used for powering threshing machines and other barn machinery.

purposes; he was asked by the consulting engineer to the local drainage commissioners to build a steam engine and boiler on a wooden frame with wheels to drive pumps for land drainage. This impressed a local farmer, who asked Howden to build him one for his thrashing machine. This pioneering engine aroused much interest when shown at the Lincolnshire agricultural show at Wrangle, near Boston, in 1842. His third engine was also built for a local farmer, in 1842, and is said to have worked until 1870. It is the precursor of all subsequent portable engines. William Tuxford, who ran the Phoenix Foundry, also in Boston, completed a prototype thrasher with portable engine in 1842. The engine was exhibited at the RASE Bristol Show in that year, and gave Tuxford national publicity. But he was not in the business of large-scale manufacture, going on to produce a total of only seven of this type of engine, and a further twelve of a modified version. Claytons of Lincoln produced their first portable engine in 1845, and Burrells in 1846. Marshalls of Gainsborough, who began as an independent business in 1848, were for some years occupied with their new threshing machines, but began making portables in 1857.[20]

While Ransomes did not exhibit another steam engine until 1849, this was a crucial year, since by then the engines of Clayton of Lincoln, Garrett of Leiston and Burrell of Thetford had appeared at the RASE trials (Garrett in 1848 and 1849; Burrell in 1848). In its immediate impact and astounding growth the portable steam engine was the Ford Model T of its day. Although only producing their first portable in 1845, Claytons had sold 2,200 by 1856. There must have been many other small firms who produced their own designs of engine, or copied existing designs from other firms. One such was the small firm of Henry Elvin, of Castle Acre, Norfolk, which began business in 1845 as a whitesmith, and produced small single cylinder portables. A larger Norfolk example is Holmes of Norwich, which produced its first portable in 1855. Garretts first exhibited a portable engine at the Royal Show in 1848, although it was not entirely home-made. As was the common procedure of the time, some components were bought in – in this case the boiler, flywheel and crankshaft.[21]

The arrival of the portable steam engine, which could be used to drive all sorts of barn machinery such as cake crushers, winnowers, and chaff cutters, as well as the modestly sized threshing machines of the mid-nineteenth century, proved a boost to the whole industry, and attracted into it a host of engineers who had so far been less committed to farm machinery. But independently of this, there had been a continual expansion in the range of machinery available. This was seen clearly at the Great Exhibition at the Crystal Palace', London, in 1851. A listing of the products exhibited at the exhibition by Garretts gives some idea of the range available:

> Seed and manure drill; Drill for turnips and manure; Lever corn and seed drill; Drill for small farms; Barrow hand drill; Hand lever

> drill; Patent horse-hoe; Bolting threshing-machine; Patent threshing machine for barn work; Patent portable threshing machine; Horse power driving machine (six horses); Portable steam engine; Barley hummeller; Corn dressing machine; Chaff cutter for horse or steam power; Portable stone mill; Linseed, malt or oat mill; Rape and linseed cake crusher; Iron plough; Hand chaff-cutter; Corn-dressing-machine for hand operation; Reaping machine [an American Hussey-type reaper*]; Models of agricultural machinery; Set of iron harrows.†

Interestingly, the employees of Garretts were treated *en masse* to a stay in London, being towed up the Thames on board two schooners, mooring at Millbank, and living free on board for a week while attending the show. In spite of a specialisation in farm machinery, Garretts were still doing general engineering work and iron-founding. The town water pump at Aldeburgh, and several railway and road bridges were all products of the firm. The firm employed about 60 in 1837, rising to around 300 by 1851.[22]

By the middle of the nineteenth century, the industry was quite different from what it had been in 1815. There was much more specialisation, although it was unlikely that any of the firms were entirely agricultural specialists; there were more firms; and the product range was much wider. The greatest contribution to farming probably came from the barn machinery, including a variety of machines to thresh grain, produce clean seed, and prepare animal feed. Meanwhile in field operations there was a range of metal ploughs to suit a wide variety of soil types and farming conditions, as well as soil breakers, seed drilling machines, haymakers and drainage machines. To the rising numbers of farm horses could be added the new power being provided by the 'portable' steam engine, which could drive all the barn machinery, including the threshing machine, objection to which amongst the labour force seems to have evaporated. The late 1850s saw a boom in portable engines; in the years 1855–59, the five leading East Anglian makers were said to have supplied engines to the value of 40,000 hp to farmers.[23]

Transport developments

The position of the industry by the early 1830s was thus that it was emerging slowly into a position of specialism, as the smiths, wheelwrights, and carpenters were developing into, or were replaced by, more specialist producers. As Blake remarks,‡ of the period from 1815 to 1835:

---

* This was probably a plagiarisation of the Hussey reaper, rather than one made under licence: see Whitehead, R. A., *Garretts of Leiston* (1964), p. 67.

† Quoted in Whitehead, *Garretts of Leiston*, pp. 52–5.

‡ Blake, thesis, p. 81.

> The whole of Britain still had a network of craftsmen producing implements in response to local demands, interspersed amongst which were the more specialist units. These were gaining in size, expertise and production capacity, and, consequently, were able to extend the size of their market areas.

However, there were limitations on the size of the market. Although the improvements in river and canal transport had clearly benefited the industry, as they had the economy generally, they were now played out, and were to be superseded by the next great transport innovation, the railway. The first recognisably modern British railway was the Liverpool and Manchester, which began operations in 1830. This was a great success, both in the carriage of freight and passengers, and in profitability. Within two years of its opening the company's shares had doubled in value. There followed a national boom in railway construction, which peaked in 1840. By 1843, when this first railway boom was really over, there were 2,043 miles of public rail track in the UK (including Ireland), and the outlay so far had amounted to about £54 million.[24] There followed a much stronger boom during the 1840s, aptly dubbed 'the railway mania'. This was at its height in the years 1844–48, during which Parliament passed about 1,720 Railway Acts. Not all the lines projected were built, but by the time the constructional phase had worked itself out, in 1852, the total network was 6,628 miles in Britain and a further 708 in Ireland.[25] By this date the basic network of the British rail system had been laid down; all the major towns and cities had been linked, and the main body of the country was criss-crossed with trunk and secondary lines. The framework had been laid for the extension of the system to smaller settlements and suburban areas, which was to take place in the 1860s and 1870s.

The coming of the railway had enormous implications for society as a whole. Personal travel became cheap and easy; the transport of building materials, raw materials, fuel and manufactured products was made much quicker and cheaper; towns and cities were enabled to grow much more rapidly, as internal migration quickened, and the towns found their building materials (brick, wood, slate and glass) from specialist suppliers further afield.[26]

Not only did the railway accelerate the movement of industry away from the small workshop to the factory; it also provided opportunities for firms to benefit from sub-contracting to the great railway promoters and entrepreneurs. The agricultural engineering industry was part of this. Although not much documentation has survived, what remains in the archive of Ransomes, which was still by far the largest firm, suggests that railway work was the mainstay of the firm until well after 1850. This may be linked to the coming to the firm of Charles May, also a Quaker, in 1836. In many ways he filled the vacuum left by William Cubitt, who had left in 1826. One or two pieces of evidence show that a great proportion of the trade was with the great railway contractors such as Peto, Betts, and Thomas Brassey. The latter is known to have contracted with Ransomes for 2,400 railway wagons on one single occasion. The balance sheet for 1851 shows agricultural work valued at only £35,000 for the year, as against nearly £87,000 for railway and general engineering work. There is evidence that the firm was best known for its chairs and their fastenings, used to secure the rails to the sleepers. These were based on patents taken out in 1841 by James Ransome and Charles May. The item produced in the greatest quantity was the 'trenail', which was an oak pin, compressed to about three-quarters of its volume in the factory. On being driven through the 'chair', an iron clip which linked the rail to the sleeper, the trenail expanded with the taking up of atmospheric moisture, so that the rail remained immovably fixed to the sleeper. The manufacture of trenails was done on a mass production basis in the newly constructed Orwell Works, and railway work was mainly responsible for the growth of the firm in 1848–50. Railway work went on for a long period after this, being transferred to the new separate firm of Ransomes & Rapier in 1869. In

Method of tipping railway trenails (wooden spikes securing the chairs to the sleepers) at Ransomes' Orwell Works, Ipswich, from *The Engineer*, 16 January 1857. (*Museum of English Rural Life*)

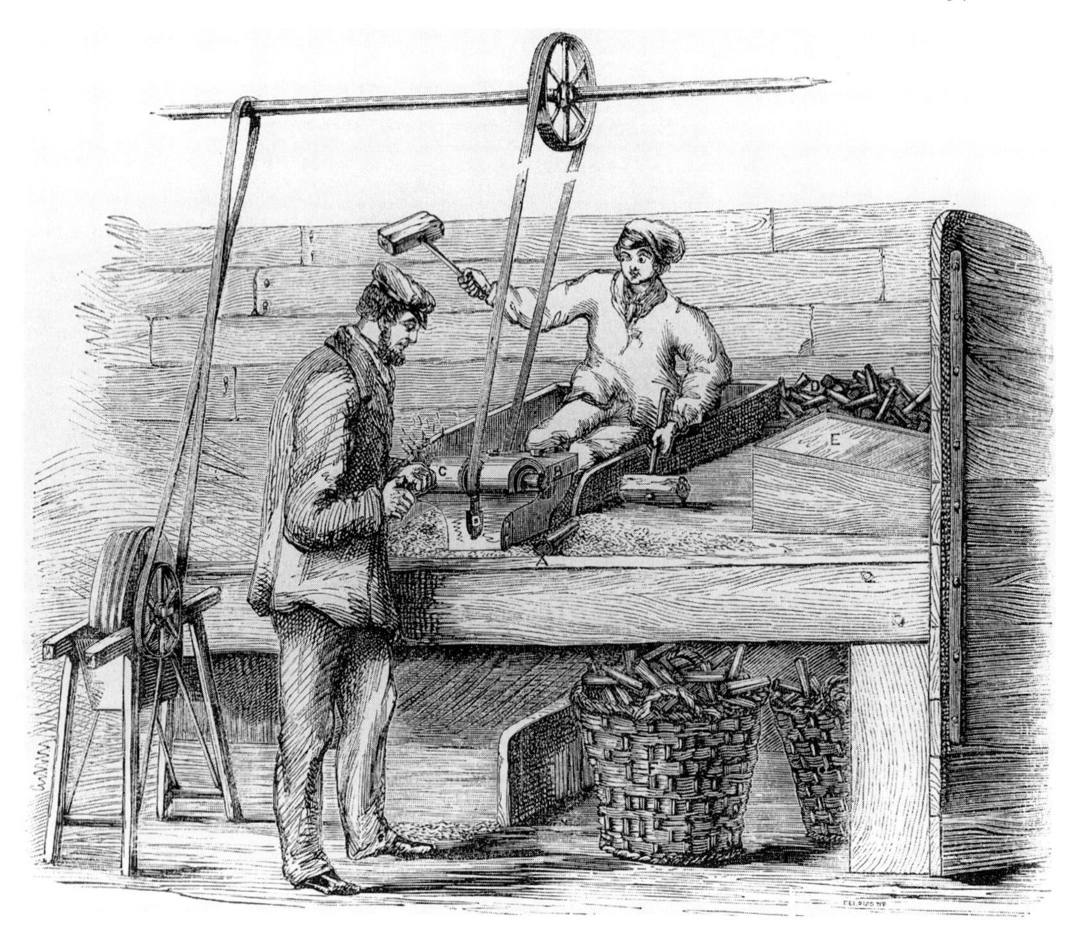

1866 the daily castings of railway chairs amounted to 50 tons, and by then the firm had supplied 'upwards of 115,000,000 Patent compressed Railway Keys and Trenails representing a money value of £700,000'.[27]

The use of the railway extended the marketing radius of the industry enormously. From the mid-century, firms usually quoted the price of transport by rail in their catalogues. Samuelson of Banbury listed the price of his patent digging machine for delivery at the works, and also at London, Bristol and Liverpool. Woods (Stowmarket) stated that delivery was free at London, Hull and Gainsborough by sea, and at all East Anglian railway stations on the Eastern Counties and East Anglian Railway. Garretts stated that implements would be carried free 'to any station on the line of the Eastern Counties and Eastern Union Railways; or 30 miles by land, or by water as far as Newcastle-upon-Tyne'. By 1854–57, Hornsby's threshers and Reeves' drills were being sold up to 200 miles from their works; the Reeves' market was concentrated in a broad belt reaching from Lincolnshire down to Dorset.[28]

Sales and marketing

From the first, successful firms showed an acute awareness of the importance of marketing. The advertisement by Ransomes in the Norwich press in 1785 has been described already. There is of course no doubt that personal recommendation and personal contacts were, and are, of great importance in selling products of the industry – there is one anecdote of Garretts in the 1840s, in which Richard Garrett sold a threshing machine to a guest at his dinner table. But the methods used in selling were modernised from an early stage. The most interesting example of this is the use of agents. In 1795 Ransome used a network of 47 agents in Norfolk and Suffolk to extend the effective market area for his plough shares and parts.[29] Ransomes was an exceptional firm at this point, and other agent networks have left few traces. Two which can be traced are those of Hanford & Davenport of Hathern in Leicestershire in 1809, and the network of 22 agents used by Bentalls of Maldon between 1846 and 1854 for the distribution of the Broadshare plough. Little other evidence of agents has survived in any detail.

The agents involved with Ransomes were not professional agents, but probably consisted of the village tradesmen – general store owners, smiths, ironmongers or carters. Of the agents used by Hanford & Davenport, 28 were publicans, 3 were warehouse companies, and 1 was a village smith. The use of public houses as depots for implements and publicans as forwarding agents was a common practice. For instance, Smyth of Peasenhall received, on 14 September 1847, an order for a drill for John Challis of Borough Green, Newmarket, which was to be sent to the Star Inn, Newmarket. By that time, it was becoming customary to send machines by rail, but usually no further than the station itself. In the case of Bentalls, the first reference to an agent comes in 1837, with a firm of dairy engineers and agents in Shropshire. But this was exceptional, and most of Bentall's agents worked in East Anglia.[30]

Manufacturers' agents must have taken their opportunities to meet farmers at village and town markets. For many farmers, attendance at a weekly market was essential as a way of selling their produce, and getting in farming equipment, seeds, fertilisers, and household supplies. At a time when personal contacts were essential, some importance also attached to the role played by innovators such as Coke of Holkham, who encouraged exhibitions of machines at his annual 'sheep shearings'. Thus in 1804 Burrell's threshing machine was shown at the Holkham shearing; the following year drills and a Norfolk plough by Balls of Holt; in 1807 a threshing machine by Cordwell & Brewster of Norwich.[31]

Marketing the products of the industry in general, and of individual firms in particular, received a substantial boost with the formation of the Royal Agricultural Society of England in 1838. Not only was the Society's annual show an opportunity for firms to display their latest products, and for the public to compare the products of different manufacturers, but its peripatetic nature meant that the industry over time reached a large part of the country. The first show in 1839 was dominated by Ransomes, which sent six tons of implements to the show, and was awarded a gold medal 'for their excellent display of implements'. The leading firms, and many others, took advantage of the Royal and other shows such as that of the Bath & West, and the Lincolnshire Agricultural Society, to promote their wares.

Some examples of this process come from the records of the North Lincolnshire Agricultural Society (founded at Brigg in 1836). At the 1848 show, held at Lincoln, consternation was caused by the late arrival of implements on the railways; several consignments sent by rail, from as far afield as Reading, Bedford, Exeter and Grantham, had only got as far as Nottingham when the show commenced. An entry had also been received from Crosskills of Beverley, East Yorkshire. At the 1849 show, held at Brigg, the Beverley firm of Barrett Aston & Co. exhibited 34 items, including a steam engine and two threshing machines. Dinner was held in the railway warehouse, the NLAS having been granted five days' free use of it by the railway company. Special excursion trains had been laid on from Hull, Grimsby, Louth and Lincoln. At the Boston Show of 1855 there was, according to the *Farmer's Journal*: 'a vast array of first class Agricultural Implements amounting to 400 entries that might vie with any local exhibition in the kingdom.'[32]

The apogee of this process came at the Great Exhibition of 1851, at which 'more than 300 agricultural engineering firms exhibited'.[33] The catalogue of 'Agricultural and Horticultural Machines and Implements' had entries from 291 firms. The major manufacturers were all represented, this being a marketing opportunity which it would have been most unwise to miss. Of the major firms, or soon-to-be major firms, the only absentee was Marshalls, which had only been founded two years previously. Hornsby was there, as were Ransome, Garrett, Bentall, Howard, Clayton & Shuttleworth, Crosskill, Barrett, Exall & Andrewes, and Mary Wedlake. The two last were destined to fade from the scene, but they were well written up in the catalogue, with impressive illustrations, the Barrett entry being blessed with the most illustrations, and the most pages – no fewer than four whole pages to itself. Both the Barrett and Wedlake stands presumably benefited from their proximity on the exhibition floor to Ransomes & May. The illustrations may have all been the work of Jabez Hare, although only the 1½ pages of the Hornsby entry were signed 'Hare'.

Using the catalogue of the Exhibition as a guide to the industry has obvious difficulties, since it was a self-selected sample. Confining the list to those who made machines or implements, which means excluding the inventors who did not seem also to be manufacturers, the horticultural entries (there were a lot of improved beehives), the makers of tile drainpipe-making machines (a recent but short-lived vogue), and simple devices like harness and whippletrees, or milk churns, reduces the number of firms to 93: 26 major ones (as far as can be judged) and 67 minor. Of the major ones, those which were clearly on any reckoning among the very top of the profession may be reduced to no more than eleven: Crosskill, Burrell, Ransomes, Wedlake, Barrett Exall & Andrewes, Garrett, Samuelson, Bentall, Hornsby, Howard, Clayton & Shuttleworth However, Burrell may not have quite reached the top rank, since at the time it employed only about 50 men.*

The exhibition catalogue is full of tantalising glimpses and suggestions. What can one make of the entry for Lord Willoughby De Eresby (no. 195) and his description of a roundabout ploughing system (without illustration)? Was the Earl of Enniskillen (no. 232) ever in business as a manufacturer? Did Crump of Derby (no. 160) find that the exhibition boosted the sales of their 'instrument for singeing horses with gas'? More importantly, what is the significance, if any, of the widely varying amount of space given to firms? The inventor who supplied a prolix description of his ideas may have had half a page to himself. Ransomes & May, the exhibitor with the most floor space, got only seven lines, without illustrations. But did it mean anything when Crosskill got two pages, well illustrated, Hill of Dudley (no. 140) got almost two illustrated pages, Deane, Dray & Deane (no. 180) got 1½ well-illustrated pages, and Williams of Bedford (no. 151) got five of his products illustrated? The three latter firms disappear from the record thereafter, and the only indication that they may have been of significance in the agricultural engineering industry of 1851 comes from the pages of the Great Exhibition catalogue itself.[34]

---

* These lists form an Appendix to the Great Exhibition Catalogue.

Manufacturers' display stands at the Great Exhibition in the Crystal Palace, Hyde Park, London 1851. Prominent at the front of the picture are those of Ransomes & May, Mary Wedlake, and Barrett, Exall & Andrewes. Garrett & Sons is also visible. (*Museum of English Rural Life*)

However, the exhibition and its catalogue was an excellent promotion for the industry. Illustrations of the agricultural machinery section, showing the area occupied by Ransomes, Barrett, and Wedlake, appeared in the press, and made a big impact. Finally, the illustrations provided a showcase for the leading-edge products of the industry, and demonstrated how far best agricultural practice had come since the end of the French wars. The products of Garrett, Crosskill, Wedlake, Barrett, Hornsby were well and clearly delineated to a wide public in 1851.

CHAPTER THREE

At the works around 1850

THE SHAPE AND SIZE of the works in the agricultural engineering industry were determined by the principal, basic functions which have to be carried out within them. These are the casting of pig-iron shapes; the finishing of those shapes by grinding and fettling; the transformation of pig into wrought iron; and the working and shaping of wrought iron by beating and hammering it in a smithery. Since many machines still had many wooden components, carpentry was also necessary. Then the various pieces had to be put together and fitted. Finally machines had to be checked, and paint applied, before being despatched. A fully fledged factory which encompassed all of these processes would therefore comprise a foundry, a grinding and fettling shop, lathes and other metal working machinery, a smithery, a carpenters' shop, a painting shed, an assembly area, and a despatch area.

Not all firms fulfilled all of these functions. Many rural foundries survived into the twentieth century without being much involved in the business of agricultural machinery, and in the early years manufacturers did not necessarily have their own foundries. The growth of businesses with their own foundries begins in the 1780s with Robert Ransome. Bentall followed shortly afterwards. It was sometimes the case that the early entrepreneurs took over an existing foundry, and used it for a variety of purposes, only beginning to specialise in farm machinery after the 1820s. This was the case for Richard Hornsby, and later Foster of Lincoln and Marshall of Gainsborough.

The growth of the factories

As the industry entered upon its most rapid growth period, after 1830, so even small firms saw the advantages of having a foundry on the premises. The case of Taskers of Andover, who had so far been essentially blacksmiths, but who established their foundry in 1813, has been referred to above. Another small firm, Lampittt & Co. of Banbury was established in 1835, and established new premises at the Vulcan Foundry, in Foundry Square. The firm produced traction and portable steam engines, and millworks.

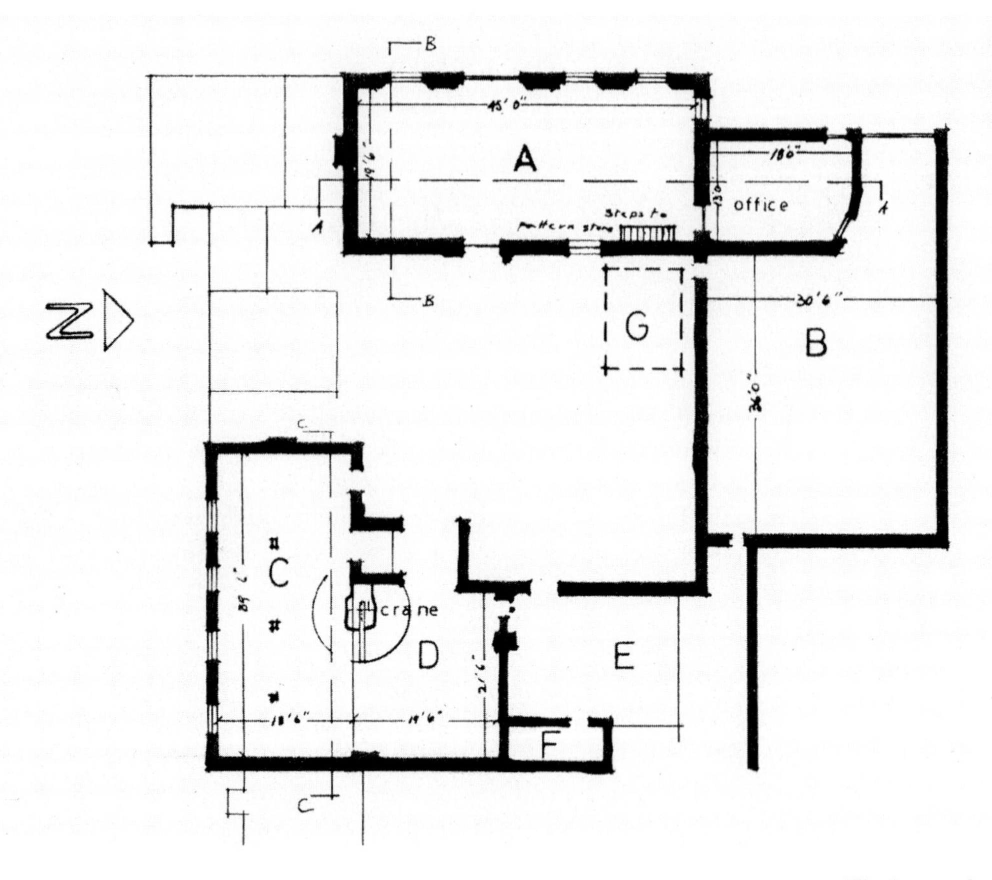

The layout of a small works: Lampitt's Vulcan Foundry, Banbury. The firm was established in 1835; Hartland, G. C. J. 'The Vulcan Foundry, Banbury', *Cake and Cockhorse*, vol. 3, no. 12 (1968).

The plan of the Vulcan Foundry above shows machine shops (A–C), moulding room (D), core room and fettling shop (E), kiln (F), and engine house (G). There was a crane, still extant in 1967, its radius indicated by the circle between D and E. There was probably a cupola furnace in the square room behind the crane. The main points to note are the compact nature of the site layout, with the foundry as its core activity. This was a purpose-built works, but with little in the way of mechanisation, apart from the power provided by the steam engine.

An impression of the working environment in a small- to medium-sized foundry at any time between the mid-nineteenth and late twentieth centuries may be had by looking at the mid-twentieth-century photographs of the foundry of Wm Elder & Sons at their Vulcan Foundry at Low Greens, Berwick-upon-Tweed (*opposite*). They show a visit to the foundry, probably in 1965, by the Mayor of Berwick, escorted by the Chairman of the company, William Elder.[1]

At the top end of the industry, the prosperity of the post-1830 period enabled some firms to construct new works on spacious sites, exhibiting the latest techniques of production and layout, with up-to-date methods of materials and product handling. The acme of this movement, and the

above
The foundry of William Elder & Sons, Berwick-upon-Tweed.
(*Berwick-upon-Tweed Record Office*)

Pouring molten metal at Elder's foundry.
(*Berwick-upon-Tweed Record Office*)

showpiece of the industry as a whole for many years, was the new works constructed by James and Frederick Howard at Bedford in 1856–59. This was sited on 15 acres of land adjoining the Midland Railway and the River Ouse. The firm had begun in the agricultural implement business in a small way in Bedford in 1813. James Howard had joined the firm in 1837, and had developed an interest in the manufacture of ploughs. He was an engineering genius, who took out more than 70 patents during his lifetime. Two of his designs for ploughs, in 1839 and 1841, had laid the basis for the rapid expansion of the firm in the 1840s. By 1850 Howards was one of the three leading plough makers, along with Ransomes & May, and Richard Hornsby. Their new Britannia Works was a superlative expression of the most modern, purpose-built factory design, with layout on the most rational, flow-line principles of the time. It was also very big, instantly rivalling the size of Ransomes' works at Ipswich. While there seem to be no photographs of the works from the outside, there are line illustrations made for poster and catalogue illustrations.[2]

The firm's catalogue of 1864 contains a bird's-eye view of the Britannia works (*below*). The core of the works is the six-bay central building,

The works of James Howard, Bedford: a bird's eye view drawn for the firm's 1864 catalogue. (*Bedford Record Office*)

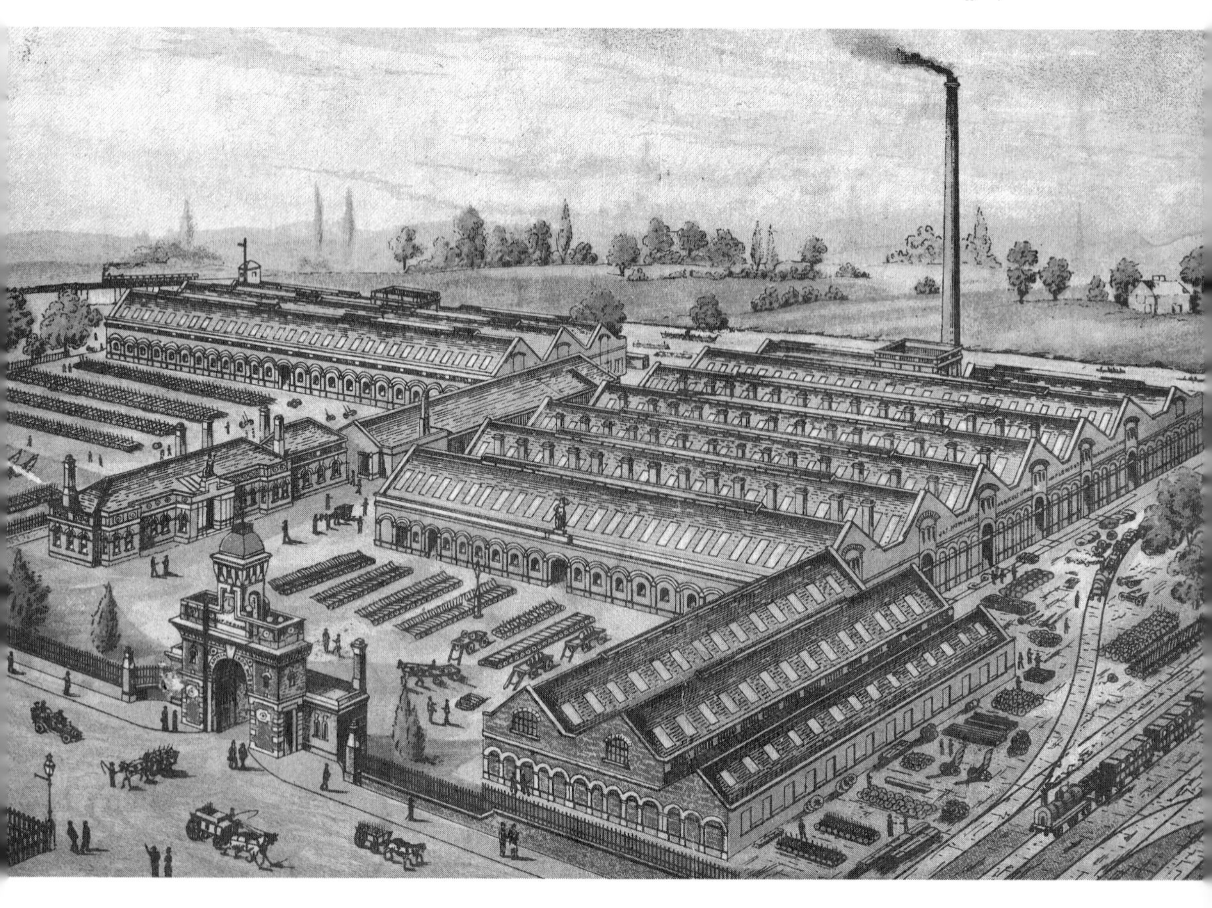

THE SMITHERY OF THE BRITANNIA IRON WORKS, BEDFORD.

The smithery of Howards' Britannia Iron Works, Bedford, 1864. (*Bedford Record Office*)

containing the foundry (furthest from the eye), fitting shops, forge, and finishing shop. The offices are on the left of the picture, and the timber shop to the right of the offices. The sidings of the Midland Railway, with contemporary locomotives, are also to be seen, terminating in the forwarding warehouse, where the goods were packed for despatch.

The catalogue also showed an internal view of the works (*above*). The internal plan of the works was printed in *The Engineer* on 20 July 1860 (*overleaf*); it seems to have been unchanged since the works were constructed. It exemplifies the best practice of the day, in terms of internal layout and equipment.

The entire works was planned so as to utilise flow-line principles. The various 'shops' were arranged in sequence, and materials and implements were transported on tramways laid around the whole works, each shop having a turntable with which to redirect the tramcar on its way. Pig-iron and coals were landed from the River Ouse, from where the iron was broken up, brought in on the tramway, and raised into cupolas by a steam-lift. After smelting, the molten iron was then served out to the moulders from a large truck ladle, running on the tramway. The iron

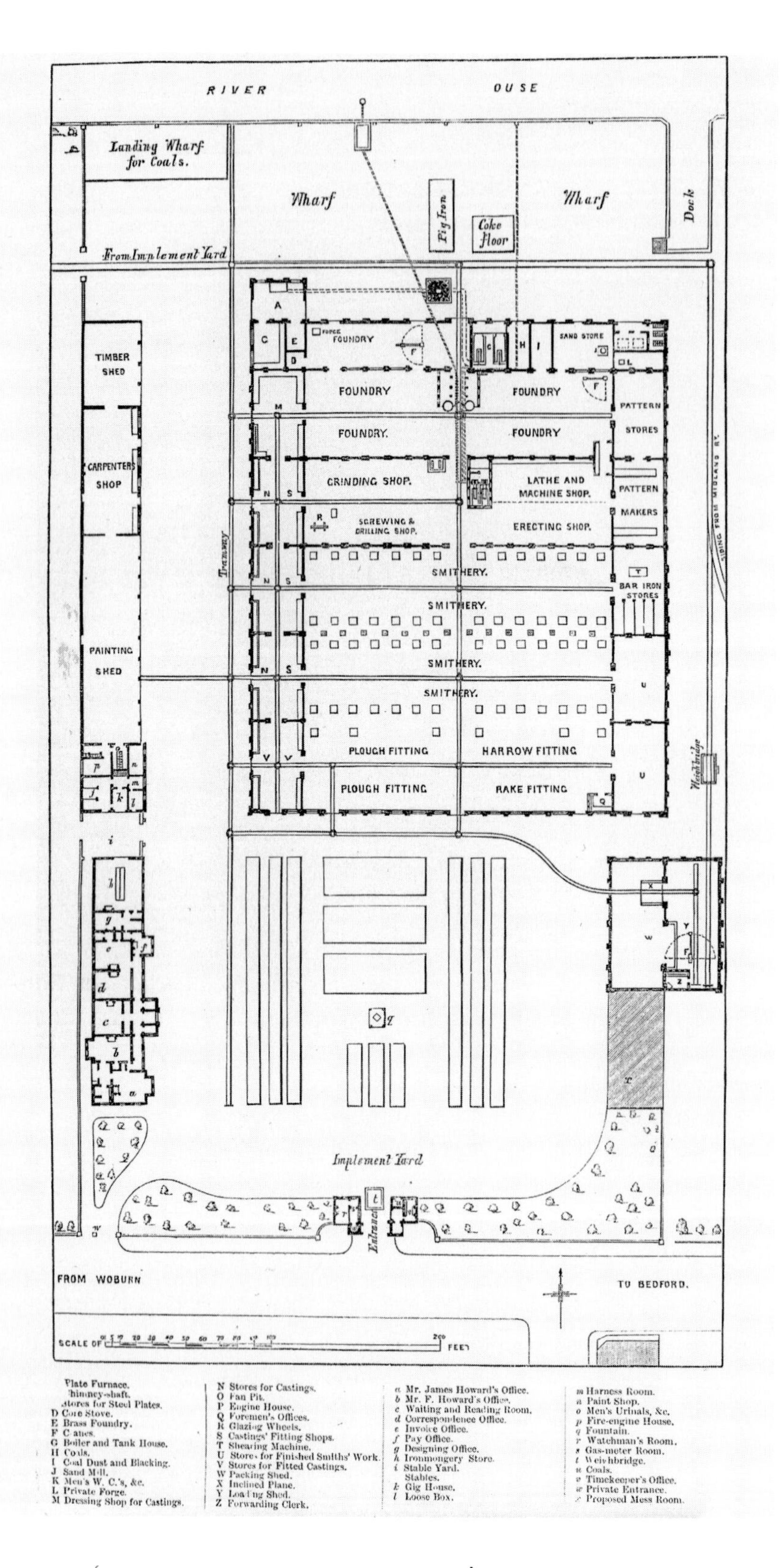

The internal plan of Howard's works, 1860. (*Museum of English Rural Life*)

which had been converted into bar-iron was delivered to the door of the bar-iron stores by the tram. A shearing machine cut the iron bars to the required lengths, from whence they were conveyed to the forges in the smithery.

Everything about the works was on an enormous scale. The forging shop ('smithery' on the plan) contained 57 hearths, the blast for which was supplied from a fan driven, as was all the machinery in the fitting shop, by a compound beam engine of 40 hp. It was a feature of the works that important processes were as mechanised as possible. The most notable piece of mechanisation was the process, patented by James Howard, of making moulds for the casting of the pig iron by machinery. By this process, which 'combine[s] perfect truthfulness and rapidity of production', unskilled labourers could mass-produce castings accurately. About ten tons of castings daily were thus produced. The process was proving so successful that it had been licensed to the Coalbrookdale Company, Ransomes, 'and other eminent founders'. Labour was also saved in the paint shop, where the colours required for the paint were ground by machine.

Work in the factory

The conditions of work in the Howard factory were those of most factories of the time. The working week did not cease until 2 o'clock on a Saturday. Piecework was used extensively; the furnacemen were paid a fixed price per ton for the metal produced. Physical comforts for the labourers seem to have been exiguous. While there was provision of urinals and WCs, there was no mess-room for meals, although one was proposed to be built (at the shaded area marked × in the plan). Office workers had better quarters – 'strong rooms are provided for principals and clerks; indeed the offices are thoroughly well arranged, spacious and complete.' It was a feature of the times that large draughtsmen's offices were not required, since the principal employers themselves designed and patented many of the products. The offices of John and of Frederick Howard (a and b) were noticeably larger than the 'designing office' (g).

It may be asked whether the artistic representations of such works exaggerated their size. Bearing in mind that each bay of the Britannia works was 52 feet across, it is likely that the relative dimensions of the buildings in the external view are correct. However, when it came to internal views, the artist exaggerated the size of the works by the simple expedient of drawing the human figures small. This may be seen by comparing the drawing of the smithery in 1864 with a photograph of the same smithery, taken in the 1920s (*overleaf*)[3]

A firm which was probably about the size of Howards in the mid-1850s, and which, along with Ransomes, comprised the three leading firms in the industry, was that of Richard Hornsby & Co., of Grantham. Although

Howards' smithery in the 1920s. (*Bedford Record Office*)

there is no contemporary ground plan to refer to, there is a description of the premises in the *Official Illustrated Guide of the Great Northern Railway* in 1857, which makes it clear that the firm possessed much the same set of shops as did Howards. A foundry is not mentioned, but it is known that Hornsby had a foundry from 1823:

> … The whole of the operations in this establishment are on a great scale and will astonish the uninitiated visitor. The plant is estimated at £100,000. The number of hands employed is 500. There are rooms and yards for every department of implement manufacture – for carpenter's work; for testing engines; a smithy with thirty forges; a lathe room; draughtsman's room; four joiners' shops; sawing room, with six saws at work; immense quantities of wood [oak] and iron lie about. The timber yard contains a stock of average worth of £6,000. Machines are held in readiness to be sent to all parts of the world, especially to New Zealand and Australia, Sweden, Austria, France and South America.*

A view of life in an agricultural engineering factory in the middle of the nineteenth century is afforded by the reminiscences of Charles Hopewell,

---

* M. Pointer, *Hornsbys of Grantham, 1815–1918* (Grantham, Bygone Grantham, 1978), p. 13.

written in 1909. Born in Newark in 1846, he was employed briefly at Clayton & Shuttleworth, before he joined the firm of Ruston & Proctor in Lincoln as an apprentice in June 1860. At the time the firm was still only modest in size; his works number was 70, and it seems that this may have been the total number of employees. He worked in the boiler-making shop. The main shop included the erecting, fitting and turning departments. There was also a lagging and painting shop, a time office, and the wood works. This was on the first floor of the building, and completed machines had to be lowered out of it by a large side door. This seems to have reflected the limited space on the site. He described the foundry as 'a very small affair', and such was the constraint on space that the smiths (of which there were only two) were also accommodated in the boiler shop. In Hopewell's boiler shop there were only about 20 men, with little equipment. He remarked:

> At first sight, it seemed very difficult to me, how the various parts had been shaped, because there was really no machinery in the shop, only one punching press with Shears at the top. There were no drilling machines, and no Rollers for bending ... The method of bending the Barrels,* was to heat the plates in the furnace, summon all the men in the shop, then in great excitement, pull one of the Plates on to the Floor, cramp another Barrel, already rolled, at the end of the hot plates, and with the aid of side dogs, pull same over to the other end, using mallets to bend the plate more quickly ... I remember that the first two Rivetters were John Winter, (still living) and John Bell, who had all to do, as at that time of day [sic] Hydraulic Rivetters were not even thought about.†

At the time (1863) weekly wages were 32 shillings for smiths, 30 shillings for platers, and 28 shillings for riveters. However, as a 16-year-old apprentice Hopewell received only 5*s.* 6*d.* He resented this, considering his work at the age of 16 to be as good as that of an adult riveter. The lack of machinery meant hard manual labour: 'Take, as an example, the cutting out of a Manhole. There was nothing to do, but take the hand-hammer and Cross cut, and cut away a hole 14″ × 10″ into the plates 3/8″ thick. I can assure you, this was solid hard work.' Hours of work were 58½ a week, and work did not finish until 6 o'clock in the evening. When 'some agitation was made in the North of England, and the nine hours [a day] was conceded' in 1872, the labourers in the Lincoln factories went on strike, and after this the Lincoln firms conceded a week of 54 hours, work to cease at 5 o'clock in the evening, except on Fridays. Hopewell had nothing but praise for the management of what was still not a large firm. Joseph Ruston, one of the two principals, was: '... a gentleman.

* The curved side plates of a boiler.

† The first hydraulic riveters were used in Rustons (the first in Lincoln) in 1875.

Always approachable, and one that was not satisfied without knowing what was going on, and what his apprentices were doing. If there were any trouble between us, we were sent for and after hearing both sides, passed judgement, to the point and which was generally right.' The other principal, James Proctor, '... was very kind to all the lads, and summing up his character, I should say, that he was a very kind, humble minded and christian Gentleman.' The works manager was Samuel Rawlinson, who managed with few foremen, '... as Mr Rawlinson was a hard-working man, and took most of the responsibility upon himself'.[4]

A view of work at one of the three largest firms in the industry may be had from a report in *Bell's Weekly Messenger* of 1859. This was of Clayton & Shuttleworth of the Stamp End Works in Lincoln. This site, which occupied more than nine acres, had been built up over a period of time, the buildings 'somewhat irregular in their structure, but most of them exceedingly commodious for the work to be done in them'. In the winter gas was produced on site for 'upwards of one thousand lights'. Branches of both the Great Northern and of the Midland Railway ran into the works, and there was a canal link to the river Witham. The firm employed around 800 men and specialised in producing portable steam engines and threshing machines. The labour was highly specialised, and most workers were paid piecework. Hours of work were from 6 o'clock in the morning to 8.30 at night. On Saturdays work finished at 12 o'clock. It was noted that 'some of the best and most dexterous hands earn as much has 60 shillings per week', but since the total wage bill was around £1,000 per week for 800 men there must have been very few of these.

The factory organisation at Claytons was devolved into individual workshops, each one run by a foreman, who could raise wages or discharge men, the latter being subject to an appeal to one of the principal employers. Each acting principal (Nathaniel Clayton or Joseph Shuttleworth) kept a set of books, detailing each week's proceedings in each department. Thus errors in materials, pay, or work were quickly detected. The main workshops were for 'moulding' (i.e. the foundry), forging, and fitting. There were also major shops for boiler work and threshing machine erection. The most mechanised of all these was the fitting shop, which had 108 machines, for planing, slotting, drilling and shaping metal. These were powered by a series of leather bands running on wheels on huge shafts, which were worked from a 16 hp steam engine. The smiths' forging shop had 30 forges, each one with a blast supplied by a steam engine, and each one manned by a smith and a 'striker'. The moulding shop could cope with castings up to 10 tons, and employed 36 men, most of whom seemed to work by time rather than by the piece. The boiler shop had some machinery – punching, shearing, drilling and rolling machines, but as yet the hammering in of rivets was still done by hand; there was a perpetual din of rivets being hammered in with sledgehammers by the 150 men, who earned on average between 26 and 30 shillings a week. The threshing machine shop used thousands of pounds' worth of timber annually; mainly oak and ash, both of which were seasoned for two years. The building was made of corrugated iron. Portable engines and threshing machines accounted for most of the output, and the firm was the biggest manufacturer of these in the world. Other products such as corn mills were produced, but they were only a small part of total production.[5]

There is a later description of the Clayton works, written by a visiting journalist ('The Druid') in 1870. By then, the works had grown to twelve acres. The river and canal links of the site were praised for allowing the import of: 'pig-iron from Scotland, deals [timber] from the Baltic, etc.' The visitor entered the turning, fitting and erecting department, 'filled with lathes and slotting and drilling machines in great variety'. In the stores were found the component parts and sub-assemblies of the steam engines, each set of fittings being ticketed, so that it was known which workman assembled them. In the engine-shed, finished engines were pressure-tested (with cold water) to double their intended working pressure. In the great forge house (180 × 80 feet), there were 52 furnaces and their attendant smiths. The boiler shop, the largest of all, at 255 × 190 feet, was filled with the clangour of metal punching and shearing machines. The wood stacks in the drying shops were filled with large quantities of oak (for wheel spokes), ash (for wagon axle-beds), and mahogany (for corn riddles). In all, it was an impressive tribute to large-scale organisation.[6]

Conclusion

Advances in technology and the development of products had progressed very rapidly since the early years of the nineteenth century. They were now paralleled by developments in the layout and equipment of the factories, which were now becoming better organised. Many small blacksmiths and small foundries still survived, of course, as some would into the twentieth century. The showpiece Howard works was not entirely typical, and most of the small- to medium-sized firms such as Lampitts, Elders, Reeves, had less advanced equipment and practices, often squeezed into cramped sites. By improving factory layout and equipment, larger firms were able to gain a competitive edge, and few of the smaller firms after the 1850s had the opportunity or resources to catch up these businesses which had already achieved a measure of dominance in the industry.

CHAPTER FOUR

A brief supremacy, 1850–1875

The expansion of the home market

The two decades or so after the Great Exhibition in 1851 are recognised as a period of prosperity for the British farmer. R. E. Prothero (Lord Ernle) saw the period from 1837 to 1874 as one of 'high farming', in which farmers and landlords made heavy investments in order to raise productivity and profits.[1] Since Ernle's time, research has shown that higher rates of investment actually did little to lift yields per acre or raise the weight of animals in the 1850s and 1860s: the results of technical improvement had actually been greater in the 'agricultural revolution' of the seventeenth and eighteenth centuries than in the several decades after 1851. The most recent work of E. J. T. Collins, in the Cambridge *Agrarian History of England and Wales*, covering the period 1850–1914, suggests that the yearly rate of growth of the output of agriculture was lower in the third quarter of the nineteenth century than it had been in the previous few decades. Cereal yields per acre probably stagnated from the late 1850s; the supply of meat was inadequate for the rising population, and led to rising prices; the performance of the dairy sector, which had experienced less technical change than the arable and meat sectors, was disappointing. Overall, productivity rose much faster in the first half of Lord Ernle's 'high farming' period, from the accession of Queen Victoria in 1837, to about 1858, than in the second half, until the onset of the 'great depression' in English farming in the mid-1870s.[2]

Yet this was a period of high incomes and profitability for farmer and landowner alike. Recent research puts the rise in farm rents between the early 1850s and the late 1870s at between one-third and one-quarter, and on many estates much more. It is difficult to ascertain the level of profits; few farmers kept accounts, and in the absence of direct information for each farm it was assumed by the Inland Revenue that a farmer's income was equivalent to his rent. This was certainly a low estimate, and there is no doubt that the income tax assessments lagged behind the rise in farm incomes. But even the income tax schedules show a rise in farm profits of about 25 per cent in the period from 1851/52 to 1876/77. The rise was greatest for the corn/mixed farms, and lowest for the dairying farms. Contemporary stories of the prosperity of the larger farmers were rife, and form the nub of the work of writers such as Richard Jefferies, who was writing about the 'gentrification' of the farming community in downland Wiltshire. In clayland Suffolk in the 1860s it was remembered that: 'The breath was no sooner out of poor George Wilson's body before men were into their gigs and racing, which could get first to his lordship's agent and say "let me have that farm never mind the rent".' At Orsett in Essex the village schoolmaster remembered the larger tenant farmers as, 'a very cultured class, far removed from the popular Farmer Hodge, as depicted in *Punch*. They lived well in substantial houses, rode to hounds, shot in winter and entertained lavishly. When they were jaded they went off to Brighton for a week.'[3] William Wood, who grew up on the low Weald of Sussex, wrote of his father:

> For such men as this, with ample capital, farming was a good business through the major part of the nineteenth century, up to about 1876. They lived in the style of country gentlemen, were very hospitable, entertained largely, kept a good table and a good cellar, and enjoyed what sport there was going – enjoyed life to the full.*

This underlying prosperity helped the agricultural machinery industry, whose output increased enormously in this period. This was further stimulated by a growing labour shortage in the countryside as more and more workers migrated to the towns. The population census of 1861 was the first that recorded a fall in the farm labouring population, and between the censuses of 1851 and 1871, the numbers of agricultural labourers (who lived off the farm) in England and Wales fell from 908,678 to 764,574, the number of farm servants (who lived on the farm) falling from 189,116 to 134,157. Labourers left the land because they saw better opportunities elsewhere, and were becoming aware that their wages had become distinctly lower than the wages in towns. Many farmers were now buying improved implements and machinery in an effort to make up for the loss of their labourers.[4]

Technical and product change

> In fitness for the urgent hour
> Unlimited, untiring power,
> Precision, promptitude command
> The infant wills the giant hand.
> Steam, mighty steam, ascends the throne
> And reigns Lord Paramount alone!†

---

* W. Wood, *A Sussex Farmer* (Jonathan Cape, 1938), p. 36.

† William Harrison, quoted on the frontispiece of R. H. Clark, *The Steam Engine Builders of Norfolk* (Sparkford: Haynes, 1988).

To contemporaries, nothing was more noteworthy and striking than the successful use of the steam engine on the farm. Agriculture was the last great industry to take advantage of this, the greatest technical advance of the nineteenth century. The steam engine was now the prime source of industrial power; the steam railway was now firmly established; shipping had moved decisively into the steam era. Farming was now to follow.

Steam Power I: Evolution of the portable engine

The main way in which steam came to the farm was by the use of the movable, or 'portable' steam engine. This was to be used chiefly to drive threshing machines and other machines such as chaff cutters, cake breakers, corn mills, drainage pumps and saw benches.

The 'portable' engine was so named because it could be moved from place to place, rather than remain fixed in one position, as were those used for industrial power in the factories. In this, its evolution mirrors the earlier development of the threshing machine, from the fixed installations of the Napoleonic war period, to the wheeled machines that were adopted from the 1830s. The typical portable consisted in a steam engine with a large horizontal boiler, the firebox at one end, along the lines of a railway locomotive, with a single cylinder on top of the boiler, at the firebox end. This drove a piston which was connected to a crankshaft leading to the front end of the machine, where it was used to drive a large flywheel located on one side of the smokestack. There might also be a smaller flywheel on the other side, to provide low-geared power. There would be some form of pressure safety valve or governor. At the front end was the smokestack, usually very tall, in order to carry sparks and smoke away from the operations, and to increase the force of the draught. The smokestack was often hinged, so that it could be lowered for passing through barn doorways. The whole machine was mounted on iron wheels, and at its front end was a pair of shafts, so that the machine could be pulled to new locations by a horse.

The form of the portable steam engine had become established by the end of the 1840s and remained largely unchanged throughout the steam era. Clayton & Shuttleworth's first portable, of 1845, was typical. It had both a horizontal boiler and a horizontal cylinder (although others, such as Burrell's first machine, had vertical boilers), and inclined cylinders. The crankshaft was located next to the chimney. By the end of the 1840s engines were operating at around 30 psi,* which was accepted as the high pressure norm of the period, increasing to around 40 psi in the 1860s. The power supplied would be up to about 8 nhp.† The power rating was raised, although considerations of cost, weight and convenience meant that most portables remained at between 6 and 10 nhp. Fuel economy was much

* psi, pounds per square inch, a measure of pressure.
† nhp, nominal horse power.

A typical 'portable' steam engine of the 1850s, by Marshalls of Gainsborough. (*Museum of English Rural Life*)

improved over time; the prize engine shown at the Royal Show in 1849 burned 11.5 lbs of coal per horsepower-hour, and this had been reduced to 3.5 lbs by 1856, and to 2.8 lb in 1872, although the engines supplied to farmers were not as economical as the 'racers' which were demonstrated at farming shows. Safety was improved with the addition of gauges for boiler pressure and water level. Steam jackets were introduced for the cylinders, which kept the cylinder hot, thus avoiding problems of steam condensation, as well as saving fuel. Boiler-making techniques improved, with rivet holes being drilled instead of punched. Plates were cast with flanges instead of having to be joined by angle irons; the first Garrett boiler which was machine-flanged throughout was made in 1876, and the plates were more precisely formed so as to fit snugly together instead of having to be forced together by cramps.* For over two decades, the only material considered suitable for boiler and fireboxes was wrought iron, such as from Low Moor or Bowling in West Yorkshire, or from Staffordshire. It was only replaced by mild steel in the 1870s. *The Engineer*

* See above, p. 47, for the reminiscences of Charles Hopewell.

commented favourably in 1877 on Garrett's use of steel for their traction engine boilers.[5]

By the end of the 1840s, steam and metal engineering had advanced far enough to allow the mass production of portable engines, and the market demand was strong enough to induce many manufacturers to enter the business. In 1851, Clayton & Shuttleworth exhibited their portable engine and thrashing machine at the Great Exhibition in Hyde Park, and that year sold a total of 126 engines. The portable had come of age, and the market was hungry for it. In 1851, ten years after the first portable appeared at a Royal Show, it was estimated that there were 8,000 in use in the UK. This was probably an overestimate, but Clayton & Shuttleworth, founded in 1842, had sold over 200 portables by 1851, and by 1857 had made a total of about 2,400. By the end of the development process, there were some 90 firms manufacturing steam engines. Claytons, one of the largest in the field, reckoned that in the thirty-five years between 1849 and 1884 they had made 21,000 engines (portables and traction engines) and 19,000 portable threshing machines, although many of these were for export.[6] In the next 70 years or so, Clayton, Marshall, Ransome, Ruston and many other firms made an enormous number of portables for the home and overseas markets.

The scramble of manufacturers to get into the business of portable engine manufacture began to resemble the gathering of the Gadarene swine. Very few of the larger manufacturers remained aloof, and the movement drew into it a host of smaller firms, many of whom made very few engines, and remained small firms until they finished in business. By 1851 the major firms of Ransomes, Clayton & Shuttleworth, Garrett, Hornsby, and Howard were all involved, and also made portables for other firms. The smaller firms must have been numerous, although little is known of most of them. In Norfolk, where the firms have been minutely chronicled by the late Ronald Clark, there were in the period up to *c.* 1875 Baker of Kings Lynn, Barnes of Great Yarmouth, Crowe of Gaywood, Cubitt of North Walsham (1878), Elvin of Castle Acre, The Farmer's Foundry Co. of Great Ryburgh, Garood of Fakenham, Hambling of East Dereham, Holmes of Norwich, Lefevre of Norwich, Pertwee & Back of Great Yarmouth, Powell of Brancaster, Riches & Watts of Norwich, Sabberton of Norwich, Savages of King's Lynn (a rather larger firm), Soame of Marsham, Sparke of Norwich, Sturgess & Towlson of Norwich, Waterson of North Walsham, and Youngs of Diss. This does not exhaust the roll of Norfolk firms making steam engines, but some of them either entered the business of agricultural portables after 1875, or made other types, such as marine or fairground engines.[7]

The technical skill required in the manufacture of steam engines meant that it was undertaken for the most part by existing firms. Among the smaller firms in the above list from Norfolk was Hambling of East Dereham. This was in business as early as 1836, when Thomas and

Robert Hambling were described as 'Brassfounders'. By 1845 they had added Millwrights to their title, and in 1854 Robert was described as 'Whitesmith, Bell-Hanger etc. and manufacturer of steam engines for agricultural purposes'. The firm ceased trading in 1877. It is recorded that they were responsible in their later years for several single-cylinder portable engines. The firm of Holmes of Norwich was established as agricultural and mechanical engineers in 1827. John Holmes produced his first portable in 1855, and this led to a series, of from 4 to 12 nhp. Later versions went up to 25 nhp. By 1867, when the firm was moving into making traction engines, they described themselves as 'engineers and machine manufacturers', and also made a wide range of agricultural machines.[8]

The origins of the much larger firm of Clayton & Shuttleworth are more exotic. Nathaniel Clayton Jnr had been apprenticed at the famous Butterley Ironworks in Derbyshire, later becoming a packet boat captain. In 1842 he joined his brother-in-law Joseph Shuttleworth, who owned a small boatyard on Lincoln's waterside. Together they established the Stamp End Works, starting as general engineers and ironfounders, with twelve men, two forges and one lathe. Earlier, Clayton had visited Howden's foundry and observed its operation. Their first important contract was for the supply of iron pipes to the Boston Waterworks at Miningsby. Iron girders for roofs and bridges were also produced. In 1845 they produced their first portable engine. Their next portable, in 1848, was claimed to save 25 per cent of the costs of thrashing by horses. By then the firm employed 100 men, and were in need of fresh capital. They were fortunate in securing the assistance of several wealthy local citizens as sleeping partners, the most important being Charles Seeley, who was a miller and corn merchant, with valuable coal-mining interests in Derbyshire. Seeley's firm contributed not cash, but a bank guarantee, which permitted continued expansion. At the Lewes Royal Show in 1852 Claytons took the prize for a movable, steam-powered thrashing machine in the face of competition from Garretts, Tuxford, Hornsby and Ransomes. By the time the Royal Show came to Lincoln in 1854 they employed 520 men and 80 boys.[9]

Hornsby's had been in business since 1815, but did not enter the steam engine market until Richard Hornsby's three sons had entered the business, producing their first portables and thrashing machines in 1849. The firm of Burton & Proctor had been established in Lincoln since about 1841, but did not enter the steam engine business until joined in 1856 by the dynamic Joseph Ruston, who had served his apprenticeship at a firm of cutlers in Sheffield. The firm was renamed Ruston, Burton & Proctor, and immediately began making steam engines – to the consternation of Mr Burton, without having orders in place. But such was the demand that Ruston was 'assembling portable steam engines as fast as he could buy or make the parts' (some parts were still being bought in from other makers, rather than made by the firm itself) and making threshing machines to go with them, and they sold. By 1860 the firm had won awards for their engines and threshing machines as far away as Gothenburg and St Petersburg.[10]

The portable engine had been developed mainly for farm use, pulled around the farm by horses. To make engines move under their own steam, more power, as well as a reliable transmission system, was needed. The resulting 'traction engine' was being developed in the 1860s, though widespread only after the 1870s, by which time it was used for a wide variety of purposes. The first traction engine built by Tasker of Andover, in 1869, was typical of the type. Named 'Hero', it required a man to sit at the front to steer the wheels, an engine operator and a man walking in front with a red flag (for legal reasons). It was of 12 hp power, with a single speed. On its first, 30-mile journey to Southampton Show on 31 May, it towed a four-wheeled water-cart, and was followed by a horse and cart carrying additional coal. The sight was a novel one: 'The loud rumbling and grinding sound of Hero's iron wheels on the rough road surface of chalk and flint heralded her coming and summoned little groups to cottage doorways or village street corners to see her pass.' The journey, with one slip into a ditch, and one accident involving the drive chain to the wheels breaking, took two days. Fortunately, the engine's journey was captured in photographs as it occurred. This pioneering effort was not immediately followed up. By 1878, out of 212 steam engines built by the firm only 18 were traction engines.[11]

The early evolution of the traction engine: Tasker's 'Hero' type, from their 1872 catalogue.

The spread of the threshing machine

The use of portable threshing machines also expanded enormously. The machines could now be driven by the new portable steam engines, and produce finished corn. In the RASE trials at Cambridge in 1841, eight machines had been entered, but only two had finished. But reliability improved, and the economic security felt by cereal farmers in the 1850s and 1860s bolstered demand. All the larger firms got into the business. In the late 1860s the RASE organised a series of threshing trials. The society's last great threshing machine trials were at Cardiff in 1872, when 29 machines from 13 manufacturers were entered.[12]

Garretts simplified their machines: instead of two fans, it had just one, with ducting directing the blast into other parts of the machine. Effective drum guards were introduced in 1874 around the sheaf feeder mechanism to prevent the awful accident of an operative getting a limb drawn into the threshing drum. A version of the thresher was developed in 1879 for southern European conditions. This roughly bruised the straw ejected from the bottom of the machine, reducing the dry brittle straw of southern Europe to animal fodder.[13] By 1865 Marshalls of Gainsborough were producing four types of threshing machine, their type 'A' winning the Gold Medal at the Cardiff RASE Show in 1872. Earlier, in 1868, they had been gratified to be awarded a Gold Medal at the Paris Universal Exhibition for their combination of 8 nhp portable engines and 4 foot (i.e 4 ft wide drum) threshing machine.[14] The Lincoln firm of William Foster, although not producing many threshing machines (31 in 1877; 52 in 1878), produced a type which may stand as the exemplar of the product by the late 1870s. The firm's No.1 Finisher/Thrashing machine threshed the grain, then riddled, winnowed and cleaned or awned it, following which it was given a second riddling and passed through a second blowing apparatus, then through a rotary separating screen to remove the small corn and seeds, before being finally delivered into sacks. The machine frames were made of English oak. The drum shaft and other wearing parts were made of steel, and the screen and concave sieve of wrought iron. Additional apparatus was developed to thrash other crops such as beans, peas, and rice, and finer-acting threshers were developed to act as clover hullers.[15]

The advance of the steam-driven thresher was at the expense of the threshers driven by horse gearing. In 1867 the judges at the Bury St Edmunds Royal Show had proposed that horse-worked machines should not be given trials in future, since they were old-fashioned. Although few horse-driven sets were to be sold by the 1870s, some farmers retained them, particularly in the North East. There, fixed threshing machines had always been popular, and there were times during the year when the need to thresh a small crop for animal feed would not have justified the raising of steam. Also, the cost of the horse-gears was small, and they were very durable, so that it cost little for them to be kept idle most of the time.[16]

Mechanical threshing by steam power had advanced very rapidly from its start in the 1830s. By the late 1870s, about three-quarters of the cereal threshing in England and Wales was done by machine.[17]

Steam Power II: The steam plough

Many people had given thought as to how to use the power offered by the steam engine for cultivating of land. Early attempts were made to use the steam engine for land drainage. The most spectacular experiment was that of John Heathcoat in 1837. He had developed an enormous steam engine, designed to be moved down the centre of a field, at the same time operating an endless flexible band made of iron strips, which operated two draining ploughs, one on each side of it. In total, with fuel, the engine weighed about 30 tons. Completed in 1834, it was tested by the *Highland and Agricultural Society* at Red Moss, in Lancashire, in 1836, and at Lochar Moss, Dumfriesshire in 1837. After the latter trials, which seem to have been a failure, for reasons which are unknown, the machine disappeared from the record. Indeed, it disappeared from view entirely, and it seems not unlikely that it was left in the field to rot, eventually sinking under its own weight into the soft ground. This early and most expensive failure probably cost Heathcote some £12,000.

MR. HEATHCOAT'S STEAM-PLOUGH.

Heathcoat's Steam Plough (1835). This worked on the principle of indirect traction. The ploughs were hauled across the field by an endless flexible band composed of linked strips of iron, wound onto a winding drum powered by the enormous engine, which remained on the headland. The engine and its apparatus weighed around 30 tons. (*Museum of English Rural Life*)

The failure of Heathcoat's engine demonstrates the principal difficulty in applying steam power to field operations: so heavy were steam engines that they risked either getting bogged down or compacting the soil unduly. The latter effect would lead either to the soil becoming impermeable and waterlogged, or to the formation of a hard 'pan' under the topsoil, which would inhibit drainage and prevent effective cultivation and sowing. The alternative to running the engines directly on the land was to devise some sort of cable system whereby the power of the engine could be applied to the land from a distance – from the sidelines, as it were. The two competing modes – direct and indirect traction – fought it out in the fields.

For some time, proponents of the direct system seemed to have found a solution to the weight problem. The most promising for some years was the system developed by Boydell, which was applied to engines made by various manufacturers, but mainly by the firm of Burrell. The result was the Burrell-Boydell engine. The system was patented by Boydell in 1846, and exhibited by him at the Great Exhibition five years later. It consisted in attaching a set of large horizontal hinged plates, made of wrought iron and wood, to the rim of the driving wheel of the engine. These would revolve with the wheel, descending into contact with the ground in advance of the wheel itself, which would thus never touch the ground. The weight of the engine would thus be spread onto the plates, reducing the pressure on the soil – the same principle eventually utilised by 'caterpillar' tracks.

This, Boydell's 'Endless Railway' system, was designed primarily for use on the poor roads of the time, but was quickly adapted to direct field cultivation. First applied in 1854, to an engine made by Bach, a Birmingham engineer, the Bach-Boydell system was exhibited at the 1855 RASE Show at Carlisle. In the following year Boydell exhibited his system on an engine built by Garretts of Leiston. But the most spectacular show was at Louth in 1857, when the *Illustrated London News* printed a fine double-page engraving of the Burrell-Boydell engine ploughing.

The appearance of the next Burrell-Boydell engine, at the Chester RASE Show in 1858, was a failure, since mechanical problems led to the engine being withdrawn, and it was on this occasion that the superiority of the competing cable ploughing system by John Fowler was demonstrated. But the Boydell system had already received much useful publicity, and there was something of a brief 'endless railway' mania, with a number of patents being filed. The endless railway wheels were applied to steam engines by some other makers, notable Clayton & Shuttleworth and Tuxford. But in all, between 1856 and 1862, Burrell, the largest maker, manufactured no more than about twenty of these engines, and, although they had some success overseas and in military wagon-hauling applications, they proved to be a dead end as far as agricultural field work was concerned. High cost, severe wear and tear, and sheer bulkiness were

Steam ploughing demonstration with the Boydell-Burrell engine, at Louth, Lincolnshire, in 1857. (*Illustrated London News*, 15 August, 1857)

NGINE, NEAR LOUTH.—(SEE NEXT PAGE.

probably the most important considerations. Direct ploughing would have to wait for the arrival of the tractor in the twentieth century.[18]

A different approach to the problem of applying steam power to direct ploughing made its appearance at this time, though with only limited success. This proceeded from the assertion that ploughing itself was inherently inefficient, producing a seedbed too deep for optimum plant growth, and that the action of the mouldboard as it was drawn repeatedly through the soil smoothed the surface of the subsoil, leading to the creation of a hard 'pan' underneath the topsoil, impenetrable to roots, thus inhibiting plant growth. It would be preferable to abandon the plough and develop a machine that could replicate the action of a man digging with spade or fork, or use some rotating digger to stir and aerate the soil. The most popular exponent of this view was the agricultural journalist and farmer, Chandos Wren Hoskyns. In 1849 he commented in his *Inquiry into the History of Agriculture* that: 'The application of steam power to tillage – will probably be vastly more successful if made to imitate the action of a spade'. In his later book, *Talpa: or the Chronicles of a Clay Farm* (1853), he poured scorn on the idea of horse ploughing, maintaining that the horse was only another form of power used to tow a plough horizontally – and, in his view, inefficiently.[19]

Hoskyns backed up his view by submitting a patent for a steam-powered rotary ploughing machine which operated by means of a toothed vertical wheel at the front of the plough. His machine was never constructed, but there was sufficient interest in this point of view, in the period before the problems of indirect steam ploughing were settled, for a number of makers to experiment with digging or rotary ploughing steam engines. The most spectacular machines used their power to dig the ground and move themselves sideways across the field as the digging progressed. These 'broadsiders' were produced in the 1870s by Thomas Darby of Pleshey, Essex. His machines were also made by the Agricultural and General Engineering Co. of London, by McLarens, and by Savages. These enormous machines were first produced by Darby in 1877, and he produced improved versions in 1878 and 1879.

In spite of good reports of the quality of the work done, the enormous cost of the machines, and their slowness in transport from field to field, militated against further development. The total number of digging engines made is not known, but estimates range from 18 to 23. Only four or five were sold direct to farmers, the rest being hired out for contract work. But Darby and his family persevered, and turned in the 1890s to a new system of direct digging, in which the steam engine propelled itself normally across the field on wheels, and used a system of horizontally rotating digging forks, hung at the rear of the engine, to do the ploughing. These 'tender diggers' had some success, and were made by various firms up until 1913. They were the ancestors of the 'gyrotiller' made by Fowlers in the 1930s, but made little impact on the farming scene before 1914.[20]

An expensive technological dead end: the Darby 'Pedestrian Broadside Digger'. This dug the land with the oscillating forks along its flanks, at the same time dragging itself sideways across the field. Various versions were made from 1877 to the 1890s. This, possibly the version built by J. & H. McLaren, would have weighed over 15 tons. (*Museum of English Rural Life*)

Experiments were also being made with 'indirect' ploughing, which in the longer run proved to be the solution to the problem of the application of steam power to the plough. This transmitted the power of the engine to the plough via a windlass, operating an endless cable, which pulled the plough from one side of the field to the other. The engine remained at the edge of the field, being moved along as more of the field was ploughed. Moveable pulleys at the corners of the field were necessary, and rope 'porters' to suspend the iron wire above the ground, so that it could move freely. At first of hemp, the ropes were soon being made of iron wire, and later of steel. The system required quite a lot of labour, but it worked. Many variations on this 'roundabout' system were made from its inception in 1850, when a farmer named Hannam got Barrett & Exall of Reading to make the first windlass to his order. One notable variation was invented by another farmer, William Smith of Little Woolston in Buckinghamshire. This made the windlass as two contra-rotating drums, by means of which the plough could be pulled to and fro across the field, dispensing with most of the pulleys. Smith's system was put into

production, mainly by Howards of Bedford, and sold by them for about ten years. In 1858 at least 39 sets were in use, and by 1862 at least 200 sets. Trials in 1858 showed that at least 3.5 acres per day could be broken up, although the system did not plough the land, using a three and five-point scarifier (termed by Smith his 'smasher') to stir the soil, rather than actually ploughing it.[21]

The greatest success in steam ploughing, which led to substantial machinery manufacture, was that of John Fowler. An engineer, he formed a partnership in Bristol with Albert Fry. Fowler & Fry made carts, wagons and railway rolling stock, and Fowler moved it into the agricultural field, developing horse-powered mole draining machines. He first showed his steam-powered mole draining plough system at the RASE Show in 1854 at Lincoln. In that same year the RASE offered a prize of £500 for, 'the Steam Cultivator which shall in the most efficient manner turn over the soil and be an economical substitute for the plough or the spade'. The following year, Fowler produced his first steam cultivation system.

Later improvements in 1856 were associated with the firm of Ransomes & Sims of Ipswich, who made a system to Fowler's order. This employed the first balance plough made by the firm, capable in trials of ploughing an acre an hour. Further development followed, and at the Chester meeting of the RASE in 1858, Fowler was awarded the Society's £500 prize. Three other entrants competed (Boydell; Crowley of Newport Pagnell; and Smith's system, as manufactured by Howard of Bedford).

William Smith's ploughing system, as manufactured by Howards of Bedford. This, the 'roundabout' system, was invented by a farmer, William Smith. This cable-hauled system needed only one portable engine, but it required much labour to reset the heavy guide pulleys after the plough completed each furrow, and so lost favour to the double engine system developed by Fowler. (*Museum of English Rural Life*)

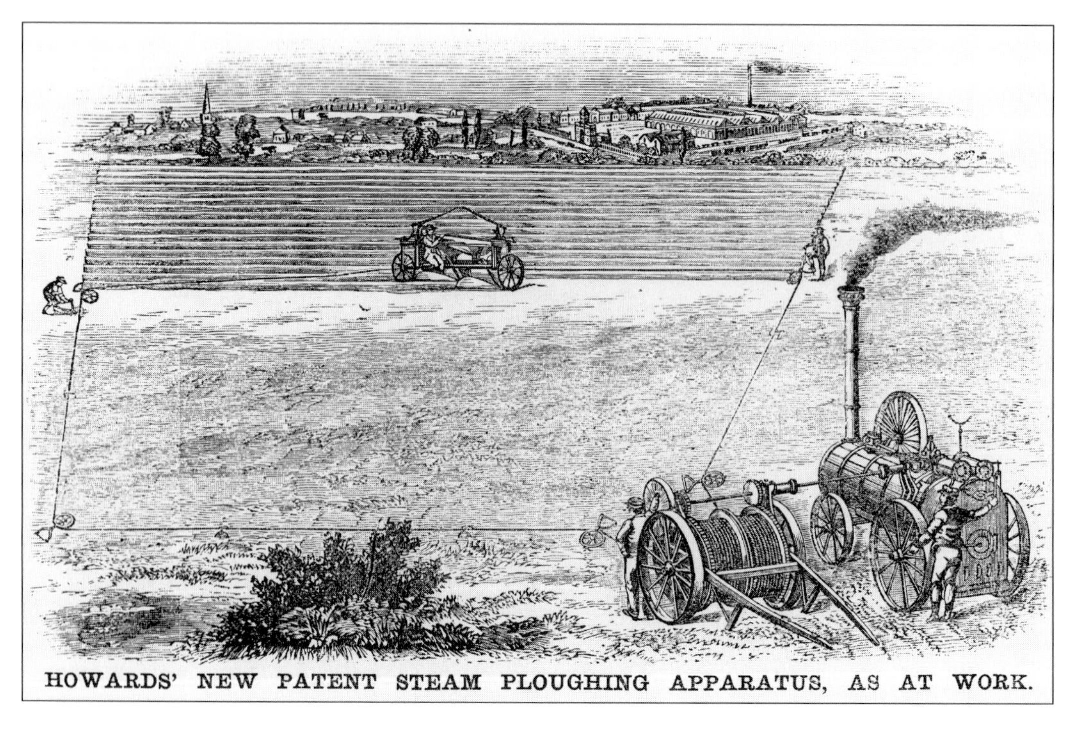
HOWARDS' NEW PATENT STEAM PLOUGHING APPARATUS, AS AT WORK.

Fowler's first ploughing system (1862), as manufactured by Kitson & Hewitson of Leeds. Like Smith's, it was a 'roundabout' system, worked by one stationary steam engine. This is one of the earliest known photographs of steam ploughing. (*Museum of English Rural Life*)

The Boydell failed mechanically; the Crowley omitted to supply its own motive power, and was disqualified; the Smith/Howard and Fowler systems both did good ploughing work, but Fowler was adjudged to have won on grounds of economy.

Following the success of his designs, Fowler formed a company in 1859 to supervise the construction of his system under licence – the Steam Plough Royalty Company, with a nominal capital of £16,000. The firm of locomotive builders of Kitson & Hewitson of Leeds was contracted to build Fowler ploughing engines; Robert Stephenson & Co. and Ransomes continued to supply other parts of the tackle. In 1860 Fowler began to build his own engines and equipment, on a site in Leeds, under a new partnership styled Fowler & Hewitson. The first engines made in the new Steam Plough Works were delivered in November 1862, to the Wakefield Steam Ploughing Company. By then Fowler listed three types of ploughing engine in his catalogue (two at 14 nhp, and one at 12 nhp). He collaborated with major firms (Clayton & Shuttleworth, Burrell, Ransomes), collected royalties, and refined his system, filing further patents.

By 1864 the Fowler system was clearly ahead of the field. In that year he showed a 14 nhp set and the innovation of a double-engined set of two 7 nhp engines at the RASE Show at Newcastle. The double-engine system used the power of both engines, drawing the plough in turn with the aid of a large horizontal winding wheel on the underside of each engine. It was an early attempt at what became later one of the firm's main products. The single engine set was priced at £875 and the double engine set at £1,066. At these prices only the largest farmers or landowners could purchase a

Fowler's final system: the 'double-engine' system, of the 1880s. Two large traction engines on opposite sides of the field draw between them a double-ended balance plough, whose shares are alternately raised and lowered as the direction of travel is reversed. The engines move along the headlands as the ploughing progresses. (*Museum of English Rural Life*)

ploughing engine, but many were sold to ploughing contractors. In the fifteen years to 1875, over 100 hiring companies were formed. By the end of the 1860s the Steam Plough Works was producing about 150–160 sets a year, about a third for export.[22] Fowler's double-engined system became the main way in which steam was used for ploughing in the last quarter of the nineteenth century, its virtues apparent from an engraving prepared by Jabez Hare for Fowler's catalogues of the 1880s (above).

The system required less heavy manual labour than the roundabout system, with its back-breaking task of moving the heavy guide pulleys as the plough completed each furrow. The drawbacks of the system were the high cost, which could in total be about £2,000, and the fact that not many farms possessed the large, flat, rectangular fields for which the system was most suited. An enquiry by the RASE in 1867 found that about 200,000 acres were being cultivated by steam power. This was an impressive achievement in such a short time after its introduction, but still a very small part of the nearly 14 million acres of tillage in Great Britain in that year.[23]

Reaping machines

Harvest time has always been hard, and expensive in labour. Until the middle of the nineteenth century, cereal crops were reaped by hand, usually by knife or sickle. In the first half of the century, the bagging hook and the scythe were increasingly adopted, and these made possible some economy of labour,[24] but on the eve of the Great Exhibition hand reaping was by far the most usual way of cutting crops.

Attempts to mechanise the harvest went back centuries. Pliny mentioned a type of cart, pushed into the crop by cattle, with a row of serrated teeth on its leading edge, which tore the upper stems of the corn from the stalks. Although it had no moving parts, this was similar to a technique used in the twentieth century by the makers of the Australian 'stripper-harvester'. Later machines oscillated between push and pull types, and between the use of cutting knives and rotating cutting wheels. Among these may be mentioned Gladstone's reaper of 1811, Smith of Deanston's reaper of 1831, and Mann's reaping machine of the early 1830s. The most promising British development was Bell's reaper in 1831. This contained many of the features which would be found on the successful reapers after 1850. But Bell's secrecy militated against further development, and his ideas were not taken up by any manufacturer.[25]

Development had proceeded more rapidly in the USA. McCormick had marketed his improved reaper with success since 1845, and both he and another American, Hussey, exhibited their machines at the Great Exhibition. These, and especially the McCormick, caused a sensation. The MP and agricultural writer Philip Pusey wrote a brief special report on it for the RASE in 1851, in which he observed:

> The machine, drawn by two horses, and carrying two men, a driver and a raker, cut the wheat about eight inches from the ground with the utmost regularity. The horses found the work light, though the machine was cutting at the rate of 1½ acres per hour, making fifteen acres in a day of 10 hours. The raker, standing behind the driver to rake the cut wheat from the platform certainly had to exert himself; but it is obvious that he and the driver, who only has to sit on the dicky, might well exchange places from time to time ... This trial was witnessed by many farmers, and no fault was found with the work ... [he observed that the machine worked less well on uneven ground] ... But, on this level land, it was wonderful to see a new implement working so smoothly, so truly, and in such a masterly manner ... the most important addition to farming machinery since the threshing-machine first took the place of the flail.*

McCormick followed up this triumph by sending his machine on a demonstration tour of the country. On 11 September 1851 it was tried on a farm at Much Wymondley, near Hitchin, Hertfordshire, where: 'Great numbers attended to witness its operations. The sodden state of the ground presented considerable difficulties, but the energies of the American conductors overcame all, and the giant reaper pursued its way triumphantly up the side of the field ... Of the general success of the experiment there can, however, be no doubt.'[26]

The other American competitor to McCormick's reaper in 1851 was

* P. Pusey, 'On Mr McCormick's Reaping-Machine', *JRASE* xii (1851), p. 160.

The beginning of transatlantic competition for the British industry: McCormick's reaper. This is the 1862 model, which threw out the cut corn to one side. The version shown at the Great Exhibition in 1851 had deposited the cut corn behind it. (*Museum of English Rural Life*)

Hussey's. This was a simpler affair, without a reel to hold the corn up to the cutting bar, and with only one man operating it, who combined the functions of driver and raker. It was cheaper (at £30) than the McCormick (£42), and this may account for the fact that it was taken up more rapidly than the latter. Only in 1857 did McCormick agree with the firm of Burgess & Key, of Brentwood, Essex, to manufacture his reaper under licence. More Burgess & Key reapers were sold in the UK than any other make between 1852 and 1862 – perhaps in all about 2,800. This contract was the making of Burgess & Key, who had established their works at Brentwood in 1856, and the company gained a national reputation for reapers and mowers in the 1860s. But in the 1860s they fell into dispute with McCormick, who was then forced to import most of his machines direct from Chicago. This made them uncompetitive with the British domestic makes of Cranstoun, Samuelson, as well as the imported Walter A. Wood machines.

The Hussey reaper was, however, taken up by the firm of Wm Dray in the USA, and by several British makers: Wray & Son; Robert Cuthbert

Reaping machine, 1851, by McCormick

The grain reaper had been long in evolution, and evolved continuously in the second half of the nineteenth century, but McCormick's design of 1848 is generally held to be the first really successful machine, the prototype of most subsequent developments. It was the one exhibited at the Great Exhibition at the Crystal Palace in 1851, and between 1855 and 1859 the company sold 15,000 machines.

The 1851 McCormick, from Pusey's report to the Prince Consort, *JRASE* (1851).

The main points of interest of the machine were two; the serrated, saw-edge cutter, which was a long reciprocating knife working through protruding fingers, and the revolving reel to bring the standing crop up against the cutter. A particular feature of the cutter knife was that its serrated edge had teeth inclining alternately to left and right, so that the stalks would be caught in the teeth on each sideways movement of the knife. The gearing for the machine was driven from the main travelling wheel. The gearing shown in this illustration of the 1851 model is exposed to view; later models covered it up, as the gears were vulnerable to the dust and dirt of the reaping field. Unlike Bell's earlier machine, which was pushed into the crop by the horses, McCormick's was pulled by two horses. It was reported that the McCormick reaper shown at the Great Exhibition could easily cut fifteen acres in ten hours. The original design, like that of the Hussey reaper, used manual labour to sweep the cut stalks off the platform. The Hussey machine was a one-man effort, with the driver also operating a hand rake to sweep the cut stalks from the machine, but the 1851 McCormick had two men; the driver, and a man standing at the back of the reaper, sweeping the stalks off with a hand rake. By 1862 McCormick had modified his machine to sweep the cut crop off the machine to one side.

& Co.; Gardner & Lindsay, and others, doing particularly well in Scotland. Apart from McCormick, the only other competition from the USA was from the firm of Walter A. Wood, who were said to have sold over 2,000 of their reapers in the four years from 1858 to 1862. Finally Samuelsons of Banbury took up the manufacture of another American reaper, the Dorsey. This had the innovation of a revolving rake, which swept the cut stalks off the platform at intervals, leaving them lying on the ground in separate groups, conveniently placed for tying into sheaves by the following labourers, and this type of reaper became the standard type henceforth. Samuelsons became one of the largest domestic manufacturers, making 1,200 self-rake reapers by 1868.[27]

The significance of the reapers of 1851 was not lost on British manufacturers. At the next year's RASE Show, reapers were exhibited by Thompson, Woods, Holmes, Ransome, Mason, Burgess, Crosskill, Samuelson, Garret and Howard. Continuing product improvement was made, notably in the adoption of side-delivery of the cut corn, and other manufacturers entered the field, notably Bamlett of Thirsk, Yorkshire. His machines dated from 1859, and were at first made by Picksley, Sims & Co. of Leigh, Lancashire, and by Kearsley of Ripon, but a little later he established his own works. In 1872, the firm of Harrison, McGregor (also of Leigh) entered the business. By the 1880s the dominant firms were Burgess & Key, Harrison McGregor, Hornsby, Dray, McCormick, and Hussey. The first three were British, the latter American. Mechanical reaping had spread rapidly to a dominant position; by the late 1870s, at least two-thirds of the cereal harvest was got by machine.[28]

REAPING MACHINES.

First Prize Awarded this Year!

WILLIAM DRAY & Co. have again obtained the First Prize for their PATENT IMPROVED HUSSEY REAPING MACHINE, at the Meeting of the Bath and West of England Agricultural Society.

Prizes awarded in 1854 by—

The Royal Agricultural Society of England;
The Bath and West of England Agricultural Society;
The North Lancashire Agricultural Society; and
The Stirling Agricultural Society.

Numerous prizes have also been awarded to the same in previous years.

A Descriptive Catalogue may be had on application to the Manufacturers,

WILLIAM DRAY & Co.,
AGRICULTURAL ENGINEERS
SWAN LANE, LONDON.

The Hussey reaper, made and imported into England by William Dray, in 1855. A simple and cheaper alternative to the McCormick reaper. (*Museum of English Rural Life*)

Mowing machines

Haymaking was just as hard work as cutting corn, as anyone who has had to carry hay bales can testify. However, its mechanisation lagged behind that of corn harvesting. Although the principle was similar, haymaking was more complex, involving cutting the grass, scattering it to dry, gathering it in lines (windrows), and then into cocks, and finally collecting it into the haystack.

The mechanisation of haymaking involved the development of three separate types of horse-drawn machine: a mower; a haymaker, or 'tedder', to spread the grass around for drying in the field; and a rake to push the grass into windrows and also rake it up into large mounds for carting. In 1814 Robert Salmon patented a mechanical hay-tosser which in essentials remained the type for the remainder of the century. A workable horse-drawn rake was available around London by 1831 (Middlesex was recognised as the model district for haymaking), being manufactured by Wedlake of Hornchurch, Essex. Further rakes were patented by Garrett in 1842 and by Ransome in 1843. In 1846 the RASE held its first trials of haymaking machines. At the society's 1857 show haymakers by Nicholson, Barrett Exall & Andrewes, Smith & Ashby, H. A. Thompson, Ransome, Wyatt, Samuelson, and Lane, were shown. Mowing machines had been much improved, following the experience of making reapers. One of the newer entrants in the field was Bamletts of Uttoxeter, which had been founded in the 1830s as manufacturers of stoves, but turned to reapers and mowers, and showed its paces at the RASE in 1860. By this time mower technology was more or less perfected, as was the horse rake, of which 26 types were exhibited at the Royal Show at Taunton in 1875. By the 1870s the farmer had the choice of many different models of mower, haymaker, and rake, and the technology did not alter much for the next twenty years, in the view of the judges at the Darlington Show in 1895.

By the end of the 1870s, the leading mower makers were Wedlake (Romford); Barrett, Exall & Andrewes (Reading); Smith (Stamford);

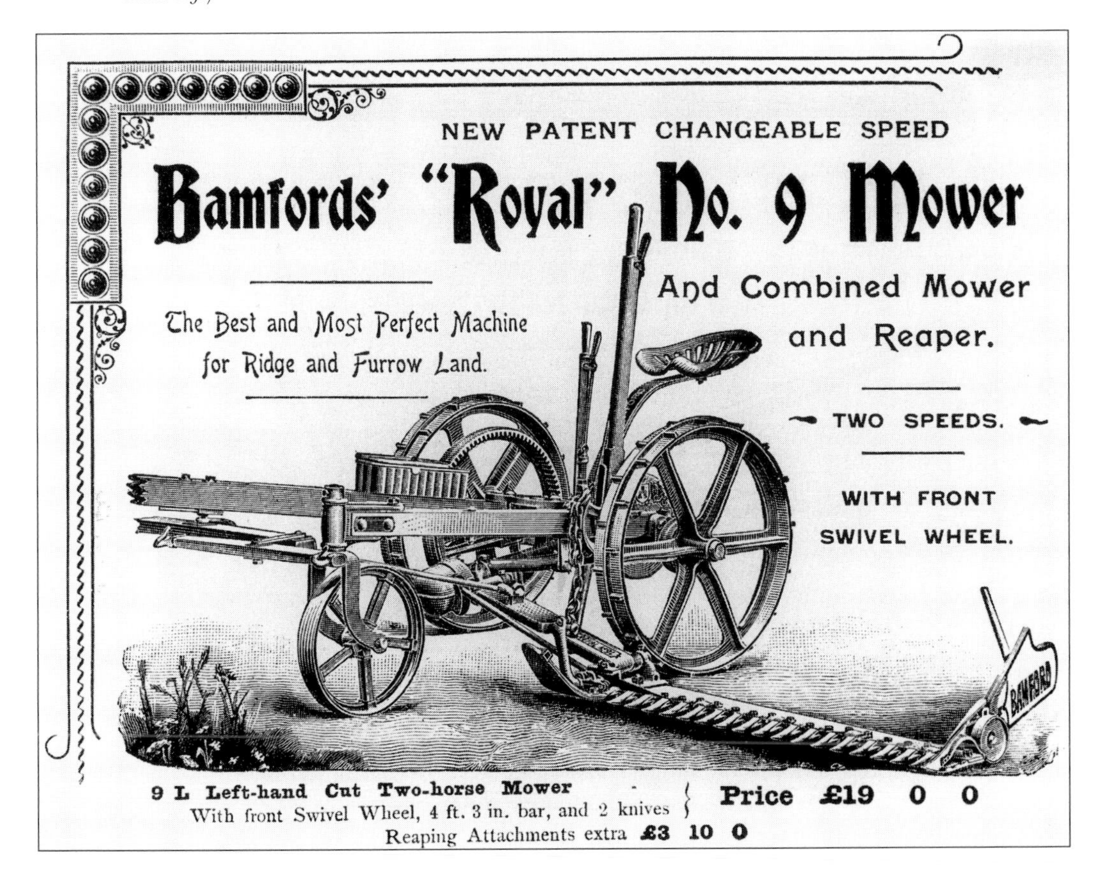

NEW PATENT CHANGEABLE SPEED
Bamfords' "Royal" No. 9 Mower
And Combined Mower and Reaper.
The Best and Most Perfect Machine for Ridge and Furrow Land.
TWO SPEEDS.
WITH FRONT SWIVEL WHEEL.
BAMFORD
9 L Left-hand Cut Two-horse Mower
With front Swivel Wheel, 4 ft. 3 in. bar, and 2 knives
Price £19 0 0
Reaping Attachments extra £3 10 0

Bamfords 'Royal' grass mower, from an 1897 catalogue. (*Museum of English Rural Life*)

Horse-drawn mowing machine, 1860

By 1860 grass cutting had been revolutionised by the use of the horse-drawn mower. Instead of using dozens of men, advancing in line across the hayfield cutting the crop with scythes, one man sitting on a mowing machine drawn by two horses could work for about six hours a day, cutting about an acre an hour. The only drawbacks were that a laid crop slowed the work, and it was heavy work for the horses to pull the machine through a standing crop of grass, so that it was not uncommon to bring along a relief horse, to rest one of the others.

The machine illustrated on page 71 shows the principles of construction and operation. Inside one of the land wheels, a circular ratchet wheel turns as the mower moves along, and engages an enclosed pinion wheel. From this, a drive shaft rotates to operate the cutting mechanism. This is enclosed in the cutter bar, which is suspended to one side of the mower. A common length of bar was 4 feet 6 inches. At the front of the bar, short fixed 'fingers' protrude forward. Inside these, there is a knife blade running the full length of the bar. The blade consists of a thin flat strip, to which are riveted V-shaped sections. The knife is free to oscillate to and fro inside the fingers, driven by the shaft from the land wheel. The resulting motion gives a reciprocating cutting action similar to that of scissors. The cutter bar could be raised out of work by a lever, attached to the cutter bar by a chain; this could be adjusted so as to change the height of the cutter bar above the ground, to accommodate different sorts of crop. At the far end of the cutter bar, a divider called a track board steered the outer edge of the crop into the path of the cutter bar, and a small land wheel kept the cutter bar slightly above ground level.

Workable mowers dated from the 1850s; one of the most successful firms, A.C. Bamlett of Thirsk, was founded in 1857. A feature of their mowers was the addition of a front wheel to the mower, which gave a steadier motion. Rapid improvement in mower construction gave a machine which could withstand the difficult conditions of mowing on uneven ground, using moving parts close to the soil, and with great stresses on the cutter bar as it was forced into heavy, damp or laid crops. Two-speed gearing became available (the lower speed being used for clovers and easy work), and the crank shaft and gearing could in some models be enclosed in an oil bath; without this, frequent oiling was essential in the dirty and dusty conditions of the hayfield.

Although cutting grass with the scythe survived until the First World War, it was a residual technique. Mowing by horse-drawn machine dominated the hayfield until the advent of the tractor after the Second World War. Now the cutting mechanism of a mowing machines is attached to the back of a tractor, but it is recognisably the same as that of the 1860s.

Samuelson (Banbury); Harrison McGregor (Leigh); Bamford (Uttoxeter); Bamlett (Thirsk); Wood (imported from the USA); Howard (Bedford); Nicholson (Newark). Smaller makers were such as Brenton (Polbathic, Cornwall), and Williams (Rhuddlan). There were also some other minor imported makes such as the Eureka. By the 1880s, there was general agreement that haymaking by hand was little more than an archaic survival, and the larger firms were undoubtedly making many mowers each year. The firm of Harrison McGregor, formed in 1872 principally for

The mechanisation of haymaking: Bamfords 'Lion' horse hay rake, from an 1897 catalogue. (*Museum of English Rural Life*)

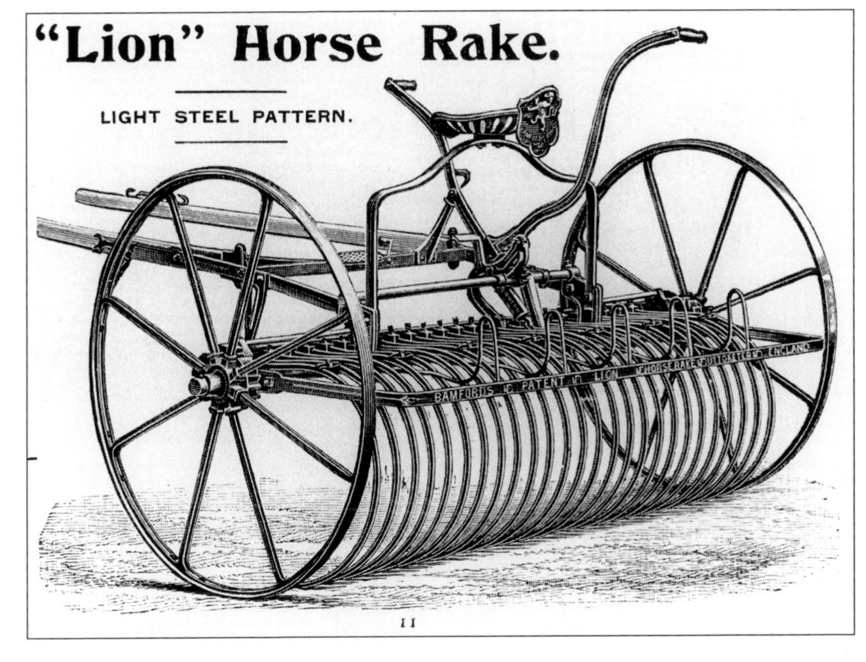

Taskers Elevator, from their 1872 catalogue. A further stage in the mechanisation of building corn stacks and hay ricks. This model folds down for travelling.

the manufacture of mowers, recorded sales of £28,222 in the calendar year 1875. At around £40 a mower, this meant an output of 705 a year.[29]

The only other innovation of note before the 1880s, was the hay or straw elevator, which began to appear in the 1860s. Powered by a horse gear, it eased the manual labour of building a hayrick.

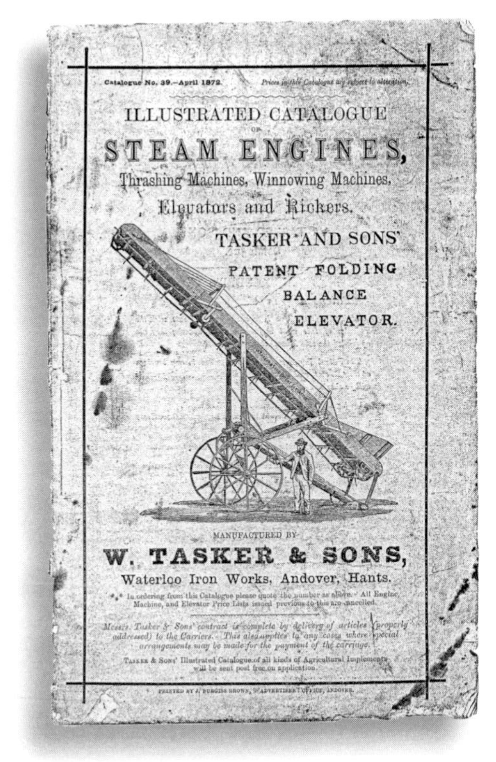

Catalogue No. 39.–April 1872. Prices in this Catalogue are subject to alteration.

ILLUSTRATED CATALOGUE OF STEAM ENGINES, Thrashing Machines, Winnowing Machines, Elevators and Rickers.

TASKER AND SONS' PATENT FOLDING BALANCE ELEVATOR.

MANUFACTURED BY W. TASKER & SONS, Waterloo Iron Works, Andover, Hants.

*** In ordering from this Catalogue please quote the number as above. All Engine, Machine, and Elevator Price Lists issued previous to this are cancelled.

Messrs. Tasker & Sons' contract is complete by delivery of articles (properly addressed) to the Carriers. This also applies to any cases where special arrangements may be made for the payment of the carriage.

Tasker & Sons' Illustrated Catalogue of all kinds of Agricultural Implements will be sent post free on application.

PRINTED BY J. BURGISS BROWN, "ADVERTISER" OFFICE, ANDOVER.

Improved field machinery

One product which may be classed as practically new in 1850 was Crosskill's clod crusher. This was a heavy horse-drawn roller, consisting of some 23 large serrated iron rings, equipped with both vertical and horizontal teeth, revolving independently on the axle of the roller, to chew up the clods on newly ploughed land. It had appeared at the 1840 RASE Show at Cambridge, and won the Society's Gold Medal in 1846. It was immediately popular. By the end of 1850 Crosskills, of Beverley, in the East Riding of Yorkshire, had made a total of 2,478. By 1851 the firm employed 240 men, and further Crosskill specialities had come into production. But the firm was already in financial difficulties, and the founder, William Crosskill, retired from the business in 1856. The clod crusher was subsequently marketed by other firms, and became widely used, in conjunction with or as substitute for, the Cambridge roller. In 1914 an encyclopaedia of agriculture noted of the Crosskill clod crusher that: 'For Pulverizing hard clods upon heavy soils it had no equal', although it noted that since by then much of the heavy land in England had been seeded down to grass, it was not as much used as formerly. It was illustrated, and was the subject of the leading article, in Crosskill's own publicity newspaper, with extensive testimonials quoted as to its efficiency.[30]

Crosskill's clod crusher, patented by William Crosskill in 1841. (From *Crosskill's Implement Newspaper of Practical Information connected with Agriculture for extending the more general application of Machinery*, Part I, July 1848, p. 5)

Improvements to existing machines and implements continued. The seed drill had reached its peak of perfection by 1860. In the form of the 'Suffolk' drill developed by James Smyth of Peasenhall (Suffolk), it reached its ultimate level of development. By 1874 this small firm was making 600 drills a year.[31] Its principal features were a cup feed, lever lift mechanism and shoe coulters. Different sizes of machines were available, from a seven-row light-land drill, to the 18-row, top of the range 'Eclipse' drill.[32] This drill did not alter perceptibly in design before the firm ceased business in 1969.

The first patents for machines for planting potatoes had been taken out in the 1850s, but it was not until the early 1870s that workable machines appeared on the market. Scottish manufacturers were to the fore in this movement, on account of the larger proportion of arable land under potatoes in Scotland. In England, a successful model was produced by Coultas of Grantham.

Ploughs were under continuous improvement. Ransomes had a notable success with their new Model RNF, which carried off the top prizes at

PATENT

"SUFFOLK" LEVER CORN DRILL.

THE above engraving illustrates the well-known "Suffolk" Drill with which J. S. & SONS' name has been associated for the last eighty years, and which has served as a model all over the World.

These Drills are usually fitted with wood levers, but wrought-iron levers are supplied at an extra cost (when desired).

For hilly land J. S. & SONS strongly recommend their LEFT END COGGING APPARATUS, which avoids the thick and thin sowing, by applying a smaller cog wheel at one end of the cup barrel than on the other end, thus using the quicker speed when going up, and the slower speed when going down hill.

For stony land it is advisable to have coulters with wrought-iron stalks, which obviate the liability of breakage.

The cog wheels can be changed much quicker than in most Drills and with much less trouble, owing to the absence of screws and such like appliances, which are always liable to corrode, and choke with grease and dirt, and prove a source of constant annoyance to the Drillman.

All J. S. & SONS' Drills are furnished with boxes of an improved shape, which dispense with the extra tins necessary in most Drills to drill out seeds to a very small quantity.

These Drills can be had with a Fore Carriage Steerage (see remarks on page 3), but the prices quoted are simply with shafts, being now usually preferred for land ploughed in small stetches only, whereas for broadwork our "Eclipse" or "Nonpareil" are generally supplied.

Smyth's 'Suffolk' drill, which became the dominant type in British farming. Smyth, a small firm in a small village (Peasenhall, Suffolk) eventually drove a world-wide trade. It remained a one-product firm throughout.

the Newcastle RASE Show in 1862, and was known thereafter as the Newcastle plough. By attaching different shares and breasts to the frame, the plough could be adapted to a number of purposes, from deep work or cross ploughing to stubble breaking, ridging, subsoiling and paring. Eight versions of a double furrow derivative were also on offer because the savings it made possible in horses and labour were by that time attracting attention. It was assiduously promoted by the firm at shows and ploughing matches, and held the position of one of the major types of plough, being manufactured until the 1940s. Four of the versions of it from Ransomes 1869 catalogue are shown on page 78.

Howards also used their Champion plough to perform a variety of

Ransomes' all-iron 'Newcastle' plough, 1864

Manufacturers continued to make ploughs with wooden frames throughout the nineteenth century, but from about 1840 turned increasingly to all-iron ploughs. Ransomes of Ipswich began their range of all-iron ploughs in 1843, with their improved 'Rutland' design, and the famous YL (Yorkshire Light) plough. The latter became the dominant design for Ransomes' horse ploughs for the next 100 years, and was still the basis of some tractor ploughs in the 1980s. It formed the basis for the 'Newcastle' plough, with which Ransomes won four out of the six first prizes at the 1864 Royal Agricultural Society of England Show at Newcastle. The Newcastle plough is illustrated here; it contains the essential features of almost all general-purpose horse ploughs thereafter.

The main features of the plough shown on page 77 are (*from right to left*): the draught chain, for attaching the plough to the horse harness; the wheels; the skim-coulter; the knife coulter; the plough mouldboard with its pointed ploughshare at its front end; the plough handles. In work, the larger wheel runs in the furrow cut by the coulter during the previous passage of the plough, whilst the smaller runs on the unploughed land. The skim coulter is in form a sort of miniature plough, which precedes the knife coulter, turning a very small furrow of about 1 inch depth from the left side of the furrow slice to the right; the vegetation thus moved is deposited at the bottom of the furrow, when the main slice is turned, instead of being left to protrude between furrow slices, as it is apt to do otherwise. The function of the knife coulter is to make a vertical cut in the sward of vegetation in advance of the ploughshare, to begin the process of detaching the furrow slice from the soil. The share, which detaches by a horizontal cut the furrow slice from the soil below it, is only held on to the front of the mouldboard by the pressure of movement; shares often only lasted a few days in work, and were cheap and easily replaced. The mouldboard inverts the furrow slice; its form is convex, the twist in it gradually increasing towards its rear. For the furrow to be inverted unbroken, it has to be in contact with the mouldboard throughout its length.

While the ploughshare and mouldboard were of cast iron, with the share chilled to become self-sharpening (Ransomes' patent having expired, all makers now supplied chilled cast-iron shares), the frame of the plough was of wrought iron. This gave it greater strength and lightness than a wooden frame, and made it longer lasting. The use of wheels made it very easy to keep the plough working in a straight line, with minimal steerage from the ploughman; it was said that a two-wheeled plough could be operated by a boy, and would run almost without holding it. The tasks, with the standard form being capable of many variants, wheeled or non-wheeled, the individual parts for which could be ordered simply by quoting the relevant pattern numbers. In an age in which the postal and telegraphy services had become ubiquitous and highly efficient, and parts could be despatched by the makers to the customer's nearest railway station (usually free of delivery charge), these attempts to tailor the product to local conditions had obvious attractions for farmers.[33]

Finally, the 1850s and 1860s saw a great increase in the use of scarifiers, cultivators and extirpators, which are all variations of a basic type of implement designed to improve the soil tilth, either after ploughing, or instead of the conventional plough. Bentall's Broadshare plough, Coleman

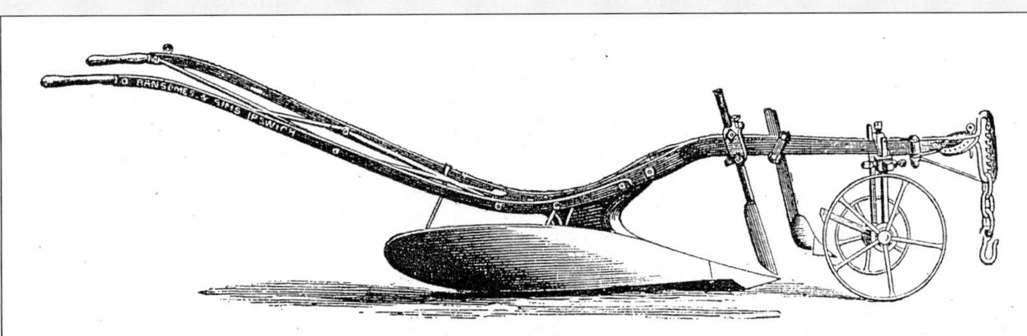

FIG. 3.—RANSOMES' NEWCASTLE FIRST PRIZE PLOUGH.

RANSOMES' NEWCASTLE FIRST PRIZE PLOUGH, fitted with long match breast, adapted for turning unbroken furrow slices, is represented by fig. 3, whilst fig. 4 is the same plough with breast adapted for general farm work, with two subsoilers behind. The fact that these ploughs have won since September, 1864, 367 All-England champion prizes, and £1300 in money prizes, may best be left to bespeak the practical merits of the implement. The subsoilers can be applied to fig. 3, so that the only difference between the two ploughs is the form of the breast. There is, however, a great variety in the form of breasts, there being several hundred patterns, including those of ploughs made for foreign use. The Newcastle ploughs are universally admitted as the most perfect ploughs now in use, and the following may be given as a summary of their advantages. (1.) They are simple in construction, strong and durable, and light both in weight and draught. (2.) Although light in weight, the beam, frames, and handles are so connected as to give the implement perfect rigidity, so that the ploughman has entire control over it in ploughing. It is otherwise when there is any want of strength and consequent twisting, for then the ploughman is apt to set the various parts of the plough so as to work one against the other and cause extra draught. (3.) The handles, as will be seen from the engraving, are braced diagonally, which not only increases strength, but further, gives the greatest power to the ploughman over his plough. (4.) They can be fitted with different breasts to suit different soils, or the same soil in different conditions. This is all-important in practice, as an additional breast is equivalent to another plough. (5.) The shares and other wearing parts are of the best quality.

Newcastle plough, *Agricultural Gazette*, July–August 1875.

knife coulter could be replaced by a disc coulter, which was very effective in cutting through thick or matted turf. As ever, this was a basic design whose components could be varied to suit different requirements. The Newcastle plough could be supplied with more than a hundred different patterns of mouldboard and ploughshare, making it suitable for all types of soil, work and local conditions.

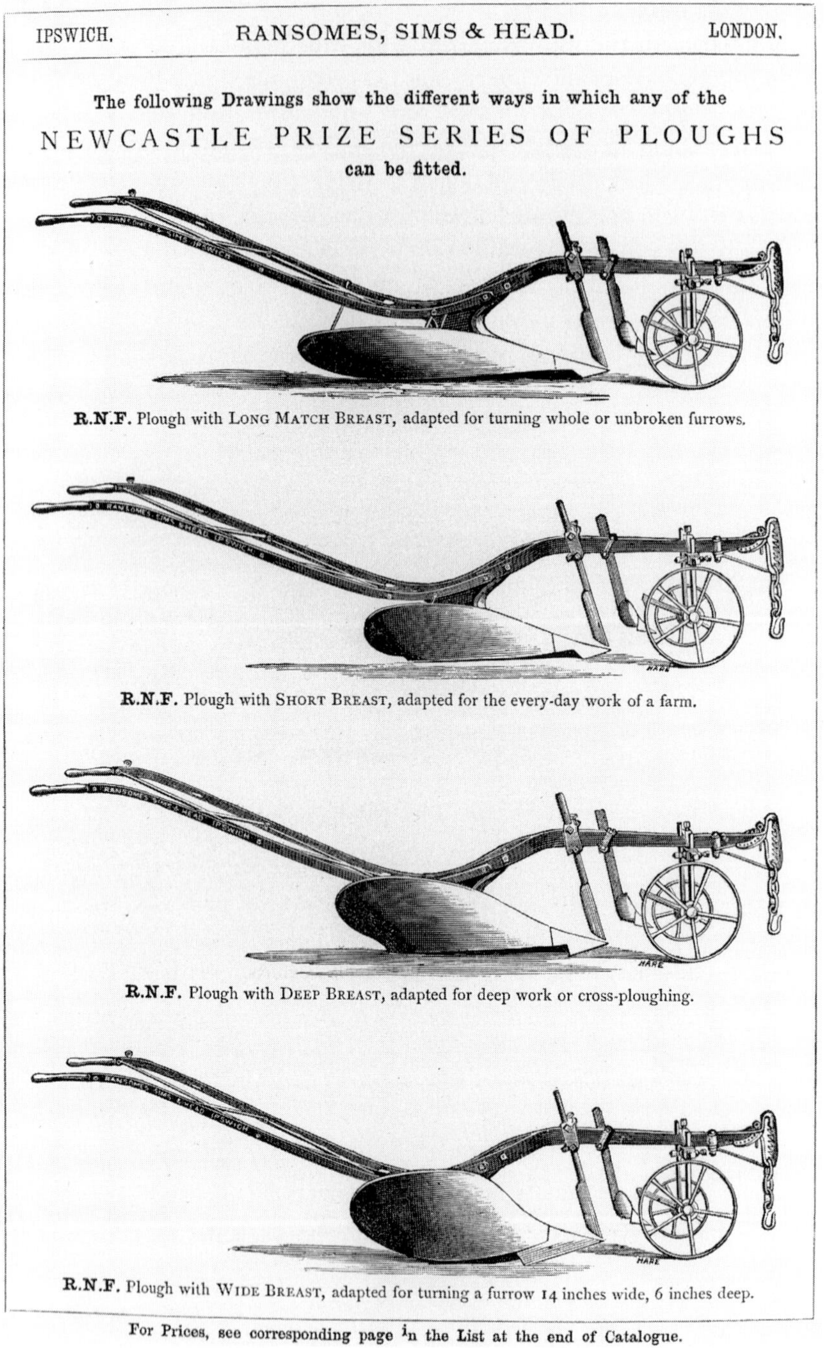

IPSWICH. RANSOMES, SIMS & HEAD. LONDON.

The following Drawings show the different ways in which any of the

NEWCASTLE PRIZE SERIES OF PLOUGHS

can be fitted.

R.N.F. Plough with LONG MATCH BREAST, adapted for turning whole or unbroken furrows.

R.N.F. Plough with SHORT BREAST, adapted for the every-day work of a farm.

R.N.F. Plough with DEEP BREAST, adapted for deep work or cross-ploughing.

R.N.F. Plough with WIDE BREAST, adapted for turning a furrow 14 inches wide, 6 inches deep.

For Prices, see corresponding page in the List at the end of Catalogue.

Ransomes' 'Newcastle' plough, named after the firm's success at the Newcastle Show of the Royal Agricultural Society in 1862. Page from the firm's 1869 catalogue. (*Museum of English Rural Life*)

& Morton's cultivator, and Tennant's grubber took the prizes at most exhibitions, although Ransome (Ipswich), Finlayson (Ayr), and Carson & Toone (Warminster) also produced well-known implements.[34]

The growth and prosperity of firms

The growth in demand, largely from the home market, led to considerable growth in the industry. A national list of manufacturers by William Kent in 1867 recorded 540 firms, to which may be added probably another 100 omissions. The *Post Office Directory of Engineers* recorded 724 for 1870 and 889 for 1890. How many of these firms were principally concerned with agricultural engineering is uncertain. In the 1870s it was probably around 700. Professor Collins suggests that in 1900 some 900 firms might have claimed a connection with agricultural engineering, although most of these were very small, and fewer than half of these were actually engaged in manufacturing.[35]

The size of the leading firms also increased. By 1868, only seven years after its inception, Fowler's Steam Plough Works in Leeds was employing about 1,000 men, the rate of growth tailing off thereafter, to reach a total of 1,200 in 1886. Sales in 1884 (a bad year) were 'less than £30,000'. In 1886 the partnership was changed into a limited liability company, with a nominal capital of £680,000, made up of 50,000 preference shares at £10 each and 18,000 ordinary shares at £10 each. Financial information is lacking from the earlier partnership era, but the first annual general meeting of the new company in 1887 revealed that it had a turnover of £157,400, though with a net profit of only £4,622. Fixed assets were valued at £216,000, with work in progress at home of £843,000, and of £79,700 at the firm's foreign depot at Magdeburg. A dividend of 2 per cent on the preference shares only was declared. This was not a very healthy financial situation, but by 1891 the turnover had risen to £275,000, and the net profits to £42,000. In only thirty years of operation, this was a truly impressive performance.[36]

Fowlers was the last of the great pre-1914 firms to enter the business of agricultural machinery. The older firms grew more slowly. Garrett's first recorded steam engine production was in 1866–67, when 98 were made. From March 1874 to April 1875 it was 299. The firm was also making large numbers of threshing machines; some 300 a year by 1874–75. The first record of the financial value of the firm's sales is in 1879, when it was £64,000. Foster of Lincoln, which had branched out into the manufacture of agricultural machinery in 1856, became a limited liability company after the death in 1876 of its founder, William Foster. The first annual report of the directors in the new company revealed sales in 1877 of £18,383, with a net profit of £3,227. It was probably smaller at that time than the lesser known firm of Harrison, McGregor, which showed sales in 1875 of £28,222.[37]

Marshalls began as a one-man firm in 1848 and was incorporated as a company in 1862, by which time it employed 21 men and 10 boys. At the first general shareholders' meeting in May 1863 it was noted that the growth of the firm had been held back by lack of production facilities, so that 'fully one half of their manufactures were purchased instead of being produced in their own premises'. A large works extension was proposed. In 1865 a new iron foundry, smithy and boiler shop were completed, and the number of employees rose to more than 100. The value of the company's assets at that first AGM was put at £11,652. This had risen to £21,211 the next year, and in 1865 the capital was increased by the issue of a further 1,500 shares at £20 each, increasing the total capital to £50,000. By 1866/67, the company's assets were valued at £67,343, and profits were £3,244. The annual dividend paid to shareholders had begun at 5% in 1863, but was 10% in each of the ensuing four years. At the 1873 AGM, the value of raw materials and work in progress was put at £71,264. The total value of the firm's capital was higher than this, since this figure did not include the value of buildings and machinery. The dividend was raised to 15%.[38] *

All these firms were dwarfed by Ransomes, which had become the leading firm before 1815, and retained its lead up to 1914. Even in 1849, it had employed well over 1,000 men. In a market which was becoming more crowded, Ransomes made more room for itself by changing its product structure. In the 1850s it had a variety of irons in the fire, including some delicate instrumentation for Greenwich Observatory. It later shed its railway work, passing it on to the new firm of Ransomes & Rapier in 1869. It also shed its food-preparation machinery business, selling this to Hunts of Earls Colne. It reshaped its lawnmower business, dropping the manufacture of Budding's mower, and acting as agents for Green's 'Silens Messor' ('silent reaper') mower. In 1866 the firm produced its own mower, the Gear Automaton, and in 1872 built a new, separate Lawnmower Works. The firm briefly manufactured harvesting machinery in the 1860s, but then wisely dropped it, in view of the fierce competition in the home market, and the efficiency of American imports. In 1871 (a bad year), its home sales alone were £90,364, and these were dwarfed by export sales of £124,377, making total sales of £214,741. By 1874 export sales had recovered to £217,534, and total sales were £306,778, although this was a peak year in the trade cycle, and sales fell in ensuing years.[39]

The earliest production records extant for Richard Hornsby of Grantham show that the firm made only 27 threshing machines in 1854. In that year only one of the machines was for export, but by 1874 production was 192, of which 54 were for export, mainly to continental Europe. Along the way, in January 1858, they had sold one to the Emperor of Austria. They were also making a large number of steam engines – 499

* The conversion of partnerships into companies makes better financial data available for the historian, since accounts were then made public at the annual shareholders' meetings. However, before 1914 the legal requirement was that the minutes and balance sheets of only the first AGM had to be published. Some companies took this literally; some published annual results, even though they were not compelled to do so.

in 1874 . The size of a smaller firm, with a largely local and some regional trade is indicated by the sales figures of Nalder & Nalder, of Challow (near Wantage, Berkshire) which were £10,660 in 1870, but rose to £21,397 in 1875 and to £21,361 in 1879, thus avoiding the mid/late-1870s slump; this perhaps reflected the firm's limited exposure to foreign markets. Its main product was the threshing machine, production of which was 33 in 1870, 97 in 1875, and 88 in 1879.[40]

Trade catalogues reflect a continuous search for new products. Even a middling-sized firm such as Humphries of Pershore, which concentrated on threshing machines, offered a wide range of goods in their 1875 catalogue – portable and fixed threshing machines, portable steam engines (though these were manufactured for the firm by Clayton & Shuttleworth), nine other types of steam engine (also manufactured by Claytons), portable clover machines and cider presses, sack-lifting barrows, horse 'works' (horse gears), riddles, and various other machines or implements of their own manufacture: Cambridge rollers, ploughs and harrows, chaff cutters, haymakers and horse rakes, horse hoes and subsoil ploughs, seed sowing machines, hand drills, and elevators A slightly smaller range was by the early 1870s offered by Taskers of Andover, who may by that time have been employing about 100 men. Their 1873 catalogue offers portable steam engines and traction engines (including a Hero-type self-moving engine, referred to above), elevators, threshing machines, winnowing machines, a range of food preparation machinery, and reaping and mowing machines.[41]

Marketing

The methods of marketing to the home market established before 1850 were retained, and were built upon. The showing of implements and machines at local and the major national agricultural shows – especially the Royal Show, but also the Bath & West, and Highland and Agricultural continued to play an important part in firms' marketing methods. The Royal Show was especially prestigious. It was also useful because it was peripatetic, reaching a wide audience. Its *Journal* had a low circulation, but an influential readership, and it reported the results of the tests on machines and implements carried out in public at each Royal Show. If favourable, and especially if accompanied by a medal from the society, this publicity would be assiduously referred to in the firm's publicity material. This was not to say that the firms were always happy with the tests as conducted by the society. Criticisms were many and vociferous, partly directed at the way the tests were conducted, partly based on the practical observation that by *c.*1850 many manufacturers' machines were so similar that it would be misleading to give prizes to any particular one, and a feeling that there was no clear correspondence between winning prizes and selling machines. After 1855 the RASE changed the style of

Clayton & Shuttleworth's portable steam engine and threshing machine, 1855. Lincoln Cathedral in the background. (*Museum of English Rural Life*)

the tests, and concentrated on particular new machines, or classes of machines, rather than testing everything offered to them. Whatever their misgivings, the manufacturers continued to make as much use as possible of recommendations, judges' comments and medals from the Royal and other shows.[42] Thus when Clayton & Shuttleworth won the award for Best Portable Threshing Machine at the RASE Show at Lincoln in 1855 they were quick to put out a flyer proclaiming the fact, with an illustration of the thresher, and extensive quotation from the remarks of the judges.

In 1875 Ransomes Sims & Head of Ipswich claimed that they had won a total of 367 champion prizes at All-England ploughing matches since 1864, due to the success of their 'Newcastle' plough.[43] The 'Prize Newcastle Ploughs' stood at the head of their full-page advertisement of the late 1860s. The prize system undoubtedly benefited both existing manufacturers and new entrants to the business::

> Although many of the leading implement houses, such as Garretts and Ransomes, were already established by 1840, the start of the Society's sequence of shows and trials can reasonably be taken as a major catalyst in the emergence of a fully fledged British agricultural engineering industry between about 1835 and 1870. Major firms such as Crosskill, Howard, Smythe, Gooch, Bentall, Wedlake, and

The front page of Ransomes & Sims' general product catalogue in the late 1860s, drawing attention to their 'Prize Newcastle ploughs'. (*Museum of English Rural Life*)

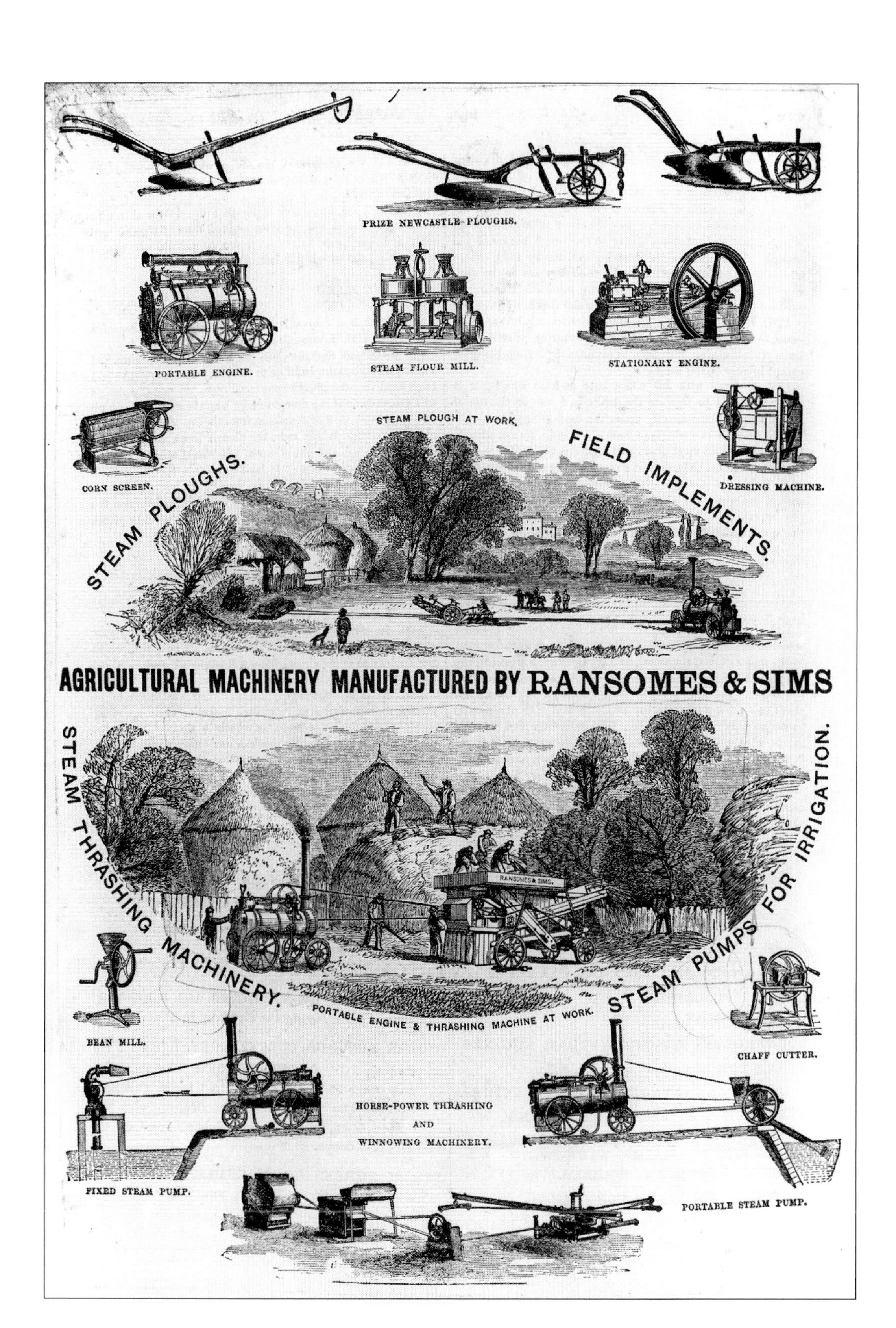
PRIZE NEWCASTLE PLOUGHS.
PORTABLE ENGINE.
STEAM FLOUR MILL.
STATIONARY ENGINE.
CORN SCREEN.
STEAM PLOUGH AT WORK.
DRESSING MACHINE.
STEAM PLOUGHS.
FIELD IMPLEMENTS.
AGRICULTURAL MACHINERY MANUFACTURED BY RANSOMES & SIMS
STEAM THRASHING MACHINERY.
STEAM PUMPS FOR IRRIGATION.
RANSOMES & SIMS
PORTABLE ENGINE & THRASHING MACHINE AT WORK.
BEAN MILL.
CHAFF CUTTER.
HORSE-POWER THRASHING AND WINNOWING MACHINERY.
FIXED STEAM PUMP.
PORTABLE STEAM PUMP.

1884. 1914.

IN CONNECTION WITH THE

Royal Agricultural Society's Show,

HELD AT SHREWSBURY

THIRTY YEARS AGO.

Three **HORNSBY BINDERS** met twenty-five Machines by the Leading Makers of the World in a week of Competitive Trials, held near Shrewsbury.

The important points adjudicated upon were:—

SIMPLICITY AND EFFICIENCY OF CONSTRUCTION;

WEIGHT, LIGHTNESS OF DRAUGHT, STRENGTH AND COST OF BINDING MATERIAL;

SECURITY OF KNOT AND EFFICIENCY IN BINDING.

Whilst during the week one or more Machines of every other maker failed to satisfy the Judges and were excluded from further Trials, **all three HORNSBY BINDERS** worked day by day, and were **all in competition to the end.**

On the last day of the Trial, August 13th, the result was announced—

First Prize of £100

To RICHARD HORNSBY & SONS, LTD., GRANTHAM.

Two years earlier, in 1882, the **HORNSBY BINDER** had won the

FIRST PRIZE OF £100

at the **Highland Show at Paisley,** with 883 points against 789 and 770 of the next highest Competitors.

At the **Australian Trials** of 1884, the **HORNSBY BINDER** competed against the Leading British and American Binders, and came off

FIRST EVERY TIME! **AND NOW IN 1914 IT IS UP-TO-DATE AND STILL ALWAYS LEADING.**

Hornsby's 1914 binder catalogue, using the publicity generated by their triumphs at Royal Shows thirty years earlier.

Hunt came into national prominence by exhibiting at the shows, participating in the trials, and winning the Society's prizes.*

The winning of medals, and publicising the fact, went hand in hand with the use of testimonials from satisfied users. This, like the use of the prize system, went on up until 1914. The Hornsby harvesting machinery catalogue of that year referred to the 'hundreds of letters of praise of the

* N. Goddard, *Harvests of Change: The Royal Agricultural Society of England, 1838–1988* (Quiller, 1988), p. 71.

celebrated British machine [their binder]' that had been received, and they made full use of the results of the RASE Shows at which they had triumphed as long ago as thirty years previously.

The promotion of products was also done by the agents to whom the firms entrusted their products. These remained an important way of marketing, especially the less expensive products such as ploughs, harrows, reapers, mowers and food preparation machinery. For the larger items, especially steam engines and threshing machines (often sold together as a package), the order books of the major large firms show that they usually dealt directly with the customer. Agents ranged from local blacksmiths upwards; frequently they were themselves manufacturers, although often in a small way. For example, John Wallis Titt of Warminster, Wilts, who began in business in 1867, advertised the firm as 'engineers and implement makers', and also acted as agents for Wallis, Haslam & Steevens, of the North Hants Iron Works. In 1875 the description had altered slightly, to 'agricultural engineer and implements agent', with a note that they were 'Special Agents' for Brown & May, of Devizes. Another Wiltshire firm, T. H. White, of Devizes, originating in the union of two former ironmongery businesses, seems to have traded almost entirely as a dealer, although there was a brief period when manufacturing was done (1855–65). The firm was still, in 2003, an important dealer in the local and regional trade in agricultural implements and machinery.[44] The balance between doing agency work and one's own manufacturing was a changing one for many small firms, who probably liked to think of themselves as manufacturers, even if they were largely reliant on selling the goods of other manufacturers.

The terms of business varied. Most manufacturers offered free delivery to the railway station of the customer's choice, with a discount for immediate payment. Thus the 1911 catalogue of the East Yorkshire and Crosskills Cart and Wagon Co., specified: 'Terms: cash on invoice, less 5%. Credit only by arrangement', but only offered free carriage to any station within 200 miles of the works. In view of the bulky nature of the product (two- and four-wheel carts), this was perhaps understandable.[45]

The variety of business with which an agency firm could be involved was considerable. Lack of Cottenham (Cambridgeshire), who were in business (as Fison & Lack) from the 1860s described themselves a little later as: 'Hydraulic and Artesian Well Engineers, Makers of Air Compressors, Hydraulic and Steam Pumps, Boiler and Deep Well Pumps. Agents for Agricultural Machines, Gas Oil and Steam Engines (Best Makes).' In practice most of their daily work seems to have been jobbing and repair work for all sorts of machines, from fire engines to reapers. It was a firm of modest size, whose nominal share capital in 1916 was £10,000, of which only £5,200 had been issued. Another Cambridgeshire firm, Hunts of Soham, who described themselves as 'millwrights, machinists and implement agents', and who began in business in 1830, seem to have been mainly concerned with work on corn and drainage mills, but also repaired agricultural machinery.[46]

There were also dealers who did not engage in manufacturing, and were specialist traders. One such was the firm of Gliddon & Squire, of Barnstaple, Devon, begun by Michael Squire as agricultural traders in 1875. The firm dealt in agricultural equipment and 'manure' (artificial fertiliser), the cake and manure stores being located on the nearby quayside at Pottington. Although they do not seem to have been manufacturers, they re-badged a Knapp corn drill as one of their own at some point in the late nineteenth century. The business obviously flourished, since in 1903 the firm acquired spacious and handsome new two-storey showrooms in Tuly Street nearby. The firm claimed to have the largest stock of implements and machines in the west of England, and seems to have dealt with all the leading manufacturers. The firm is still in business, but no longer occupies its smart Edwardian showrooms, which still stand.

Finally, the national and local press was not neglected in the search for markets; the trade even acquired a paper of its own in 1875, the *Implement Manufacturers' Review* (later styled the *Implement and Machinery Review*). This, which continued publishing until the 1960s, carried much advertising from the manufacturers, and many articles describing new machines and implements, as well as developments in the trade at home and overseas. As well as this, there were the older farming newspapers such as the *Agricultural Gazette*, which carried lengthy articles about new machines, or descriptions of new factory premises.

The spacious showroom and warehouse in Tuly Street, Barnstaple, Devon, of Gliddon & Squire (established in 1875), suppliers of agricultural machinery and fertiliser. (*Peter Dewey*)

The rise of the export trade

The British agricultural machinery industry led the world. Britain had been the first country to industrialise, and the technical superiority of the British makers was clear. Even if in places there existed some product or form of technique superior to that of the British, it was not on a sufficient scale to form a serious threat. In 1851 the British manufacturers could look out on the world with some confidence and self-satisfaction.

The export of British agricultural machinery was helped by the steady growth of the continental railway network from the 1840s (frequently built by British contractors such as Brassey or Peto). Also, there were significant improvements in ships and shipping services: the first steam ship (Brunel's *Great Britain*) crossed the Atlantic in 1843, and in the ensuing decades communications were eased by the laying of trans-oceanic telegraph cables. The transatlantic cable was successfully laid in 1866, and by the early 1880s the global telegraph network was almost entirely complete, except for a few far-flung places, and, 'Nearly every other place of real importance not in the heart of China could be reached overland or under sea.'[47]

Farm machinery exports were first returned in the British trade statistics in 1853, when £77,379 of 'agricultural implements' were recorded. This was very small in relation to total British exports of £98.9 million.[48] The largest destination for British implement exports was Australia, with £35,192. This reflected the interest at the time in emigration to Australia, and firms responded to this. Crosskills even marketed an emigrant implement kit. Their catalogue of 1851 offered an 'Emigrant Implement Box', with a plough, harrows and various accessories in kit form ready for self-assembly. In 1853 they even offered prefabricated four-room log cabins for use in Australia, an outstanding example of marketing ingenuity.[49]

Thereafter, exports rose almost without interruption, although the statistics are not completely satisfactory. The only category recognised in 1853 was 'agricultural implements'. This remained so until 1862, when 'machinery and millwork, not being steam engines: Agricultural', was added. At the same time, the statistics recognised for the first time the export of steam engines (including locomotives), but no distinction was made between those for non-agricultural use and the agricultural ones. It was only in 1882 that the category of steam engines for agricultural purposes was included. So we have figures for the export of agricultural implements from 1853, to which we can add other agricultural machines from 1862, and agricultural steam engines from 1882. At this point, the export statistics became comprehensive enough to cover all types of agricultural machinery exported. But since at the first enumeration of agricultural steam engines in 1882, their value appears as the enormous total of £1,023,985, we can presumed that their export had been building up for many years, and either went unrecorded, or was conflated with the other, non-agricultural steam engines from 1862. It is also certain that much of the export of agricultural machinery (as opposed to implements) was omitted until 1862, when it suddenly appears as worth £215,090.

With these caveats in mind, we can analyse the export trade. In 1853 it was as follows.

Table 4.1 *Exports of British manufactures of agricultural implements and tools, 1853 (£)*

	£
Russia	4,070
Hanse towns (Hamburg, Bremen, Lubeck)	5,495
France	1,201
British North America	7,370
British West Indies	2,128
USA	1,765
South America	8,471
British South Africa	5,332
Australia	35,192
Other countries	6,335
Total	**77,379**

Source: *Annual Statement of Trade*, 1853.

Clearly most of the export markets were at a considerable distance. The fact that European trade was low reflected the fact that continental countries were still heavily protectionist. It was not until the mutual reduction in tariffs in the 1850s, culminating in the 1860 Anglo-French trade treaty (the Cobden–Chevallier treaty) that a system of fairly free trade was established. In the ensuing period, the trade steadily grew. By 1860 the total value of exports was £254,481, and the biggest market was now Russia (£83,294), followed by Australia (£55,505), 'other countries' (in effect, southern Europe and Asia), £55,247, and north and west Europe (£29,770). The rest of the export market was made up of (in order) South America, British South Africa, and Canada.

After the further liberalisation of world trade in the 1860s, British exports of agricultural machinery grew apace. By 1874 they were valued at £1,051,734. This may be compared with the grand total of all British exports, which was £240 million. Comparison of these figures with those for 1853 shows that the agricultural machinery exports had risen proportionately much faster than export trade generally. They had risen also faster than the value of exports of all machinery, the total of which had been £2.0 million in 1853, and had risen to £9.8 million in 1874.[50]

By the early 1870s, the major international exhibitions (London in 1851 and 1862; Paris in 1867) had made British companies well known abroad, and they responded to the opportunities thus created. Positive efforts were made to produce foreign-language material. Ransomes had a French catalogue by 1853, adding German and Dutch by 1857 and Spanish and Russian by the early 1870s. Howards had a French catalogue by 1856, and Garretts had French, German, Danish and Spanish catalogues by the early 1860s.[51] But exporting has always been a volatile business, and trade was interrupted by the political and military upheavals of this period. Most notably there was the Crimean War (1854–56), the American Civil War (1861–65), the brief Prusso-Danish war of 1862 involving Schleswig-Holstein, the domestic upheavals in Russia following the end of the Crimean War, leading up to the Emancipation of the Serfs in 1865, the Austro-Prussian War of 1866, and the Franco-Prussian war of 1870–71.

The change in destination of exports of agricultural machinery can also be shown:

Table 4.2 *Exports of agricultural machinery, 1860 and 1874*

	North and west Europe	*Russia*	*British North America*	*South America**	*British South Africa*	*Australia*	*Other*	*Total*
1860								
(£)	29,770	83,294	3,211	17,020	10,434	55,505	55,247	254,481
%	11.7	32.7	1.3	6.7	4.1	21.8	21.7	100.0
1874								
(£)	587,767	142,309	– †	45,731	20,272	131,780	123,875	1,051,734
%	55.9	13.5	–	4.3	1.9	12.5	11.8	99.9

* including Cuba

† no data

Source: *Annual Statements of Trade.*

The largest market in 1860 had been Russia, taking over the lead from Australia. However, the Russian market failed to grow as fast as the others. The strongest growth was shown by northern and western Europe, which was taking over half the machinery exports by 1874. The American Civil War (1861–65) was responsible for an astonishing but brief rise in exports to Egypt, as cotton planters strove frantically to take advantage of the world shortage of cotton, formerly exported from the southern states of the USA. Exports of machinery to Egypt accordingly rose, from £4,874 in 1862 to £92,281 in 1865, falling back sharply, to £2,784, in 1866. By the late 1860s, north and west Europe dominated the export market and accounted for 60 per cent of the total in 1868. In the world trade boom of 1870–73, total exports soared to over £1 million. At the peak, in 1874, before the ensuing slump had affected exports, Europe's dominance was clear. The largest customer overall was Germany, which accounted for sales of £324,601, almost one-third of all exports. The Russian market had also revived, at £142,309, so that Germany and Russia together accounted for 44 per cent of total exports. South America had been a minor, if steady, market; British North America very small. Exports to the USA had always been very small, and by 1874 they had disappeared, as had those to British India, which had always been a small market.

The larger firms had paid attention to the export market from an early stage, and cultivated foreign visitors. In association with the London International Exhibition of 1862, Ransomes 'entertained a distinguished company of Foreign Visitors', who 'inspected the Works, also an outdoor display at Westerfield, of Machinery and Implements in motion'. In 1866 the firm was visited by a Japanese commission. It made 'a very important display at the Great Paris Exhibition of 1867 with a full range of their manufactures at rest and in motion'. The firm also attended the Rio exhibition of 1866, the Santiago exhibition of 1869, 'sweeping the board of the principal Medals and Awards'. Their partner, Mr Head, made annual visits to Europe. Mr R. C. Ransome made 'several journeys to America and one to the USA'.[52]

Exporting was becoming more important. This can be seen in one of the few unbroken surviving output series for an individual firm for this period, that listing the sales of threshing machines by Richard Hornbsy & Co. In 1854, the firm sold 27 machines, of which only one went for export. Ten years later total sales were 84, of which 36 were exported. In 1873, sales had risen to 208 machines, and now exports dominated, being 166 of this total. Throughout this period, the main markets were in south-east Europe, at first Vienna, and latterly Pest (the other half of what was later named Budapest) predominating.[53]

The main way in which the companies sold goods abroad was through agents. This had advantages, chiefly that of local knowledge, and of not tying up capital in depots and stock. Goods were despatched in response to orders, only occasionally being sent speculatively. Agents were responsible for import duties and freight charges, in return for a discount or commission on sales. In this way exporters minimised risk, and avoided tying up their own capital, although on occasion they did extend credit. Overseas branches were occasionally found, but usually on a limited scale. Clayton & Shuttleworth established major branches in Vienna and Budapest as early as 1858. Overseas branches would in turn deal through agents. The great exception to the agency system was that of Fowler, most of whose European business was done through their Magdeburg branch, established in 1872. Subsidiaries of the Magdeburg branch were set up at Prague and Budapest in 1887. Direct sales attracted a 10 per cent commission, with smaller commissions for any sub-agents employed. Agents often acted for a number of different firms, although they generally avoided competition in individual lines. Thus Stefan Vidats in Hungary, himself a machine maker, stocked Howards harrows, Crosskills clod crushers, Barrett Exall & Andrewes hay rakes, Garretts seed drills, Hornsbys dressing machines and Samuelsons chaff cutters. Another benefit of using agents was that they sent their orders in blocks, which allowed manufacturers to set up machine tools and production lines for long runs of the same article, although the full benefits of this were not felt until the 1880s and 1890s, as firms modernised their factories and adopted more modern machine tools.[54]

Conclusion

The third quarter of the nineteenth century was the golden age of the British agricultural machinery industry. At home, products pioneered in the preceding decades came to fruition, and were perfected. In their new, reliable and cheap form they appealed to the farming customers, who were themselves making good profits. Attracted by the new technologies on offer, they eagerly bought the new products – iron ploughs, drills, steam engines and threshing machines, reaping and mowing machines, barn machinery, and corn dressing machines. By the early 1870s, agriculture had been technically modernised out of all recognition to what it had been in the middle of the century. At the same time, the reduction and removal of import duties, which culminated in the Cobden–Chevallier Treaty of 1860 and its subsequent replications in Europe led to a new market overseas. This was eagerly seized on by the machinery manufacturers, and their exports grew much more rapidly than the overall total of British exports. Unthreatened by overseas competition, British exports acquired dominance, especially in Germany and Russia. The native producers could not as yet compete on quality or price. The only market in which the British producers exhibited weakness was the reaping and mowing market (especially the former) in Britain itself. The incursion was from the USA, which by the 1870s had a significant and growing proportion of British sales. But it may be enquired how long the market could grow at home: as agriculture became modernised, the market would increasingly become a replacement market. In that case, the only solution would be to find a new range of products for home consumption, or to pursue exports with even greater vigour. In the event, the manufacturers adopted the latter course.

CHAPTER FIVE

Exports to the rescue, 1875–1913

Problems in the home market

By the mid-1870s, arable farming in Britain had adopted many of the new mechanical aids: the iron plough was triumphant; seedbeds were prepared by iron harrows and cultivators; a great deal of cereal harvesting and hay mowing, as well as almost all threshing, was done by machine; and food for animals was also prepared using new machinery. In addition, the great mid-century drainage projects, executed using machinery and implements made by the farm machinery manufacturers, were now completed.[1] This allowed full use to be made of the new farming implements. This widespread adoption of farm machinery meant that large-scale expansion of the machinery manufacture sector could not be sustained, and manufacturers would have to look for new markets and new sources of income, either by diversifying into non-agricultural products, or finding work which was akin to their existing lines.

Apart from the mowing of grass and the preparation of animal feed, non-arable farming – the production of meat and milk and other livestock products – was less suitable for mechanisation and modernisation. This obviously presented problems for machine manufacturers, exacerbated by the 'agricultural depression' in arable farming which began in the early 1870s, and went on until the mid-1890s. A large rise in imports of cereals, especially wheat, from the USA and other newly settled territories, led many farmers to abandon cereal growing and to lay land down to permanent grass, mainly for the raising of cattle. This process was further stimulated by the growing urban demand for meat and milk, as consumers were increasingly able to afford these 'superior' foods, as against 'inferior' bread and potatoes. The result was a shift in farming away from arable and towards grassland. Between 1873 and 1894 the total arable acreage in Britain fell from 18.1 million acres to 16.2 million acres, with a corresponding rise in the amount of permanent grassland. The crop hit hardest by foreign competition, wheat, fell from 3.49 million acres to 1.93 million acres.[2]

The incomes of arable farmers fell. Lord Ernle referred to the period after 1874 as the 'Great Depression' in farming, which touched bottom in 1895, after which wheat prices began to rise again. In practice, although some former arable lands were successfully turned over to profitable use in producing meat and milk for the towns (Essex, where many Scottish farmers took over former arable farms, was a notable case of this), a lot of the former arable was merely left to tumble down to grass, and the resultant turf was of low quality. The upshot was that arable farming was in financial trouble after the early 1870s, and grassland farming could not make up the shortfall. As a result of falling prices, total agricultural income faltered, and then declined heavily. Gross farm output in England, which (at current prices) has been estimated at £132.5 million in 1873, had fallen some 25 per cent by 1894. In fact the physical quantity of output produced in 1894 was only 4 per cent less than in 1873. But this was little consolation for farmers, whose incomes and profits both fell. The result for the machinery manufacturers was a straitened home market, in which the mainstay of their growth to date, the mechanisation of arable farming, was severely undermined.[3]

As in other industry sectors, combination of firms was one response to the straitened circumstances of the early 1870s. Farm machinery manufacturers also now formed a trade association, the Agricultural Engineers Association, in 1875. Initially it concentrated on pressing for lower import duties in the industry's foreign markets, and the reduction of railway transport charges, and of charges to exhibitors at the Royal and other agricultural shows. Less publicised was the policy of attempting to maintain a standard list of agreed selling prices for machinery and parts, with pressure being brought to bear on members who failed to adhere to the list prices.[4]

A search also began for new, diversified products, sometimes spin-offs from existing lines, as in the case of the traction engine, a self-moving, high-power version of the old portable engine which found a market outside agriculture, in general industry and transport. Both Burrells of Thetford and Savages of King's Lynn, as well as making traction engines for general haulage, produced engines which were designed as showmen's haulage and power plants, for the travelling circuses and fairs which were an increasing feature of late Victorian life. Other firms produced similar engines, and experimented with oil engines in the 1890s. Taskers moved into engines for road haulage, while Garretts' engines also moved into industrial uses. Some did well by diversifying into the livestock sector, such as with the cream separator, which had been developed in Denmark in the 1880s. In the next decade Listers of Dursley (Gloucs.) made many thousands of these under licence, before developing their own model. In the 1880s Samuelsons of Banbury, one of the leading British makes of reaping machines, found themselves in difficulties, and may well have survived the worst years by turning to the production of turnip cutters.[5]

Another successful diversification was the development of the lawnmower by Ransomes, This went back as early as 1842, when Ransomes had built the machine patented by Budding under licence. Since then the firm had developed its own types, and the home market was steadily expanding, as the population and towns grew. In 1876, Ransomes sold £6,258 worth of lawnmowers on the home market, but only £264 worth of exports. The home sales of mowers represented about 3 per cent of the total sales of the firm. But by 1894 home sales were £20,799 (exports only £942), about 9 per cent of the total sales of all the firm's products.[6]

The shift to exports

Many manufacturers also turned to foreign markets. The export trade was liable to be more volatile than domestic markets, as when the strong trade and export boom, which peaked in 1873–74, gave way to a correspondingly severe depression. The late 1870s was a gloomy time for industry and the economy: national income fell, from £1,186 million (at current prices) in 1873 to £1,048 million in 1879, after which it began to rise again.[7] But by this time European countries were turning their backs on free trade. The newly united Imperial Germany in 1879, France in 1881, and Italy soon afterwards, all turned to protectionism. The USA and Russia were already protectionist, and their import duties rose even higher in the 1880s and 1890s. Furthermore, the general level of prices, largely stable since the late 1840s, now showed a general long-term decline, punctuated by the usual trade cycle fluctuations. This downward trend, due to increased competition from newly industrialising countries and cheaper commodities of all kinds, continued until near the end of the century. In 1873 the Board of Trade index of wholesale prices (1900 = 100) had been 145.2; it fell as far as 88.2 in 1896, and rose thereafter until the First World War.[8]

In spite of the problems associated with developing an export trade, the British manufacturers looked on it as the way forward from the early 1870s. The conventional view is that exports came to dominate the industry in the period between 1875 and 1914. As Dr Grace has written: 'Where figures are available, it is clear that the export trade dominated production in most areas of farm machinery.'[9] Since there are no satisfactory national export figures for farm machinery before 1882, one has to draw evidence from the production records of the manufacturers themselves. While few complete series of home and foreign sales for the leading firms still exist, that for the largest firm, Ransomes, has survived (table 5.1).

Contrary to the established view, these data do not show a long-term shift to exports. Rather, at the peak of the trade cycles of the early 1870s and early 1880s, exports accounted already for almost three-quarters of total sales, falling back to little over a half during the troughs. But the export peak of the early 1890s was, unusually, accompanied by a strong recovery in the home market, so that export proportion did not rise as much as in previous export booms. The overall performance of

Table 5.1 *Ransomes' home and export sales, 1871–92*

Home sales	*Export sales*	*Total*	*Export*	*Ratio*
	£	£	£	%
1871	90,364	124,377	214,741	58
1872	96,056 *	161,851	257,907	63
1873	92,976	187,228	280,204	67
1874	89,244	217,543	306,778 *	71
1875	90,342	207,690	298,032	70
1876	87,839	111,360	199,199	56
1877	89,057	100,263	189,320 †	53
1878	77,418	143,796	221,214	53
1879	64,015	138,542	202,557	68
1880	59,098 †	173,734	232,832	75
1881	66,746	197,267	264,013	75
1882	70,581	206,647	277,228 *	75
1883	78,262	188,988	267,250	71
1884	85,673	161,881	247,554	65
1885	72,955	100,668	173,623	58
1886	73,139	88,300	161,439 †	55
1887	76,734	105,970	182,704	58
1888	81,687	118,524	200,211	59
1889	103,848	125,144	228,992	55
1890	112,398	140,520	252,918	55
1891	130,983 *	119,810	250,793	48
1892	127,616	174,971	302,587 *	58

* peak; † trough

Source: Rural History Centre, University of Reading, Ransomes Sims & Jefferies archive, TR RAN AD 7/17.

the company was outstanding: over two decades which were difficult for British industry at home and abroad, Ransomes had maintained the money value of its sales. Since the general price level had declined considerably by the 1890s, this represents a very large real increase. The Board of Trade wholesale index stood at 135.6 in 1871, and had fallen to 101.1 by 1892; a fall of 25 per cent. If this ratio is applied to Ransomes' total sales in 1892, they would have been £378,234 rather than £302,587.[10] The lack of complete or consistent data elsewhere, until the first Census of Production in 1907, makes it difficult know if Ransomes were unusual: they were certainly the largest firm of all, and the first to move strongly into exports, so it may be that they were not going to move any more heavily into exports than they had already by the early 1870s.[11]

Some firms certainly did become highly export-oriented in the last quarter of the century. The greatest example here is Fowlers of Leeds, which had developed the best system for cable ploughing by steam engine. In the period 1886–99 they built a total of 737 ploughing engines, of which 720 (94%) were exported, almost all sold through the firm's depot at Magdeburg, which served the markets of the German, Austro-Hungarian, Russian and Turkish empires. Sales were also made to Egypt, Tunis, British South Africa and the Transvaal, Mozambique, Australia, Honolulu, Peru and Brazil. Between 1875 and 1894 'up to 90 per cent of Garrett's production of engines went abroad, and the engine register of Burrells shows a distinct upturn in exporting after 1875.'[12]

The increasing emphasis being place on the export market is most easily traced in agricultural engines, as most firms kept detailed registers of customers. Of those produced by Burrells between 1869 and 1913, at least 30 per cent went abroad, including all of their steam ploughing engines. Garretts sent 65 per cent of their main class of agricultural engines to foreign markets between 1858 and 1913, with periods when up to 90 per cent were exported. Over 50 per cent of Savage's 'Agriculturalist' engines (used in steam ploughing) up to 1885 went to exports. Among the smaller firms, the Wantage Engineering Co. was exporting 68 per cent of its engines in the 1880s and 1890s.[13]

To some extent, firms developed new products for distant markets. A notable development was the patenting, by Head (of Ransomes, Sims & Head) and Schemioth in 1862, of a device for utilising straw as fuel for a steam engine. It was taken up by Fowlers in 1873, and promoted in 1876–77 in Russia by their overseas sales engineer, Max Eyth. This was particularly useful in areas such as the plains of Hungary, the pampas of Argentina, or the black-soil regions of Russia, where coal was scarce and expensive. The Head and Schemioth device had been anticipated in 1871 by Paul Kotzo, who was the agent in Pest for Garretts, and an acrimonious debate on the merits of the two methods ensued in the columns of *The Engineer.* Another development by a Garrett's agent, Grimaldi of Milan, was the adaptation of their threshing machine to process the dry, brittle straw of southern Europe into animal fodder. The development of plantation machinery for tea, sugar and sisal plantations was another aspect of this. Special steam engines were developed for the colonial markets. In 1890, the manager of Fowler's Magdeburg depot developed a cable hauled implement capable of lifting sugar beet, which met with 'a very large demand'.[14]

The increasing export competition, and the development of domestic agricultural machinery manufacturers in Europe, the USA and Canada, meant that the established markets for the British makers came under strong pressure. As can be seen from the figures of British agricultural machinery exports in 1860 and 1874 (table 4.2 above), after the early domination of the Australian market, Europe had by the early 1870s become the main destination. In 1874, the north and west European sales were 56 per cent of the total, and if one added Russian sales into this, then it becomes 69 per cent. But by 1888, north and west European sales had fallen to 31 per cent of the total exports, the addition of Russia bringing this up to 43 per cent. The geographical destination of the exports had changed considerably.

Table 5.2 *British exports of agricultural implements and machinery, 1874 and 1888*

	1874		*1888*	
	£	%	£	%
North and west Europe	587,767	56	446,778	31
Russia	142,309	14	182,757	13
South America	45,731	4	275,674	19
British South Africa	20,272	2	45,807	3
Australia	131,780	13	232,972 †	16
Others	123,875	12	269,026	18
Totals *	1,051,734	101	1,453,014	99

* Totals do not add to 100 because of rounding
† 1888 figure is for Australasia
Source: *Annual Statements of Trade.*

Of the established markets, Russia held its position, buoyed up by the expansion of wheat production following the emancipation of the serfs in 1861, as large farmers and estate owners bought steam ploughs, harvesting and threshing machinery. Australia improved its position slightly, although competition from America and Canada undermined the position of British producers.[15] The great growth was in the South American market: developing rapidly as a cereal producer, as immigration rose, and land was taken into cultivation, less well developed economies offered good opportunities for entrepreneurs. Strong trading and financial links developed with Britain, so that an increasing proportion of British imports of wheat, meat, and wool came from them. The use of the pound sterling as an international currency, which became dominant after 1870, also eased the growth of trading links. So strong did the economic relationship grow between Argentina and Britain that it has been described by some as a sort of 'informal Empire'.

Marketing, agents and overseas depots

While the use of agents abroad had been growing before the 1870s, much attention had also been paid to more informal methods of marketing. Some firms had branches abroad, although these were probably repair depots combined with warehouses, rather than manufacturing branches. The only firms which made products locally were Clayton & Shuttleworth, in Vienna, and Robeys of Lincoln, in Pest. Claytons' factory in Vienna was a fully fledged manufacturing establishment, employing a workforce of 700, but this was exceptional. Senior partners and directors had made visits; younger sons of the principals of the firms had made a sort of Grand Tour to familiarise themselves with commercial conditions abroad; firms had made efforts to be represented at international exhibitions. After *c.*1870, more attention was placed on having many agents, and less on having a small number of branches/depots. Ransomes wound up all its foreign branches, except for Odessa, between 1874 and 1914, and replaced them with carefully chosen agents. Generally, their agents seem to have been well informed, and of great value to the firm in keeping up with changes in the market. They were of sufficient importance for J. E. Ransome to produce in 1897 a booklet on how to treat agents in the course of travelling on behalf of the company. Commercial travellers were another source of information, and firms seem to have used them effectively. Marshalls and Ransomes had semi-permanent representatives in Russia, and the latter sent a prominent family member on several overseas visits. Elsewhere, Ransomes used a journalist, James Drew Gay, to send back reports on prospects in India and Ceylon.

The rise of North American competition

British technology had seemed dominant at the time of the Great Exhibition of 1851, but this was not so by the end of the 1870s. The technical developments in the USA were most spectacular in the field of harvesting and mowing machinery. By the 1870s US firms were also producing the whole range of improved farm equipment pioneered in Britain, and the leading firms were enormous by British standards. As early as 1879, the capital value of the farm machinery manufacturers in the USA was estimated at $106 million, probably on a par with that of Britain. Ten years later, the USA firms' capital had trebled, to $302 million, probably far larger than that of the British industry.[16]

In the 1890s the North American industry was subject to great internal competition, reducing the number of smaller firms, and leading in 1902 to the biggest merger to date – the International Harvester Corporation, originally capitalised at $120 million. A similar story of rapid growth was evident in Canada where, as in the USA, a protectionist system had become very marked in the 1880s and 1890s. Despite American competition, Canadian firms grew rapidly, the largest being the Massey company, which merged with the Harris company in 1891 to form Massey-Harris.[17] North American firms had the benefit of a very large and rapidly expanding market. The value of farmers' capital investment in implements and machinery in the USA was estimated at $152 million in 1850, $271 million in 1870, $494 million in 1890, and $750 million (£164 million) in 1900.[18] The size of these firms, especially after the mergers of the 1890s and the early twentieth century, meant that they could reap large economies of scale, their products being sold at very low prices. This, allied to the early US superiority in harvesting machinery, meant formidable competition for British manufacturers. This was seen in the difficulties which British exporters faced in overseas markets, while at the same time US competition was becoming a factor on the British market itself. This cannot be deduced from British statistics, because before 1901 the trade returns did not enumerate imports of agricultural machinery. In that year they stood at £369,000 (£255,000 from the USA, and £55,000 from Canada). This may be compared with the value of British farm machinery exports in that year of £1,748,000, or with the British farm machinery industry as a whole, whose total production was at least £3 million. American imports can be analysed from USA trade returns:

Table 5.3 *Exports of agricultural machinery from the USA (total and to UK, 1865–69/1910–14 (5-year averages, in $000)*

Period	*Total*	*To UK*	*Mowers and reapers*	*Plows and cultivators*	*Other*
1865–69	1,082	98	–	–	–
1870–74	1,872	334	329	1	70 †
1875–79	2,441	598	391	*	207
1880–84	2,989	661	347	6	307
1885–89	2,667	426	305	18	96
1890–94	4,112	552	337	20	194
1895–99	7,175	900	611	52	236
1900–04	18,491	1,556	924	128	463
1905–09	24,450	966	518	97	351
1910–14	35,077 ‡	1,107	775	118	172

* = less than $1,000 † Includes spare parts ‡ 1910–13 only.
Source: USA, Department of Commerce, *Foreign and Domestic Commerce.*

While Britain had become a minor market for the US manufacturers by 1914, most of the US exports went to Canada and South America, with some to European markets. The main products sold in Britain were harvesting machines (reapers and later reaper-binders) and mowers. Imports became substantial in relation to the total size of the British market soon after the Civil War, and grew rapidly in the early 1870s, although there was a period of slow growth (and near-stasis or decline at some points) from then on until the mid-1890s. Thereafter they resumed rapid growth, and after 1900 averaged at least $1 million (£200,000) a year. While this was not a large sum in relation to the total production of the British industry, its concentration in the harvest and hay field made it formidable. In fact, the British market for harvesting machinery was firmly in the hands of American manufacturers. By 1913, only 30 per cent of British home demand for self-binding reapers was met by home production, and some of this was by an American firm with a factory in Britain – Walter A. Wood, whose output in Britain rose from 500 *per annum* in 1853 to almost 21,000 in 1873. By comparison, the peak production of what was probably the largest British producer,

Horse-drawn reaper-binder, 1870

The improvement of the horse-drawn reaper reached its final stage with the reaper-binder, which not only cut the crop and delivered it in bundles onto the field, but also tied the sheaves with thick twine, so that they could be easily gathered and 'stooked' together in the field and built into ricks before threshing. Prior to this, the labourers had had to follow the reaper, tying each bundle of cut stalks together with bands of straw. The binder saved much labour, and loss of corn from handling in the field. It dominated the harvest field from the 1890s until the arrival of the combine harvester in the 1950s.

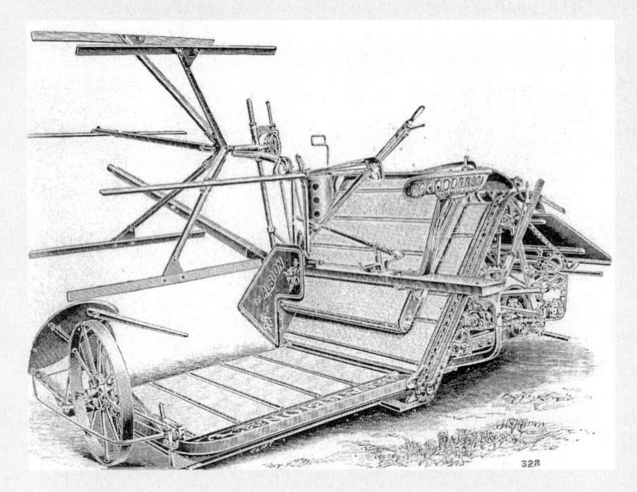

The reaper-binder was built upon the by-now standard reaping mechanism of a reciprocating cutter bar. Pushed onto the cutter bar by the revolving reel, the corn was cut, and fell back on to a platform behind the cutter bar. Over the platform an endless canvas conveyor belt, strengthened with wooden slats, travelled sideways, moving the cut corn towards the centre of the machine. There it passed up an incline (taking it over the machine's travelling wheel), sandwiched between an upper and a lower travelling conveyor belt. Being delivered by the canvasses at the highest point of the machine, the corn slid down an inclined deck on the side of the machine furthest from the cutter bar. Here it was packed together by rapidly moving arms, whilst the butt ends of the straw were straightened by a board. The fall of the material down the incline was arrested by an arm, and when sufficient weight had collected the pressure tripped the tying mechanism into action. This was the most complex and delicate of the machine's moving parts. The twine began its passage around the sheaf by being held in a retainer, from which it passed on the outward and undersides of the sheaf being formed, below the deck, and through the eye of a large needle. When the trip set the needle in motion, it carried the string round the inward and upper sides of the sheaf back to the retainer, where it was caught, and the encircling of the sheaf was complete. At this moment two short knotter bills caught the now doubled twine near the retainer, and by performing a quick revolution, formed a loop in the twine, the bills then opening and seizing the strings. The sheaf was then thrown clear of the machine by three delivery arms, and the weight of the sheaf drew the loop from the bills over the ends held by them, thus making the knot. The strings between the bills and the retainer were pressed against a knife, which cut them, and the needle receded below the deck, leaving the end of the string in the retainer, whilst the cut short end fell out. The sheaf then fell to the ground, awaiting collection by a labourer, and stooking.

All the mechanism of the binder was driven by gearing from the main travelling wheel running in the centre of the machine. This required a lot of effort, which increased the tendency of the travelling wheel to skid. When the ground was soft or wet, considerable delays could result. The binder was also a very heavy machine, having a cut of five or six feet, and requiring to be pulled by three good horses. These could only be expected to work for about five hours. A common practice on all except very small farms was to use six horses in two shifts, when 10–12 acres a day could be covered.

Samuelson, was no more than 18,000 *per annum* for both home and export markets. Harrison McGregor of Leigh (Lancashire) geared itself up for mass production of its 'Albion' reapers in the 1880s, but by 1906 annual production was still only 6,409. By then only two British firms were left producing self-binding reapers.[19]

Apart from harvesting and mowing machines, and some of the plough trade, the threat posed by foreign competition on the home market was not large. In 1913, when the total output of the British farm machinery industry was estimated at £6.5 million, the total of imports was £1,285,000. However, from this should be deducted re-exports amounting to £592,000, leaving £693,000 of retained imports, equivalent to about 10 per cent of the output of the British industry.

The last export boom

By 1913 exports accounted for some 57 per cent of output of the British industry. In the first decade or so of the twentieth century, there occurred the last great export boom to be enjoyed by the industry before the First World War. This was also true of British exports in general. For the agricultural machinery industry, the rise in export values was startling. Between 1901 and 1913 the value of exports more than doubled, from £1,708,000 in 1901 to £3,735,000 in 1913 (137 per cent). To some extent, this is due to comparing the depressed year of 1901 with the boom year of 1913. Some of it is also due to price inflation, as the index of export prices rose by about 11 per cent between these two years. But most of it was due to a rise of the real volume of exports.

The main destinations showed a strong continuation of the tendencies apparent from the 1870s. For the peak period 1909–13, the main export markets were:

Table 5.4 *British agricultural machinery export markets, 1909–13 (£, annual averages)*

	*Engines**	*Machinery*	*Implements*	*Totals*
North and west Europe	128,751	426,896	16,150	571,797
Russia	373,558	351,072	3,809	728,439
British South Africa†	31,107	53,307	64,076	148,490
British India	37,113	23,731	117,835	178,679
Australasia	60,549	63,378	77,400	201,327
South America	232,473	239,585	216,566	688,624
Other	345,014	352,161	115,427	812,602
Totals	**1,208,565**	**1,510,130**	**611,263**	**3,329,958**

* Described as 'Prime Movers' in the trade statistics. Includes all types of engines, but consists mainly of steam engines.
† = Cape of Good Hope and Natal.
Source: *Annual Statements of Trade*.

By the eve of the First World War, the major markets were Russia (about one-fifth of total exports), South America, and north and west Europe (about one-sixth each). These markets had developed rapidly since the late 1880s. The Imperial and colonial markets (British South Africa, British India, Australasia), although large, were less important. The Russian trade was clearly the largest single market, and was a great consumer of 'prime movers' (mainly steam engines) and machinery. But the trade at its peak was even larger. Total exports of all engines, machinery and implements peaked in 1913, at £3,735,168. At that point the Russian import of 'prime movers' was £619,324, and of machinery £484,059. The rise to dominance of the Russian trade was a considerable achievement. Imperial Russia had raised its protective tariffs in 1887, although in the following year they were relaxed, with 'complex' machines such as steam engines, binders and reapers being allowed into the country free of duty.[20] British manufacturers took full advantage of this concession. A further stimulus came as a result of the 1905 Revolution in Russia: the ensuing agrarian settlement allowed farmers to take their scattered strips out of the open-field system and create consolidated holdings. The resultant boom in farming and in particular grain production sucked in imports of machinery, benefiting British exporters.

The next great market was in South America. Here, although the exporters made great efforts across the sub-continent, the trade was dominated by Brazil and the Argentine Republic. Thus in 1913, out of total implements exported worth £745,883, Brazil alone took £217,785. The Argentine trade was more in machinery, and sales of machinery to Argentina amounted to £150,900 out of a total export to all countries of £1,628,241 in 1913.

In both Russia and South America, the market was buoyed up governments which encouraged immigration and frontier settlement. British exporters, banks and shipping concerns also had strong links with the buyers, and put a lot of effort into selling in these regions. In the case of Argentina, such connections were encouraged by the strong import link, in the form of frozen and chilled beef. None of these favourable factors was at work in the markets of north and west Europe. The return to protectionism on the continent after 1879 and the rise of European manufacturers did affect British exports, clearly, and the European trade no longer dominated exports as it had done in the 1870s, but exports to Europe were still important. Most of these went to France and to Germany. In 1913 the main national destinations of all exports were: Russia (£1,105,665), Germany (£303,143), Argentina (£236,456), Brazil (£241,171), and France (£165,394). There were also small but not insignificant trades with Portugal and Italy.

Conclusions

The maturing of the home market by the mid-1870s posed a potential threat to the continued growth of the industry. This was accompanied by greater import penetration into the UK, chiefly of North American reaping and mowing machines. Some British manufacturers diversified into non-agricultural products, but most, even if already strongly committed to exports, looked to overseas markets. The older, European markets were more difficult to maintain, as protectionism revived and as domestic manufacturers began production. More success was had in the further-off markets of the Russian empire and South America, in spite of the distances involved. The success of the last great export effort is not in doubt, with exports, at the historic 1913 peak, accounting for over half the industry's output. Nor was the home market by any means dominated by imports, which accounted for only about one-fifth of the total UK machinery market. On the other hand, the export trade, like export trades everywhere, was more volatile than home trade, and there were particular risks associated with the Russian market. Above all, the continuing prosperity dependent on exports required the maintenance of international peace and security; yet 1913 was to be the last full year of peace before the outbreak of the First World War.

CHAPTER SIX

A mature industry, 1875–1913

The rise of some firms and the fall of others

The population of Britain virtually doubled, from 20.8 million to 40.8 million in 1911, and demand for food grew enormously, a demand that was actually being met by fewer agricultural workers, as from the middle of the century there had begun an enormous exodus of farm labourers from the countryside. The number of agricultural labourers and farm servants in England and Wales had fallen from 1,097,794 to 622,279 (males) and from 143,147 to 13,214 (females) between 1851 and 1911. The farm labour force which remained on the land had at its disposal both new machines and new sources of power. The number of horses available for agricultural work in England and Wales rose from 720,000 in 1840 to 815,000 in 1910, and increasing amounts of power was also now supplied by portable steam engines and a smaller amount from the new oil, petrol or electric motors on farms. In all, it is estimated that the total supply of power (stationary, draught and human muscle) available to English and Welsh farmers had risen from about 688,000 hp in 1840 to about 1,002,000 hp in 1910. In terms of power per worker, the rise was greater – about 60 per cent. In effect, human effort in farming was being replaced by mechanical and equine power.[1]

Some of the principal beneficiaries of this process were the farm machinery manufacturers. Victorian Britain was a generally favourable environment in which to engage in business, and on the whole it was a period of success for enterprising individuals and companies. Indeed, there are few documented cases of commercial failure in the agricultural machinery industry before 1914. So what were the particular circumstances and reasons for company decline or collapse in this period of general growth and optimism?

Company failures do not often yield a great deal of documentary evidence. One case, however, where evidence has been examined is that of the Reading firm of Barrett, Exall & Andrewes. In 1851 it was one of the three or four largest firms in the industry, employing some 250 people. At the Great Exhibition of that year, it occupied one of the four prime sites on the Crystal Palace's ground floor, its near neighbours being Ransomes (800 employees), Garrets (300) and Crosskills of Beverley (240). It had a

The stands of the major manufacturers at the Great Exhibition of 1851 in the Crystal Palace, Hyde Park, London. From left: Ransomes & May, Barrett Exall & Andrewes, Mary Wedlake. (*Museum of English Rural Life*)

prominent position in the subsequent illustration of the industry's stands which appeared in the *Illustrated London News*.

The antecedents of the firm went back to the first decade of the nineteenth century, but it did not begin to grow rapidly until it was joined by William Exall, who had considerable energy and inventiveness, and took out an impressive number of patents. The early growth of the firm was based on sub-contracting work in the 1830s for Brunel's Great Western Railway line, which ran through Reading. In the late 1830s it moved into agricultural machinery, and won a prize for its plough at the RASE Show in 1841. By making what its Great Exhibition catalogue desfribed as 'a peculiar study' of ploughs, the firm forged a national reputation. While still undertaking local foundry work, the firm moved into the more complex business of making agricultural machinery. It had developed a hand-powered threshing machine which was exhibited at the 1843 RASE Show; the following year this was recognised by the judges as the most original on display, so much so that several other entries had apparently been copied from it. A horse-powered version was also developed.

In design the firm's machines were both simple and original. In its 1847 catalogue, the firm claimed that its machines were to be found in every county in England, and in 1858 stated that they had sold 'upwards of 2,500' threshing machines in total, a large number for a firm so new to the business. The firm progressively improved the machines, and made changes to make them more attractive to colonial markets. In 1849 Exall had constructed the earliest set of equipment for ploughing on the 'roundabout' system, and shortly afterwards went in for the manufacture of steam engines. In 1851 their portable engine won a prize at the Great Exhibition. In 1852 at the RASE Show, one of its engines had won first prize, Ransomes coming second.

In the late 1850s the firm seemed to be at the top of its profession. However, its technical ingenuity was not seemingly matched by managerial efficiency, and its research costs were not being recouped sufficiently. In 1849 one of the original partners, George Barrett, who had been an able manager, retired. In the early 1850s the energies of William Exall and of C.J. Andrewes were diverted into local politics, both serving as mayors of Reading. Growth did continue, and the number of employees reached about 350 in 1864, peaking at 400 in 1870. Yet a cash-flow crisis loomed: research and show expenses were high; it had inadequate capital of just £10,000, partners who lived in some style, and who may have preferred to draw out profits rather than re-invest them in the business. When another manager, George A. Barrett, retired in 1863, the opportunity was taken to pay him his share of the business by means of a capital reconstruction, and the firm was incorporated as a limited liability company. The capital structure of the new company was on the optimistic side. At a time when turnover was just under £70,000, the authorised share capital was set at £200,000, of which the first issue was to be £100,000, although in fact only a little over £70,000 worth of shares was issued. However, an additional £20,000 in debentures was issued, and the paying of dividends on them would give a hostage to fortune should earnings fail to be maintained. The old partnership's assets were valued at only £50,000, to which was added £20,000 for goodwill. The prospectus of the new company pointed to the need to raise more capital to underwrite the continuing expansion of their recent growth in business to Europe, which, now that free trade had been established, could be expected to grow. At the first general meeting (the only one which the law at the time required to be publicly reported, and the only one whose records have survived) in 1864, a 7½ per cent dividend was declared, and total distributed profits have been estimated at £2,000.

Yet the firm was slipping. In spite of the technical excellence of its steam engines (made famous by a trial against the engine of Clayton & Shuttleworth at the RASE steam trials in Cardiff in 1872), the market for them was becoming overcrowded, and the firm did not pay enough attention to cheapness. By the early 1870s death or retirement had taken its toll of several key directors, and the inventive genius of William Exall waned after the tragic death of his son in 1868. In 1875 C.J. Andrewes took over as Managing Director, but does not seem to have been up to the task (his nickname among his subordinates was 'Major Pomposity'). None of the surviving sons of the founders were able to enter the firm, and by the 1870s there were clear managerial, financial and commercial weaknesses in the firm. The decline of grain farming found it unprepared to respond creatively. By 1880 its losses were large enough to cause a special meeting of the shareholders to be convened. The value of the issued capital was brought down from £80,780 (up to which it had crept) to £70,020. Yet the debenture debt had continued to rise, being £29,500 by 1885, and representing a prior charge of £1,400 a year in interest. The company now had few new products to offer, and although they had some success with a patent self-acting horse rake, they pinned most of their hopes on a line of wrought-iron pulleys. These did not sell well, and by 1887 the workforce was down to 200. At an extraordinary meeting of shareholders in March of that year, the company went into voluntary liquidation.

This case study is a story of early technical success – with William Exell prominent – later undermined by senior management failures which included insufficient cost control and reinvestment. A failure to maintain continual innovation in new and improved products led to faltering sales, while the firm's drastic financial reorganisation in 1864 was unable to cope as profits turned into losses. Then, crucially, the company ran out of cash.[2]

Some of these factors can also be discerned in the story of William Crosskills of Beverley. William Crosskill had been born in 1799, and was only 12 when his father, a whitesmith, died. The next year William became apprenticed to his mother, and by 1825 was taking on his own apprentices. In 1827 he moved to newly built foundry premises, initially concentrating on general castings, such as railings and lamp standards. He seems to have had a keen eye for business, and got the contract to re-hang the bells in Beverley Minster in 1829. By the late 1830s he had moved into agricultural machinery, and in particular his clod crusher (see page 74), by which his name was to be chiefly remembered. This was a heavy roller with serrated iron rings, which pulverised the newly ploughed field before sowing. It was shown at the RASE Cambridge Show in 1840, and won prizes at various other events, culminating in the RASE Gold Medal in 1846. By the end of 1850 a total of 2,478 clod crushers had been manufactured. In 1851 the foundry employed 250 men, and further Crosskill specialities were being produced, including his portable farm railway to facilitate field work in wet weather, ploughs, harrows and threshing machines. The Crimean war provided another opportunity, and he manufactured shells, mortars and over 3,000 carts and wagons for the army. At one point there were over 800 workmen, although this had fallen to 500 by 1861. The firm seems to have been in sole ownership, and not incorporated. The Crosskill works now covered some seven acres, and were notable for the use of steam-driven machine tools to mass-produce patent wheels and axles for carts and wagons. This feature excited the particular admiration of the British Association members when they visited the works during their 1853 meeting at nearby Hull. Considerable attention was paid to the export business, and this probably accounts for the opening of a branch in Liverpool in 1849 for the sale and repair of wheels. An earlier, more extensive foreign venture had occurred in

1844, when he and two other businessmen fulfilled a contract to set up a gas-works and street lighting system in Hamburg, whose extent can be judged by the fact that another Beverley ironmonger was sub-contracted to supply 3,000 copper lanterns for the project. That the firm was to the fore in the latest ideas may be seen from the fact that it produced one of the earliest direct steam ploughing machines, Robert Romaine's steam cultivator, in 1857, although this did not proceed further.

However, much of this business success was achieved against a background of financial difficulty, caused apparently by over-expansion with insufficient working capital. As early as 1847 the business was mortgaged to the East Riding Bank. In 1855 the bank foreclosed, and there was drawn up a deed of assignment by which the whole property of the firm passed over to this bank, as trustees on behalf of Crosskill's creditors; the debt came to about £73,000. At about the same time, Crosskill obtained the choice government sinecure of Distributor of Stamps for the East Riding, in reward for political services (Crosskill had served on Beverley town council for over twenty years, and was mayor in 1848). Having his sinecure, he virtually retired from the business, from which presumably he received nothing, and was, apart from the government job, virtually bankrupt. The 'Old Foundry' carried on under its new ownership as the 'Beverley Iron and Waggon Company', led by Sir Henry Edwards, the Conservative MP for the town. It suffered heavy losses in the depression

Romaine's direct-action ploughing engine, as made by Crosskills of Beverley, 1857. (*Museum of English Rural Life*)

of the 1870s and was compulsorily wound up in 1878 with the loss of 200–300 jobs, the land, stock, plant and buildings being sold off by auction. The fact that the firm had debts of £24,000 when it closed may be interpreted to mean that Edwards had formerly subsidised it, but had ceased to do so. The patterns and drawings were bought by Crosskill's sons, who had started their own business as 'William Crosskill & Sons' in 1864. They do not seem to have been active product innovators. Their 1867 catalogue listing (25 pages) was headed by the Crosskill clod crusher, the remainder being mainly carts and feed preparation machinery (not all of it their own make), although they were also marketing improved models of haymaking machines and lawn mowers, which seem to have been of their own make. This firm survived in various guises and under various owners until the 1920s, but was never as important as that of the elder Crosskill had been.[3]

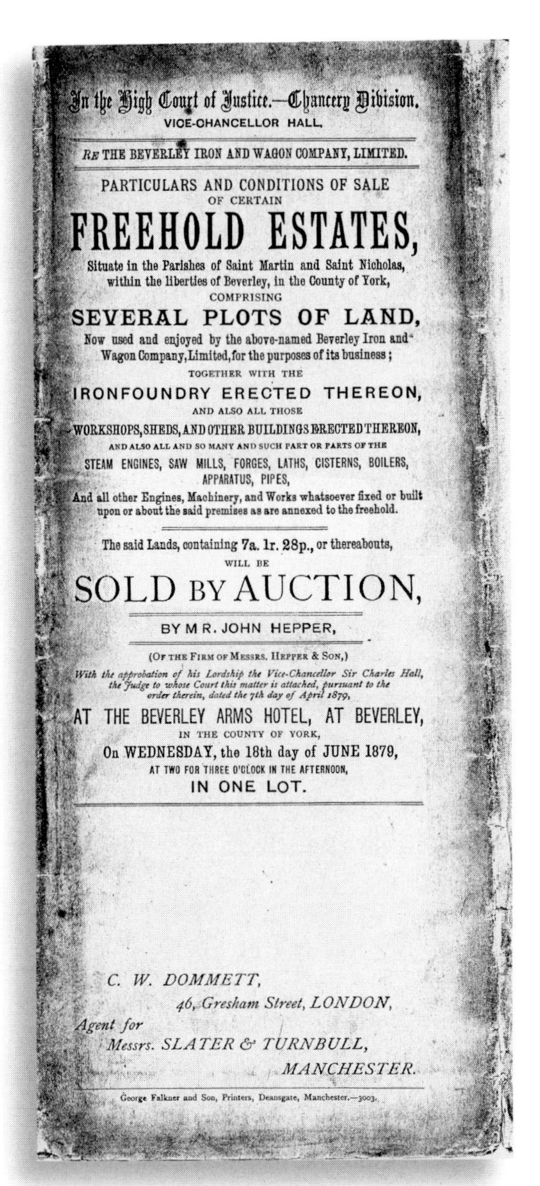
In the High Court of Justice.—Chancery Division.
VICE-CHANCELLOR HALL.

Re THE BEVERLEY IRON AND WAGON COMPANY, LIMITED.

PARTICULARS AND CONDITIONS OF SALE
OF CERTAIN
FREEHOLD ESTATES,
Situate in the Parishes of Saint Martin and Saint Nicholas,
within the liberties of Beverley, in the County of York,
COMPRISING
SEVERAL PLOTS OF LAND,
Now used and enjoyed by the above-named Beverley Iron and
Wagon Company, Limited, for the purposes of its business;
TOGETHER WITH THE
IRONFOUNDRY ERECTED THEREON,
AND ALSO ALL THOSE
WORKSHOPS, SHEDS, AND OTHER BUILDINGS ERECTED THEREON,
AND ALSO ALL AND SO MANY AND SUCH PART OR PARTS OF THE
STEAM ENGINES, SAW MILLS, FORGES, LATHS, CISTERNS, BOILERS,
APPARATUS, PIPES,
And all other Engines, Machinery, and Works whatsoever fixed or built
upon or about the said premises as are annexed to the freehold.

The said Lands, containing 7a. 1r. 28p., or thereabouts,
WILL BE
SOLD BY AUCTION,

BY MR. JOHN HEPPER,

(Of the Firm of Messrs. Hepper & Son,)
With the approbation of his Lordship the Vice-Chancellor Sir Charles Hall, the Judge to whose Court this matter is attached, pursuant to the order therein, dated the 7th day of April 1879,
AT THE BEVERLEY ARMS HOTEL, AT BEVERLEY,
IN THE COUNTY OF YORK,
On WEDNESDAY, the 18th day of JUNE 1879,
AT TWO FOR THREE O'CLOCK IN THE AFTERNOON,
IN ONE LOT.

C. W. DOMMETT,
46, Gresham Street, LONDON,
Agent for
Messrs. SLATER & TURNBULL,
MANCHESTER.

George Falkner and Son, Printers, Deansgate, Manchester.—3003.

Sale notice for Crosskills business, 1879. One of the few recorded examples of business failure in the industry before 1914. *(East Riding of Yorkshire Record Office)*

Like the firm of Barrett, Exall & Andrewes, Crosskills relied for their growth on the technical innovation, largely of one individual, of which Crosskill's has lasted longer in agricultural use. There is evidence of lack of working capital in both firms. It may also be that local politics diverted the energy of the key men to the detriment of their business. That Crosskills lasted through a long period of financial difficulty may be due to William Crosskill's perceived business acumen; that the bank found it necessary (or opportune) to foreclose during the Crimean War may mean that the military contracts had been unprofitable, or that profitable contracts were coming to an end; in the absence of accounts we cannot be sure.

The problems facing Barretts and Crosskill proved in the end fatal. Another business, Samuelson of Banbury, is an example of a firm experiencing the problems widely felt during the 'Great Depression' from the early 1870s until the later 1890s. Bernhard Samuelson had taken over James Gardner's agricultural machinery works in 1849. In 1851 he was appointed by Burgess & Key, the agents for McCormick, to make the McCormick reaper, and Samuelsons also made the Owen Dorsey reel-rakes. The firm became famous for its lines of mowing machines and reapers. By 1859 it had risen from two or three dozen men to a payroll of almost 300. By 1865 the firm was one of the largest makers of reapers in England. In 1869 it made between 1,600 and 2,000 self-raking reaping machines. However, in the 1870s competition in the mowing and reaping machine market became severe, as new firms entered the reaper market, and imports increased. Falling prices for wheat led to a decline in farm incomes. The worst year was 1879, when summer rain led to a low yield of cereals, and import competition prevented prices rising in consequence of the poor English crop. Farmers were in low straits, and so was Samuelsons. Bernhard Samuelson, by this time an MP, had to explain to the electorate in March 1880 why the works had put all its men on short time in order to deal with the depression. The firm faced major financial difficulties and had to diversify. The resulting move to turnip cutters and flour milling machinery probably ensured the firm's survival, and in 1888 it was incorporated as a joint-stock company. The Ordinary General Meeting of the company (presumably the first and possibly the last to be publicly recorded) showed a board which was firmly dominated by the Samuelson family (4 out of the 8 directors), and was still headed by Sir Bernhard. The meeting was informed that trading results were satisfactory due to the partial revival of trade, and to the hearty cooperation of employees and employed. A tax-free dividend of 10% was announced. The balance sheet showed an authorised capital of £150,000, of which £100,000 had been issued as shares. Profit to the end of September last had been £14,031, and £10,000 of this would be paid out as dividend, the rest going to the reserves. Samuelsons was to survive until 1929. Like the above cases, the firm depended very much on the energy of one man, the founder-manager, and it was its good fortune that trade revived in the 1890s up until 1914.[4]

The demise of Barrett, Exall & Andrews was one of the two largest failures in the industry before 1914. The other was that of Brown & May, of Devizes. This firm began in the business of making steam engines (chiefly portables) in 1854, and lasted until the Official Receiver was appointed in March 1913. Ironically, its two founders, William Brown and Charles Neale-May, had both served their apprenticeships at Barrett, Exall & Andrews. Starting with a fairly conventional design of portable, the firm made continuous refinements and improvements, eventually offering a range of engines from 1½ hp to 20 hp. About 80 per cent of the output was portables, but the firm also developed traction engines in 6–8 hp sizes, some steam road vehicles, and moved into building gas, petrol and oil engines after 1880. It also manufactured a wide range of minor products, such as saw-benches, chaff-cutters and winding gear, and briefly made its own threshing drums. By 1909 it claimed to have made 10,000 engines, and employed 350 men, the largest employer in the district.

The road to failure for Brown & May can be traced back to 1895, when the two founders, having reached the age of 65, decided to retire and enjoy their fortunes. Accordingly the firm was converted into a limited liability company; the founders remained on the board, but left the firm completely in 1901 and 1903. The new owners seem to have lived a long way from Devizes, and had many other interests. It did not take long for the fairly respectable profits of the 1890s (around £4,500 a year) to fall to about £2,000 a year by 1900. It seems that the output of the firm was too small for a firm of that capacity, the range of products too wide, and that the experiments with ICE engines had used up financial reserves without leading to the development of substantial new product lines. The firm finally failed in 1913, with net debts of £15,000. Running out of entrepreneurial energy and (probably) lack of managerial competence were key elements in this failure.[5]

The final example of a firm facing major difficulties before the First World War is that of Taskers of Andover. Here the problems were familiar: a shortage of working capital and managerial or administrative weakness. By the end of the century, Taskers was a medium-sized firm, specialising in steam engines. Although Henry Tasker was in charge, he was only one of three partners with equal financial stakes, the other two taking no interest in the running of the firm. This difficult situation was ended when Henry Tasker purchased the other partners' shares, at a cost of £6,620, in 1890. While this might have made control easier, it probably also exacerbated the shortage of working capital. Further, Tasker's only son died in infancy and he had virtually (although not formally) adopted George Hoare, the son of his coachman, and trained him as his successor. Hoare, although technically very competent, was probably not as strong on the managerial and financial side. The difficulties of the firm came to a head in 1896, when it was decided to form a limited liability company to take over the business. The prospectus of W. Tasker & Sons Ltd announced the issue of £50,000 of ordinary shares, plus £35,000 of debentures. Since the freehold properties were valued at £25,522, and the stock in trade and accounts due at £41,578, this implies a valuation of the goodwill at £17,900. Whether this was a fair valuation cannot be decided, since most of the financial records of the firm did not survive (a common occurrence on the failure of a firm – as this one finally did in 1932). But shortage of working capital must have continued. Contributory factors were the opening of offices in Coventry and London, and the firm's entry into the business of hiring out its machinery. This was an overcrowded field, and the move was financially disastrous. The firm was put into liquidation in August 1903. It continued in the receiver's hands for four years. During this time Henry Tasker resigned his connection with the firm in favour of George Hoare. In 1907 a new firm of the same name was started. The nominal share capital was £30,000, divided equally into £1 preference and ordinary shares, and debentures to the value of £16,350 were issued. Thus the firm was worth much less than its predecessor. The new financial structure benefited George Hoare, who was now the managing director.

He was described as the vendor, and received 2,400 debenture and 1,000 preference shares in part consideration of the purchase price, and £21,500 in cash, making a total of £39,900.

This was not the end of the financial troubles of the firm. The new company continued the policy of hiring out its machinery, and this may have been a contributory factor. A subsidiary, the Tasker Hiring Company, was set up in 1907 to purchase traction engines and other machinery from the parent company. This was a financial disaster. A member of the firm's accountants was appointed manager on behalf of the debenture holders, while George Hoare continued to be responsible for the technical direction and daily management of the business. This administrative structure continued for the next nineteen years, until the next voluntary liquidation. It was ironic that Tasker's troubles came to a head just as they were doing well with a new product – the 'Little Giant' traction engine, and in 1909 they went into the expanding business of building steam wagons for road transport. But the managerial decision-making obviously left something to be desired, and this was compounded by the problem of managerial succession.[6]

What lessons may be drawn from this history of difficulty and (sometimes) outright failure? It is clear that product innovation and inventive activity lay at the heart of all these businesses. The more successful ones demonstrated an unceasing flow of new ideas, and new patents. The failed firms were often those with a diminishing flow of inventive activity. But this was not enough. An equally close eye had to be kept on market developments, and firms had to be willing to change their product lines at short notice. Those which failed to innovate, or which were unwilling to change their product mix, soon found themselves running out of the profits with which to make the necessary technical readjustments, and were then locked into a path of failure. Finally, a high degree of commercial acumen was indispensable. This may have been incompatible with the desire of successful businessmen to play a part in local politics. In the case of Henry Tasker, religious activity was the counter-attraction: he would have preferred to have become a missionary, and was only in the business because he saw it as his duty following the death of his brothers. Constant and undivided attention to the business was a prerequisite of survival in the long run.

Specialisation in steam

While the virtues of steam engines had been demonstrated convincingly by the 1870s, the next forty years saw considerable further technical development, and steam technology came to play an increasing role in the industry. At the first Census of Production, taken in 1907, the output of the industry was classified with that of general engineering, and of the implement and tool trades, with this result:

Table 6.1 *Output of agricultural engineering products in the UK in 1907 (£000)*

	England and Wales	*Scotland*	*Ireland*	*Total*
General engineering				
Steam engines and parts	*	*	*	1,283
Agricultural machinery	1,009	75	60	1,144
Agricultural implements and tools and parts thereof:	1,138	112	27	1,277
Implement and tool trades				
Agricultural implements and tools and parts thereof:†	842	48	18	908
Total				**4,612**

* No data

† Includes 'spades, shovels, hoes, hayforks, ploughs, harrows, pickaxes, &c.'

Source: Census of Production, 1907, *Final Report*, Cd. 6320, 1912, pp. 190, 209.

This total of almost £5 million's worth of goods was impressive, although there is some over-estimate through the inclusion of the non-agricultural products of the hand-tool trades under 'implement and tool trades'. Nevertheless, the figures indicate the importance of the steam sector within the business as a whole, accounting for some 28 per cent of the total production of the industry. The manufacture of steam equipment had been a significant part of the industry since the 1850s, and includes portable steam engines; engines for ploughing; for traction or road use; and stationary engines and power plant.

Portable engines

The portable engine continued to develop technically, and more uses were found for it in agriculture and other industries. In farming, it was used most generally for threshing, although it could be put to a variety of uses: it could drive elevators, when making corn ricks or hay stacks; it could pump water, a very useful attribute in the days when few farms had a piped water supply; it could be used to drive all sorts of barn machinery, such as for feed preparation; and it could be used to power saw mills, being widely used in forestry. The construction industry also used it widely for pumping water, mixing mortar, moulding bricks and hoisting stones. In later years it was also used for driving tarmac and asphalt mixers and stone-crushers. It was also the main power plant in many small factories. Such was the cheapness of portables, that in applications where it was only moved every few months, such as in forestry or construction, it made economic sense to use a portable rather than the later self-moving and more powerful traction engine. Consequently portables continued to be made alongside traction engines, and remained in production on a

Ransomes' portable steam engine, 1879

The manufacture of the 'portable' (i.e. moveable) steam engine for agricultural use had begun in the 1840s. By 1880, there were at a conservative estimate some 14,000 engaged in agricultural work in England and Wales, undertaking a wide variety of tasks. The range of portables made by Ransomes in 1879 was recommended for driving steam threshing machines, and other farm machinery such as corn mills, chaff cutters, bruising mills (for preparing animal feed), root pulpers, for working circular saws, pumps, and 'for contractors' purposes'. They were available in nine sizes, from a single-cylinder 4 HP (price £150) to a double-cylinder 20 hp engine (price £450). These ratings are in 'indicated horse power', which measured the power developed in the cylinder or on the piston. Here, it was calculated at a pressure of 35 lbs per square inch, but these boilers would take a pressure of up to 80 lbs, in which case the indicated horse power would be almost three times that, at 35 lbs.

The illustration, from Ransomes' 1879 catalogue, shows the basic model, to which various other features could be added: expansion gear, for economising on fuel and for tasks requiring much variation in the amount of power delivered; extra large fire boxes, for burning wood and refuse; a special automatic governor, for uses where the load was varying constantly, such as in wood sawing, crushing mills and winding gear.

The boiler was lagged with wood and sheet-iron; the cylinder was surrounded by a steam jacket, which kept it hot, and thus did not waste fuel and effort. The engine's fuel economy was praised (a consumption of 5 cwt of good coal per day of 10 hours for the 8 hp engine); wearing parts were case-hardened where applicable; the incoming fresh water supply was pre-heated by the exhaust steam; the water pump which delivered the supply of fresh water to the boiler ran continuously, so that the level of water in the boiler was always maintained; the chimney was fitted with a steam blast pipe, which blew steam up the chimney, increasing the air draught in the firebox, and thus raising steam more quickly.

In all, this was a much more sophisticated and complex machine than the first Ransomes portable of 1841, and such machines, and the self-moving traction engines developed from them, dominated the threshing yard and barn work until the Second World War.

AUGUST, 1879. No. 1391, H.

CATALOGUE OF
STEAM ENGINES
MANUFACTURED BY
RANSOMES, SIMS & HEAD,
ORWELL WORKS, IPSWICH.
LONDON OFFICE: 9, GRACECHURCH STREET.

INCLUDING
PORTABLE & TRACTION ENGINES,
VERTICAL STEAM ENGINES,
STATIONARY ENGINES, BOILERS, ETC.

RANSOMES, SIMS & HEAD'S factory, the "Orwell Works," established in 1785, is situated at Ipswich, distant from London 69 miles, and accessible by frequent trains on the Great Eastern Railway in two hours.

The factory covers more than twelve acres of land.

RANSOMES, SIMS & HEAD
ALSO MANUFACTURE
STEAM THRASHING MACHINES,
HORSE PLOUGHS OF ALL KINDS AND SIZES,
HORSE RAKES, AND HAYMAKERS.

COMPLETE ILLUSTRATED CATALOGUES FREE ON APPLICATION.

Source: MERL, *Catalogue of Steam Engines Manufactured by Ransomes*, Sims & Head, Orwell Works, Ipswich (August, 1879) small scale even after traction engine production ceased. Robey & Co. of Lincoln had portables on offer until the 1960s.[7]

The technical development of the portable engine may be demonstrated by comparing an engine of the 1850s with one of the 1900s. The Marshall single-cylinder portable, illustrated on page 53, is a typical machine from the mid-nineteenth century, and was available in sizes from 1½ hp to 12 hp. Such an engine would have a safety valve, and a pressure governor (to the right of the flywheel), and this model was fitted with a spark arrestor on the top of the chimney, which was hinged to allow entry into barns or passage under bridges. By 1900 the principal improvement was that the engine now had a single compound cylinder with two chambers, allowing steam first to move the piston within the larger chamber, and then, its pressure not yet exhausted, to flow into the second chamber to perform more work. The compound engine was more powerful and more fuel efficient. The other main feature of the engine would have been that it was made of steel, and could withstand higher pressure.

The difference in the performance and specification of an early and a later type portable may be illustrated by engines from Dodman of King's Lynn.[8] The earlier one would have had a working pressure of about 80 psi, and the fireboxes were of Lowmoor iron. It was available in powers of 2½–12 nhp. The smaller version ran at 150 r.p.m., the larger at 110 rpm. They were described as being 'well adapted for the use of landed proprietors, Agriculturalists and Brickmakers'. The 2½ nhp engine cost £98 and the 12 nhp was priced at £250. A two-cylinder version was available, which would supply between 8 and 40 nhp. Like many other makes, these portables could be had with enlarged fireboxes for burning wood, straw, stalks etc. The later type of portable by Dodman is of the period 1900–10. The smallest power rating was now 4 nhp, ranging up to 14. The working pressure was now 100 psi, and it was hydraulically tested to 180 psi.

Ploughing engines

Ploughing engines and their associated systems were also subject to continuous technical improvement. The early experiments had been with a single-engine, 'roundabout' system. By the end of the 1860s the leading manufacturer, Fowlers of Leeds, also offered a double-engine system, by which the implement was pulled back and forth across the field by a pair of engines on opposite sides of the field. By then the Steam Plough Works was producing 150–160 sets a year, of 8, 10, and 14 nhp, of which one third were exported. The double-engine system, although of a higher capital cost than the single-engine one, saved labour and time in working, and was especially attractive to the overseas market, with its shortage of skilled labour. Implements included a balance plough, a mole-draining plough, and a cultivator, with living vans for the contractor's labour. By this time the engines used steel for their wire ropes, replacing the original and

The final version of the Fowler BB ploughing engine, 1917. Compound cylinders, rated at 16 nominal horse-power. (*Museum of English Rural Life*)

less reliable iron cables. Engines modified to use straw as fuel had been developed, particularly for the eastern European and Russian market. From 1881, Fowler's ploughing engines were compound, with a consequent rise in power. By the late 1880s, Fowlers offered a range of single-cylinder ploughing engines, from 4 to 20 nhp, and compound engines from 6 to 16 nhp. Later, superheaters were also added. In the period 1900–14, Fowlers built a total of 2,519 ploughing engines, of which 2,312 were exported. AA compound saturated steam engines comprised 939 of the total; designed for export, these were particularly successful in the sugar-beet growing areas of Europe and in Egypt. At first these had a working pressure of 180 psi, later raised to 200 psi.[9] As well as developments in the engines themselves, attention was paid to the implements. Fowlers offered over 200 types of ploughshare as stock items, as well as over 300 different types of plough.[10]

The development of ploughing engines was closely bound up with that of traction engines. The original foray of the steam engineer into agriculture produced the portable engine, which may be dated from Ransomes' first display of one at the 1841 RASE Show. The next step was to make the engine self-moving. The credit for this is given to Thomas Aveling, who began farming at Ruckinge in Kent, and later combined this with engineering, buying a small millwrighting shop and foundry at Rochester in 1850. In 1856 he introduced a modified steam plough, based on the Fowler and Ransome designs. Attracted by the notion of dispensing with the many horses (up to six) necessary to move a portable engine, he took out his first patent on the subject (concerned with adjusting for wear in the drive chain linking the engine to its wheels) in 1859. In 1860 he exhibited the first self-moving engine at the RASE Show at Canterbury. His '8 hp Patent Locomotive Engine' was in fact a converted Clayton & Shuttleworth portable. The price at which he offered this engine in 1860 was £315, including reversing gear, two safety valves, pressure gauge and whistle. In 1861 he exhibited the first engine of his own manufacture. The next year he was joined in business partnership by Richard Porter, trading under the name Aveling & Porter.

Aveling continued to make contributions to the development of the traction engine. Possibly his most useful one was the horn plates. These were forward extensions of the firebox side members, on which could be supported the engine crankshaft and gearing. These had formerly been bolted directly to the boiler, and thus had set up considerable stresses on it, with consequent leakages and corrosion. The firm moved into adapting its engines to road rolling, and it was not until 1871 that the RASE recognised the utility of the traction engine for farm work, when the Aveling 10 hp engine won the prize for 'best agricultural locomotive'.[11]

Self propulsion of steam engines by means of a chain drive was applied initially by a number of manufacturers apart from Aveling – including Garrett, Tuxford, Clayton & Shuttleworth, Burrell and Savage. Other makers used a more satisfactory geared system. Steering initially involved an extra man sitting at the front operating a steering wheel, but this was soon was replaced by a chain and bobbin system operated by the driver. Within ten years from the first appearance of the traction engine, its main features were in place: gear drive; steering by chain and bobbin; water tanks as an integral part of the engine; and coal carried in a bunker behind the driver.[12]

Three further developments had to occur to make the engines into their final form: compounding; road springing; and the use of differential gears. Compounding, whereby steam from a high-pressure cylinder is fed, having done its work, at a lower pressure into a low-pressure cylinder, was perfected both by Fowler and Aveling, and introduced in 1881. The great advantage of compounding was the saving on fuel. The Fowler compound, first exhibited at the Kilburn RASE Show in 1879, had the remarkably low fuel consumption of 2.75 lb per hp per hour, and claimed a water saving of 30 per cent.[13] Other manufacturers followed these basic designs. Road springing was a difficult problem, tackled in different ways. Fowler developed a spring wheel system, using a patent of Thomas Aveling and Alfred Greig (of Fowlers), by which springs were interposed between the wheel rim and the wheel hub. Others, such as Burrell, applied

A typical agricultural traction engine, by Ransomes, the 'Royal George', of 8 hp. (*Museum of English Rural Life*)

springing directly to the drive axle, and rubber tyres were used from the end of the century. The final improvement was differential gearing, by which one rear wheel could revolve at a different speed to the other when turning. This came to be universally adopted.

The basic traction engine, as evolved by the later 1880s, is illustrated above. As well as this type, there were also larger and more powerful models for road transport of heavy loads, and showmen's engines. The latter provided the means to transport the travelling fair or circus, and when stationary provided the power plant for the electric lighting and power systems being used. Some *ad hoc* orders for military or colonial purposes were undertaken, and Fowlers went into the rail locomotive building business in the 1860s, when legislative restrictions on road movement of steam engines limited the traction and road engine business. They also diversified into industrial power plant, colliery winding engines, and electricity generating engines in the 1880s and 1890s.[14]

Unbroken series of statistics for the output of individual firms is seldom available during this period. The most comprehensive is that of

Ransomes' threshing machine, 1875

The threshing machine had begun in the early years of the nineteenth century as a simple box, with a drum inside, which revolved inside a concave grid; beaters on the drum knocked the grain out of the ears of corn against the grid. Its workings are illustrated in this example, by Ransomes, Sims & Head, of Ipswich. The corn to be threshed is fed into the drum from the top of the machine, one man standing there to cut the bands from the sheaves and to feed the machine (some machines had a safety guard to prevent accidents). The drum, revolving at about 1,000 rpm consisted of eight ribbed bars, which beat out the grain against the stationary bars partially encircling it and forming the 'concave'. The gap between the concave and the drum could be adjusted to suit different sorts of grain. The straw and corn from the drum **A** are thrown upon the lower end of the rotary shaker, formed of 15 revolving drums **BB**, each armed with three curved teeth. The straw is carried upward, to be ejected out of the front of the thresher. The corn and short straw (cavings) pass down onto the caving riddle **C**, which delivers the cavings out behind the forewheels of the thresher. The chaff and grain fall onto the collecting boards **DD**, and, passing over a fine screen or dresser, have all their seeds removed. Here the blast from the fan removes the chaff, it being blown onto the revolving screw **H**, from which it is bagged. The corn, falling through **E**, is elevated by a conveyor belt **JJ** to the hummeller **K**, which removes any residual chaff adhering to the grain, from which it passes to a second blast **M**, then to the corn screen **N**, which grades the corn before ejecting it into sacks.

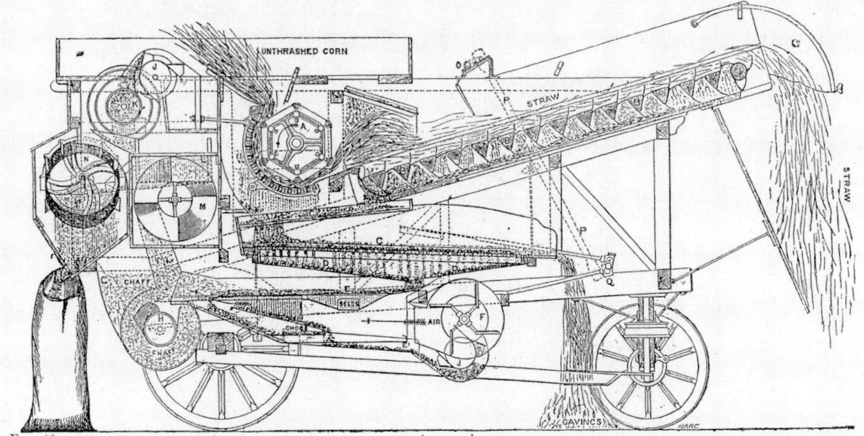

FIG. 40.—LONGITUDINAL SECTION OF MESSRS. RANSOMES, SIMS & HEAD'S FINISHING THRESHING MACHINE, WITH ROTARY SHAKERS.

A, drum; *a*, concave; B, straw shakers; *b*, inclined board; c¹, shog-board; C, caving-riddle; D, D, collecting-boards; E, dresser; E¹, corn-shoot; F, fan; G, perforated plate; H, revolving screw; J, J, elevator cups; K, hummeller; L, oscillating sieve; M, second fan; N, rotary screen; O, vibrating fork; P, connecting-rod; Q, crank-shaft.

The late nineteenth-century threshing machine and its ancillary equipment dominated the processing of cereals until the coming of the combine harvester. By 1880, about 90 per cent of the grain crop was threshed by machine. A full threshing set would comprise the threshing machine itself, the engine (usually a portable steam engine) to drive it, a chaff cutter and an elevator to build the straw stack after threshing. It took a sizeable team to operate a threshing set: engine driver; drum feeder; two or three men on the corn stack, passing the sheaves to the machine; one or two cutting the bands from the sheaves; one clearing the chaff and cavings as necessary; and one taking off the sacks, tying and weighing them. The cost of the equipment meant that only the larger farms owned their own sets, and most farms would hire them from a contractor. A team could thresh between 300 and 400 bushels of wheat or barley a day (7–10 tons), while a man with a flail could only deal with about one-fifth of a ton, and his final product would be of lower quality than the machine-threshed grain.

Ransomes, which serves to indicate the importance of steam products in the period up to 1914:

Table 6.2 *Ransomes' sales structure, 1878–82 and 1909–1913 (as % of output)*

	1878–1882	*1909–13*
Ploughs	32	23
Field implements	8	4
Threshing machines	17	23
Lawn mowers	4	9
Engines and boilers	35	37
Others	4	6
	100	**100**

Source: University of Reading, Rural History Centre, Ransomes, Sims & Jefferies archive, TR RAN AD 7/17.

The importance of the engines and boilers is clear, accounting for a little over one-third of all sales at the start and the end of the period. Ploughs and field implements were in relative decline, as presumably sales of engines and boilers were boosted by the export trade, especially to Russia. The value of sales of engines and threshing machines together amounted to more than half of all sales for most of this period, up to 1913. The association between engines and threshing machines is obvious. Steam products clearly dominated Ransomes' output in this period. The only non-steam product to show considerable growth was the lawn mower. In the late 1870s, it accounted for about 4 per cent of sales, and this had risen to about 9 per cent by 1913. Since the lawn mower trade was mainly to the home market, this reduced slightly Ransomes' dependence on exports, but not by very much.

New products

Dairy machinery

The decline of arable farming in the late nineteenth century and the rise of meat and milk production should have given some opportunities for new markets to British manufacturers of farm machinery. While this was true to some extent, the opportunities were limited, because before the First World War such opportunities were largely confined to the dairy sector, which had limited potential.

The main processes in dairying were milking, the separation of cream from milk and the manufacture of butter and cheese. Many attempts had been made to develop a workable and hygienic milking machine. By the 1890s these were reaching the stage of commercial development. In 1898 the Highland and Agricultural Society held a trial of milking machines, and but only two types were entered, one of which received a prize of

£50. Two years later the RASE also held a trial, and again only two types were entered. One of these, the Lawrence and Kennedy Universal Cow Milker, was the first really practical milking machine. In 1905 J. & R. Wallace, of Castle Douglas, showed their machine to the RASE, and in 1910 a machine by Vaccar Ltd was also on the market. But on the eve of the Great War in 1914 mechanical milking was a rarity, and another quarter of a century was to pass before the machines became reliable enough for mass use.[15]

Since milking remained unmechanised, and the making of butter and cheese had for many years been mainly done in factories rather than on farms, this left only the separation of milk from cream as a possibility for exploitation by the manufacturers of machinery. The process of separation, the essential preliminary to the making of butter and cheese, had traditionally been done by letting milk stand, so that the cream rose to the top of the mixture. This was slow and potentially unhygienic, as the milk could acquire germs from prolonged contact with air. The cream separator consisted of a metal bowl on a pedestal; milk was poured into the top of the bowl, which was then rotated at high speed (3,500–4,000 rpm), being driven by a vertical axle below it. The spinning separated the lighter cream from the heavier skimmed milk, which was forced to the upper side of the bowl, to discharge itself through a pipe. The machine could be powered by hand, horse gear, or an engine.[16]

Although known for some years on the Continent, the British market for separators began with the trial of the Swedish De Laval machine at the 1879 Kilburn show. An analysis of the results of the working of the machine by Dr Augustus Voelcker, the Consulting Chemist to the RASE, '... afforded a striking proof of the perfect manner in which the butter-forming constituents are separated from milk in passing through Laval's rotary machine'.[17] The machine proved popular, and various manufacturers entered the market. At the 1889 Bath and West Show there were five power-driven machines, and there were at least six makers producing manually operated machines.[18]

One of the most successful models of cream separator was that introduced by Lister & Co. of Dursley, Gloucestershire. Robert Ashton Lister had begun in business as an engineer in 1867, and some time after that began to specialise in agricultural machinery, producing barn machinery such as chaff cutters, grinding mills and cheese presses. In 1889 he met Mikael Pedersen, who had developed an improved cream separator. This had solved the problem of keeping the bowl spinning at a fixed speed even if the motive force varied (as it would if produced by man or horse). Lister immediately acquired the sole rights to market it, under the name of the Alexandra Cream Separator (a neat patriotic reference to Princess Alexandra, the future King Edward VII's popular Danish-born wife) in the UK and colonies, and it was a great success. It was assiduously marketed by Lister in the UK and Australasia. By the time production

VIPAN & HEADLY, Dairy Engineers, LEICESTER, ENGLAND.

The "Little Gem"

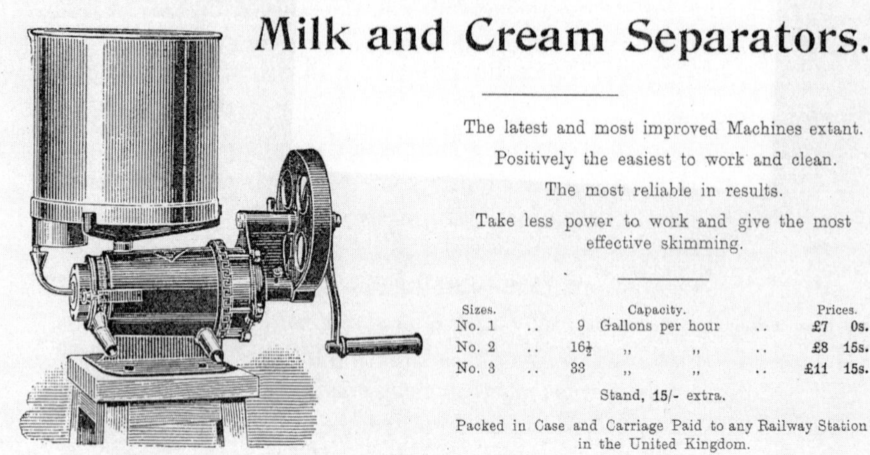

Milk and Cream Separators.

The latest and most improved Machines extant.

Positively the easiest to work and clean.

The most reliable in results.

Take less power to work and give the most effective skimming.

Sizes.	Capacity.	Prices.
No. 1 ..	9 Gallons per hour ..	£7 0s.
No. 2 ..	16½ " " ..	£8 15s.
No. 3 ..	33 " " ..	£11 15s.

Stand, 15/- extra.

Packed in Case and Carriage Paid to any Railway Station in the United Kingdom.

The "Little Gem." Can be worked by a child.

The "Little Gem." A model of simplicity and durability.

The "Little Gem." Easy and silent in motion, giving the maximum results with the minimum of wear and tear.

The "Little Gem." Has very ingenious and practical gearing arrangement.

Latest Award—GOLD MEDAL, PARIS EXHIBITION, 1900.

DESCRIPTION.

The driving gear consists of a roller gearing, combined with a spur gear and pinion, giving the bowl a velocity of 9,880 revolutions, when the crank is rotated at a speed of 65 turns a minute. All wearing parts of the roller gear are made of steel, hardened and ground. The spur pinion is hardened and ground and journalled in an exchangeable bushing of gun-metal. The gears are provided with a guard. The bowl or centrifugal vessel, the most expensive part of the Separator, has no shaft, and its only wearing part, the gun-metal bushing in the lid, can be readily exchanged by any person. The bowl, therefore, need *never wear out.*

The bowl has inside partitions, made in two parts, which are readily cleaned, and serve to increase the skimming capacity in an efficient manner, partly by preventing the inflowing new milk from mixing with the cream and milk already separated, and partly by conducting all skim milk to the periphery of the bowl, where the centrifugal force is at a maximum.

Cream separator by Vipan & Headley, Leicester, from their 1900 catalogue.

ceased in 1903, some 40,000 had been sold. By that time Listers had the reputation of being the biggest general dairy equipment maker in the UK. They followed this success by marketing their own separator, took on the agency for the Belgian Melotte separator, and diversified into other areas, notably the brief run with the eccentrically designed but high-quality Dursley-Pedersen bicycle. By 1900 the workforce of the business was about 200, and by 1914 had risen to 1,000. The Melotte sales, which were conducted through a separate limited company from 1907, were very successful, and continued until 1933. In 1907 these sales were worth

£101,000, and produced a profit of £51,000. In 1913 sales were down to £55,000, and profits down to £21,000.[19]

Other firms joined in the cream separator market, including Fullwood & Bland; Watson, Laidlow & Co. of Glasgow; the Dairy Supply Co.; and Vipan and Headley of Leicester, whose hand-powered 1900 machine is illustrated on page 122. In addition to these, there was considerable competition from imports, of which the most effective was still the De Laval machine. The trade journal *The Implement and Machinery Review* reported in 1908 that there were only three makers in the UK at that time, and welcomed the coming of a fourth, the Wolseley S.S. Machine Co.'s new factory at Birmingham,* where the Wolseley-Pedersen separator was to be manufactured, with the aid of a workforce of 170.[20]

Internal combustion engines

The other major development in new markets was that associated with the internal combustion engine, both stationary and self-propelling engines, the latter developing later into the farm tractor. But there is no doubt that most work was done on the stationary engine, and that it had a much bigger market on the eve of the First World War than the hesitant venture into tractors, whether for farm or other purposes.

The history of the early development of the internal combustion engine is well known. The earliest successful ICE was Lenoir's gas engine of 1860, although gas engines did not find much use on farms. There was more future in the agricultural market for engines fuelled by petroleum – either the cheap, heavy oils, or the lighter paraffin and petrol. The most successful type of engine derives from that of Otto, who in 1876 produced the four-stroke engine. He used gas as a fuel, but by the 1890s developments had taken place which permitted the use of petroleum. A further breakthrough was made by Ackroyd-Stuart, who developed the hot-bulb oil engine. Starting this required the use of a blowlamp to heat a bulb containing the fuel, which then vaporised, and could be ignited in the combustion cylinder. Once the engine was running, the blowlamp could be dispensed with, as ignition was maintained by the heat produced by compressing the vapour, aided by heat retained in the hot-bulb after each explosion. The Ackroyd-Stuart patent rights were bought by Hornsbys of Grantham, who produced the first 'Hornsby-Ackroyd Oil Engine' in 1891. By 1905 the Grantham works had sold over 10,000 engines, and had totally abandoned steam. In the same year they had further diversified by acquiring the Stockport firm of J. E. H. Andrew & Co. Ltd, which had developed a highly successful range of gas engines.

---

* It is not known if this machine was invented by the same Mikael Pedersen, who had come to live in Dursley. The fact that it had been on the market since 1905, and that was the year that the Dursley-Pedersen Cycle Co. was wound up, suggests that Pedersen then broke away from Listers. The initials 'S.S.' stand for 'Sheep Shearing', since Wolseleys had begun by making hand-operated sheep-shearing clippers.

The early oil engine was a pioneering effort, and very successful, partly because it ran on cheap, low-grade oil, but it was probably too large and heavy to have much of an agricultural application except on the larger farm, in spite of the makers' efforts to promote its use in agriculture. The same was true of Diesel's heavy-oil engine which was also developed, after the Hornsby-Ackroyd, in the 1890s. This relied on very high compression to start the initial ignition, and was consequently very massive and heavy, with thick cylinder walls. By 1900, however, development of the smaller four- and two-stroke petrol and paraffin engines had got to the stage where they could be fairly easily manufactured by small- or medium-sized engineering firms, and many thousands were being produced annually.[21]

Other established farm machinery firms also bought into the rising market for internal combustion engines. The modest-sized machinery manufacturers Carters of Billinghurst, West Sussex, established in 1885, had two very creative engineers, in the persons of two of the brothers of the founding family, Frank and Tod. Their 'Reliance Works' produced a variety of farm and other machines, including steam engines. In 1894 they took out their first patent on an oil engine, and followed this up with an improved patent engine in 1896. Their first engine, the 'Reliance', had caught the attention of Blackstone's representative at the Darlington RASE Show of 1895. He had been instructed by his managing director, Edward Blackstone, to 'find a good engine and somebody who knows all about them'. The upshot was an agreement in 1896 whereby the Carter brothers licensed their patents to Blackstones and themselves became employees of the firm. They went on to become works manager and chief engineer respectively, and produce many new designs for Blackstones. For Blackstones, it was the beginning of a great success story in the oil engine business. The original Carter firm stuck to making farm machinery, notably their hay and straw elevator, and was in business until at least 1955. Some of the largest firms in the business also entered the internal combustion engine market: Clayton & Shuttleworth, for example, had a large section devoted to the manufacture of heavy oil engines by 1914.[22]

Established manufacturers moved into the farm engine market with petrol engines. R. A. Lister & Co. of Dursley began to make one in 1908, to designs by Southwell & Co., of London. There were three sizes, of 2, 3½ and 5 hp, all aimed specifically at the farm market. Listers also used the opportunity to market these engines to drive their sheep shearing sets, production of which also began in 1908. In the same year Bentalls, best known for their barn machinery, brought out its own line of petrol engines, designed primarily to drive Bentall machinery, but also sold separately. The early vertical engines were supplemented by the first horizontal engine in the UK, the Pioneer, in 1912, and this was to continue in production until around 1930. The Bentall engines were simple and economical, and 'many thousands' were sold, winning medals at exhibitions in Brussels and Turin.[23]

See Stand No. 248, Smithfield Show, London, December 7, 8, 9, 10 & 11.
THE LISTER SAFETY PETROL ENGINE
(Formerly known as the "Southwell").
Specially suited for
CHAFF CUTTING,
GRINDING,
PULPING,
SAWING,
PUMPING,
DAIRIES,
ENGINEERS,
BLACKSMITHS,
WHEELWRIGHTS,
&c.
THE SOUTHWELL ENGINE
2 B.H.P.
Guaranteed Powers:
2, 3½ and 5½
BRAKE HORSE POWER.
ANY SIZE MADE PORTABLE IF DESIRED.
Extremely Simple and Durable.
Write for further particulars and terms to
R. A. LISTER & Co., Ltd., DURSLEY, GLOS.

Advertisement for Lister's stationary petrol engine, 1908. Listers were also acting as import agents for Cockshutt ploughs. From *The Implement and Machinery Review*, 2 November 1908 (*Museum of English Rural Life*)

By the early twentieth century, farmers had a considerable range of static power sources, ranging from the early portable steam engines to the more recent petrol engines made by the leading firms such as Listers and Petters. The range and total power available is indicated in the first *Census of Production* for British farming, conducted in 1908:

Table 6.3 *Power supply on British farms, 1908*

engine type	*number*	*total horse-power*
Steam	16,959	106,460
Oil	12,807	84,240
Petrol	1,347	5,041
Gas	1,855	9,504
Others/unclassified	1,482	8,280

Source: Board of Agriculture and Fisheries, *The Agricultural Output of Great Britain*, Cd. 6277 (1912), p. 21.

Not surprisingly, most of these engines were employed on the larger farms: some 89 per cent of the steam engines, and of the oil engines, 87 per cent of the petrol engines, and 57 per cent of the gas engines were employed on holdings above 50 acres.

From the census figures one can calculate the approximate average size of engines. The steam engines averaged about 6 hp, the oil engines at 7 hp, petrol at about 4 hp, and gas at 5 hp. The fact that the oil engines had about the same average power as the older stock of steam engines suggests that they were mainly the early, massive type such as Hornsby's Ackroyd-Stuart. It is very notable that the number of oil engines had almost caught up with the number of steam engines, although they had only been on the market since the early 1890s. Petrol engines were less common, partly because many farmers feared risk of fire. The low number of gas engines reflects the fact that few farms had a gas supply. The usage of internal combustion engines was to grow much more rapidly soon after this. In 1913, when the next census of agricultural production was taken, 16,284 oil or petrol engines were recorded on farms in England and Wales alone.

Tractors

Confusion can be caused by the use of the word 'tractor'. The large steam-driven traction engines of the late nineteenth century were sometimes referred to as tractors, simply meaning a self-propelled machine for pulling loads, mainly on made-up roads. These were, for the most part, too heavy to be used on cultivated farmland. The modern farm tractor may be defined as a self-propelled machine which can be used to pull field machinery such as ploughs and harrows across farm fields.

There is an ongoing debate as to who built the first motor tractor. It is probable that the first farm tractor in the world powered by an internal combustion engine started work in South Dakota in 1889, on the farm of Mr L. F. Burger. Its first job was to operate a stationary threshing machine. The tractor had been built by one John Charter, using the simple expedient of installing a large 25 hp gasoline engine on the running gear of a Rumely steam traction engine. It was sufficiently successful to have a further six made and sold to farmers in the region. Charter made no more, but by 1892 three other US makers had appeared, and the age of the farm tractor had dawned.

Early tractors were derived from steam traction engines, and in many ways they resembled each other, not least in size and weight. The US tractors of Case (from 1892), Hart-Parr (1902) and Rumely (1908) were all massively built and very heavy. This did not prevent them undertaking ploughing and other field operations on the hard, dry soils of North America; but they would have been unsuited for direct traction in the wetter climates of north Europe. In the UK the first tractor, the Hornsby-Ackroyd, was produced in 1896 by Hornsby, using the Ackroyd-Stuart engine referred to above. It was, however, still essentially a traction engine, and was too heavy to be suitable for field operations. It was exhibited at the Smithfield Show in 1896, and was afterwards sold to Mr Locke-King of Weybridge. No more were sold in Britain, although several were sold overseas, mainly in Australia.[24]

The first true farm tractor manufactured in Britain was produced by Dan Albone, a small-scale bicycle manufacturer and engineer of

The first British internal combustion engine 'tractor': the Hornsby-Ackroyd. Although referred to as a tractor it was really a traction engine. (*Museum of English Rural Life*)

Biggleswade, Bedfordshire, in 1903. Not having a background in steam engineering, he looked to the fledgling automobile industry for his design. The result was a light, manoeuvrable three-wheeled machine, which performed most farm operations, static or mobile, except rowcrop work.[25] The Ivel was small, measuring 5 ft 4 inches × 9 ft 9 inches, and light, weighing just 1.7 tons. The standard model had one speed, about 4 mph; A double-gear version gave two speeds, but it did not have reverse gear. It was rated at 18–20 hp, and could be modified to work on either paraffin or petrol.[26] It was not particularly cheap, costing £300 in 1907. The firm did not make its own engines, but bought them instead from a Coventry firm.

Albone's works manager, J. Boswell, recalled many years later that Albone, 'one of the great penny-farthing cyclists of his day', was originally the landlord of the Ivel Hotel, Biggleswade ('All the Great North Road cyclists and celebrities stopped at Dan Albone's hotel at one time or other'). The workshop at the rear of the building having been vacated by a failed cycle firm, Albone took it over, and began making cycles, moving to experiments with tractors in 1902. According to Boswell, Albone had not served an apprenticeship, but was self-taught. Unfortunately, Dan Albone died in 1906, and the tractor does not seem to have been developed further, although the firm set up to make and market it, the Ivel Motor Co., continued to operate. It is not known how many Ivels were made, although it cannot have been many. It was a small firm, its issued capital in 1910 only £8,857. By then it was making losses (the highest being £955 in 1912). The firm was probably too small to be awarded munitions work by the government during the war, and the cost of raw materials and labour were rising. In 1916–18, lack of labour prevented it making any new tractors, and it merely serviced existing ones. In spite of some success in overseas sales, of acquiring the agency for a leading US make, the Hart-Paar tractor, it made losses in every year from 1912 to 1918. Briefly in profit in 1919, and helped by loans (a total of £700) in 1918–19 from one of its directors, the famous pioneering motorist S. F. Edge, it was compelled to go into liquidation on 20 October 1920.[27]

The Ivel is often held as an example of an excellent product which could have prompted the larger manufacturers to develop a British motor tractor before 1914, thereby staving off the dominance of the Fordson after 1917. However, other manufacturers were well aware of the Ivel, and actively decided not to develop small or medium tractors. J. E. Ransome

Ransomes' petrol tractor, 1904, shown ploughing. (*Museum of English Rural Life*)

designed a motor tractor in 1903, and it was built at Ransomes' Lawn Mower Works. One was sent out to a Prussian nobleman, Prince Schoenburg-Hartenstein, who was among other things an adjutant to the Kaiser, and very rich ('fortune runs into millions', is the note which survives in the firm's archive). But if the firm had visions of selling motor tractors to large German estates, hopes were dashed, since the tractor was not strong enough for the work (deep ploughing on heavy soil), and it was sent back to Ransomes. The firm decided not to proceed with the tractor, 'owing to the great expense required to fully develop it, and the doubt as to a satisfactory demand'.[28]

The largest producer of motor tractors in Britain before 1914 was Marshalls of Gainsborough. Their product was nothing like the Ivel, being a large traction-engine type designed for use on the prairies of North America. The first one was built in 1906, with a large, 30 hp kerosene engine. Four classes of tractor were produced, and by 1912 over a hundred had been sold to North America. In that year the imposition by Canada of a large duty on imported tractors induced the company to set up a subsidiary there, but this was to make large losses during the Great War. Overall, in the period from 1908 to August 1914, over 300 of these oil tractors were produced, mainly for the overseas market.[29]

Fowlers of Leeds had early on shown an interest in the internal combustion engine. R. H. Fowler applied for a patent for an oil engine in 1904, but it is not known if this was for direct traction or cable ploughing. The first self-moving Fowler internal combustion motor appeared in 1909, and their first commercial tractor appeared at the 1911 RASE Show in Norwich. This was essentially a traction engine, designed for road haulage use, and developed the enormous power of 50 bhp. The next year, six of an improved design were built, and five more in 1913. It is clear that Fowlers was entering the road, rather than the farm motor tractor market.[30]

Other firms entered the market too. Fosters of Lincoln, seeing the large sales of US motor tractors to South America, decided to develop a similar tractor for that market. The first, fitted with a Hornsby engine, developed 60 bhp. A 40 bhp model followed, which, it was claimed, would pull fourteen disc ploughs or eight breast ploughs. It weighed 6 tons, and was first demonstrated at the 1911 Doncaster Show. Later models used Daimler engines and were known as Daimler-Foster tractors. They would play a military role after 1914, but few seem to have been used on farms.[31] Hornsby's venture into the motor trade also had a military angle. In 1908 it won a prize, offered by the War Office, for a mechanical tractor which would run for 40 miles without refuelling. In the same year the firm successfully demonstrated a caterpillar-tracked motor vehicle at Aldershot. This received wide publicity. However, the idea was not commercially successful (although a spectacular-looking steam caterpillar tractor was sold for use in the Yukon), and the rights in the tractor were sold to the Holt Caterpillar Company, later to emerge as the Caterpillar Corporation.[32]

There were three major producers of British tractors in 1914: Marshall, whose machines were designed for export; Ivel, who cannot have produced more than a few hundred machines; and Saunderson of Elstow, Bedfordshire. Although one writer considers that Saundersons was the biggest tractor producer before the war, very little is known of the firm, save that it produced a variety of popular, medium- and large-sized four-wheel tractors suitable for British conditions, from 1908, having earlier experimented with three-wheel models. In 1910 Saunderson advertised seven models. His 'Universal' 30 hp model weighed 4 tons 13 cwt; there was a smaller model of 3 tons 11 cwt. Saunderson tractors had some success, and were exported to Russia, South America and Australia. The firm sold its interests to Crossley Brothers in 1923, although some tractors were still made until the end of the decade.[33]

By 1913 there were also several small motor ploughs on the market. These had only two large wheels, and were steered by a man walking behind, or sitting above the ploughs. The Weeks-Dungey 'Simplex' and the Wyles Motor Plough made by the Crawley brothers were two such examples. Three of the Wyles were made in 1913–14 by Garretts, being referred to as the Garrett-Crawley Agrimotor, before the Crawley brothers built a factory of their own, until they closed down in 1924. Finally, the steam engine manufacturers had not entirely given up the idea of using some of their (comparatively) lighter products for direct use on fields. At the last Tractor Trials held by the RASE before the Great War, two models of Ivel and two also of Saundersons, were pitched against steam tractors by Mann, McLaren, and Wallis & Stevens; the steam tractors all weighed in the region of 4 tons.[34]

In spite of some pioneering successes by firms such as Ivel and Saunderson, most British agricultural machinery manufacturers turned away from the development of the motor tractor before 1913, and continued with steam. The most successful foray into motor tractors was that of Marshalls, and their machines were not designed for use in northern Europe. The failure to develop a British tractor on modern lines has been deplored by many British authors. Yet on commercial grounds the decision could be justified. The development time and costs were so high that only a mass-production company such as Ford could sustain it: even they, who had been experimenting with motor tractors since 1908, did not release the Fordson until 1917. None of the British companies had the economies of scale of the US giants. Finally, in 1908–9 the whole British industry was enjoying the last great export boom of the pre-war period, which was to take sales and profits to new heights. In these circumstances, it could well be argued that to do anything other than concentrate on existing lines of business would not be in the best interests of the firm or their shareholders.

The legal framework, scale of production, and profits

The only legal forms available to the early firms were the sole proprietorship or the partnership. Having only one, or several, people in charge made for clear decision making, but had disadvantages, the main one being that the capital resources of a firm were limited to the amounts which these individuals were prepared to invest. There was also a legal limit of fifty on the number of partners, although, in practice, problems of control kept the number of partners well below this. In addition, proprietors or partners were responsible for the debts of the firm, without upper limit, so could be made to pay in the case of failure 'up to their last penny and their last farthing'. The Companies Acts of 1856 allowed the creation of joint-stock companies, that is companies with more than fifty shareholders. The newly formed companies could now increase their capital by issuing shares to the public, up to the amount permitted in their Articles of Association. The remaining problem, of unlimited liability in case of failure, was resolved by the 1862 Companies Act, which permitted companies to opt for limited liability. This meant that shareholders were only liable for the debts of the firm up to the value of their own shareholding.

The 1862 Act was a revision of the earlier one, requiring among other things that when forming a company seven persons subscribe their names to a Memorandum of Association, describing the aims and financial structure of the company. Although these Acts were eagerly taken up by some, others were slow to adopt them, perhaps fearing loss of control to an enlarged body of shareholders, or simply not feeling the necessity, since they were making good profits and so had no fear of failure, as well as plenty of capital of their own to reinvest. In practice, incorporation was often not resorted to until the firm had grown to the stage at which the founding family wished to cash in on its success and sell some of the firm to outsiders, via an issue of shares. This was the case with Hornsbys of Grantham. In 1877 Richard Hornsby Junior died, and, since his two sons were not interested in continuing with the business, the family of Richard Hornsby Junior withdrew the larger part of their capital from the firm, and it was decided to turn it into a public company. In the words of the prospectus: 'The death of the late Mr Richard Hornsby, the eldest son of the founder of the business, and the consequent withdrawal of a considerable amount of capital, have induced the Vendors to convert the business into a Company.' This occurred in 1879, when the company had grown to occupy 15 acres of land (of which 10 were freehold), and the purchase price paid for the firm was £235,000, not including a sum for goodwill.[35]

Although there were periods during which there were high rates of incorporation – the two decades years of so before 1914 probably being the most active – the decision to adopt the corporate form, with or without limited liability, must have been very much *ad hoc*, depending on a variety of factors such as the size and speed of growth of the firm, the anticipated requirement for future capital, and the financial position of the founder and his family. Generalisations are very difficult to make, and some specific examples to illuminate some of the forces at work may well be more helpful.

The earliest case of incorporation on the part of one of the firms which came to dominate the industry was Marshalls of Gainsborough. Subsequent history makes it clear that the motive for incorporation was a need for capital to continue the firm's rapid rate of expansion. William Marshall acquiring his first small foundry to make threshing machines in 1848; the firm moved to larger premises in 1855, and began to make steam engines in 1857. The firm was a partnership, trading as William Marshall & Sons. By 1861 it had grown from a one-man operation to one employing 21 men and 10 boys, and, although still small, had a national reputation for quality. In June of that year the founder died. The two sons remaining in the partnership took professional advice. They viewed the prospects of the firm as very promising, but needed capital for expansion, and so in 1862 the firm became a limited company, styled as Marshall, Sons & Company Ltd. The nominal capital was fixed at £20,000, in 1,000 shares of £20 each. The shareholders were the two sons, James and Henry, and five independent shareholders. However, the two sons retained control by holding over 50 per cent of the shares. The directors accepted modest remuneration, the brothers being made joint managing directors (James being chairman and Henry company secretary) at a salary of only £300, but James lived on the premises and received £10 *per annum* for entertaining. The next few years were very prosperous, and the company usually felt able to pay dividends of 10 per cent, but still needed fresh working capital as the business grew. In 1864 the nominal capital was raised to £50,000, and in 1865 shareholders were offered debenture bonds up to £10,000. Later increases in the nominal capital occurred in 1868, 1875 and 1876, and in 1884 it reached £200,000, by which date the workforce had reached 1,190. By 1910 sales were valued at £707,266, the greater part of which (£625,903) were in export markets. By 1913 it was the largest firm in the industry, employing the largest number of men, and with the highest sales. The firm employed 4,150, and annual profit was £50,568. In 1905 Henry Marshall died, leaving a fortune of £443,523.[36]

Marshalls was the biggest firm of all. Within the second rank there were firms such as Samuelsons. Its financial records have not survived after its first shareholders' meeting as an incorporated limited company in 1889, when it was only about half as large as Marshalls, being capitalised at a nominal capital of £150,000, of which only £100,000 was issued in shares. However, the firm was making profits at a not dissimilar rate – £14,000, of which £10,000 was to go in dividends, £3,500 to the reserve fund, and the balance carried forward to the next. This may be compared with Marshall's net profits (in 1884) of £7,000 after payment of dividends, interest charges and taxes. In view of the comments above on the difficulties faced by Samuelsons in the late 1870 and early 1880s, and the fact that Samuelson family members dominated the new board, it may be surmised that, the firm having recovered from a bad period, opportunity was being taken to float the company, and bring in outside capital, thus rewarding the Samuelsons without losing control.[37]

An example of a medium- to large-sized enterprise which came late to incorporation is Harrison, McGregor & Co. of Leigh, near Wigan, Lancashire. It was founded in 1872 under this name, but in fact seems to have been a partnership. Other examples (for instance, Bentalls) can be found of this device, which may have been used to promote the company's image without the trouble and consequences of incorporation. Its early business was mowers for farm use. The earliest surviving accounts are for 1875, when sales were recorded as £28,222. Capital was £5,581, supplied by three partners (Henry Harrison, Alexander McGregor and George Rich). Profits were £2,654, divided two-fifths each to Harrison and to McGregor, and one-fifth to Rich. By 1878, when this early run of accounts ceases, capital had grown to £11,492 (from the same partners, with £610 from Mrs Harrison), sales were up to £32,613, but profits had fallen to £1,685. It is not known how the firm fared thereafter, until in 1891 it was incorporated as a limited company. At the first shareholders' meeting, in 1892, a dividend of 7½ per cent was declared. In the following year it was resolved to double the authorised capital from its existing £50,000. By 1895 gross profits were £9,224, of which £2,498 went to pay the dividend of 5 per cent, and £6,000 was put into the reserve. For the rest of the 1890s profits were around the £8,000–£10,000 level, although 1898 was much higher, with gross profits being £13,079, and the dividend 7½ per cent. In the first decade of the new century, information is limited to the dividend level, which was usually 5 per cent. In 1900 an Extraordinary General Meeting was held, at which it was agreed to raise the authorised capital to £200,000. However, not all of this was issued. When full information is next available, in 1911, the value of sales (but including increase in the value of stock in hand) was £190,551, the share capital issued £117,704, and profits £14,387. A dividend of 7½ per cent was declared, and £10,000 carried to the reserve.

This history of Harrison McGregor's finances suggests that its early policy was not one of re-investing its profits. The 1875–77 surpluses were distributed entirely to the partners, and the low level of profit in 1878, a bad year for the firm, conceals loans from the partners to the firm of almost exactly that amount (£1,682). How the firm survived the 1880s is not known, but the incorporation in 1891 was clearly an opportunity to increase the firm's capital, and thereafter trade boomed. By the First World War, it was one of the larger firms in the business, having survived lean years to ride the crest of the boom which began in the 1890s.[38]

The firm of Nalder & Nalder, Challow, Berkshire, is an example of one which was incorporated early, yet failed to grow very much, remaining at the lower end of the medium-sized company sector. The main products were threshing machines, coffee screens, malt and straw elevators, and malting and brewing machinery. The founding date is uncertain, but the first extant wages book dates from 1857, when there were 14 employees. In 1866 it was incorporated as a limited company. The earliest sales figures are for 1870, when they totalled £10,660. By 1876, when the first balance sheet is available, sales had risen to £19,704, and the capital employed is recorded as £17,225. Gross profit was £2,391, of which £1,722 was distributed in cash or shares to the shareholders, and £400 was put to the reserves, which then stood at £2,000. Sales were at a similar level until 1885, when they dropped sharply, and carried on dropping. By the time the next balance sheet is available, in 1896, sales were down to a low of £12,610. Capital was up, but not very much, at £25,000. This increase seems to have occurred as partly a result of issuing debentures, which had only been worth £1,100 in 1876, but were now worth £5,500, and thus a larger charge on the profits. After charges, net profits were only £619, of which £500 would go to the ordinary shareholders, and the balance carried forward.

The firm was obviously in trouble. When full financial information is next available, in 1907, things had improved somewhat. Sales were up to £17,456, and debenture capital had been reduced by £500 (but was still £5,000, and the reserve still only £2,024). Due to the rise in raw materials prices, which was due in turn to 'keen competition', a loss of £934 was made after charges. Things had improved by 1913, but not by much. The capital structure was unchanged. Sales had risen to £22,493. The loss outstanding from 1911 (the financial year was changed, so that the 1911 accounts were followed by those of 1912–13) was paid from the reserves, which fell accordingly, to £610. After paying 'Debenture Interest and all fixed charges, and making the usual allowance for depreciation of Plant', there was a net profit of £285. The directors commented that: 'the Works were well employed during the year, but the rise in the price of Raw Material, was not accompanied by a corresponding rise in prices of the Company's Manufactures, due to severe competition'. By the First World War, there were probably fewer than 100 employees; in 1910 the firm paid out £5,277 in wages and £585 in salaries. This level of wages would finance about 50 skilled workers (@ £2 a week) or 100 unskilled workers (@ £1 a week).

The history of Nalders leaves few clues as to why it chose to become incorporated when it did. It hardly seems to be expanding so rapidly by 1866 as to require outside capital. Nor was it the case that the firm ever got into deep trouble as a result of imprudent ventures. In fact, it seems to have reacted appropriately to market changes. The drop in sales in the mid-1880s was due to a sudden falling-off in threshing machine sales, and the firm then switched to coffee and malt screens, which by the 1900s generated most of their sales. It was Nalder's misfortune that by then competition in these sectors was too fierce to make high profits, and in the interim they had increased the debenture capital to survive, and this became a charge on profits. From the accounts one forms the impression of a very cautious firm with somewhat obsessive attention to detail. The accounts are immaculately presented, and subject to a great deal of analysis. This is the only firm in this study which went to the trouble to analyse the labour cost per machine produced, broken down further into the cost attributable to carpenters, fitters, turners and smiths. This cost-analysis might be indispensable for a large mass-production company such as Ford (or even Ransomes), working on very narrow profit margins per unit produced, but was probably a diversion of clerical energy for a modest-sized firm like Nalders.[39]

Finally, there were firms which seem to have taken incorporation as a natural progression for a family firm in the late nineteenth or early twentieth century. Such was the firm of William Elder & Sons Ltd, of Berwick-upon-Tweed. First trading as Lillie, Goodlet and Elder in Tweedmouth, in 1865, the firm manufactured agricultural implements from the outset, and won their first major award, at the Highland and Agricultural Society's show, and their first antipodean prize at the Canterbury (New Zealand) Show in 1878. They specialised in mowers, reapers and a patent machine for sowing grain and grass seeds together. By the 1880s the founder's four sons had joined the business, and it was incorporated, as a limited company, in 1904. This was followed by a move across the river Tweed to Berwick, into very splendid new premises.* The first recorded year's trading, to 31 December 1904, saw sales of £21,095,

* Ironically, in 1999, the building was the local Job Centre, though retaining the original façade.

SOME OF THE VISITORS WHO ATTENDED THE OPENING OF MESSRS. ELDER'S NEW SHOWROOMS.

Farmers gather at the opening of Elder & Son's new premises in Berwick-upon-Tweed, 1907. (*Implement & Machinery Review, 1 April 1907*) plus sales of £1,867 from a separate foundry business; net profit was only £779. By the end of 1906 sales of their own implements were worth £22,130, plus foundry sales of £1,694, to which was added agency sales of other manufacturers' goods of £8,195. In the remaining years to the First World War, sales of their own implements rose only slightly; in 1912 they were £26,909, though agency sales were now up to £12,899, and the firm had started a small subsidiary company in Glasgow in 1908 to act as a selling agency for its own and other makers' goods. The general impression up to 1914 is of a steadily growing family firm, with some original and successful product ideas, and a loyal local and regional market, yet which was careful and cautious, becoming incorporated essentially as a matter of safeguarding the family's financial position which had been slowly and steadily built up over several decades.[40]

Conclusion

By the 1870s the UK home market had exhausted the first flush of its growth. Up to the First World War the largest opportunities would henceforth lie overseas. However, both the home and the export market benefited from continued technical development and new products. In the generally favourable business environment of the nineteenth century there were few failures. The surviving records of some of the failures show a tendency to under-capitalisation, and usually a running out of new ideas. The pressure of competition in the 1880s caused some bad moments for some firms, and caused others to reassess their mix of products. The strongest technical advantage enjoyed by British manufacturers was in steam, chiefly in the form of portable engines, ploughing and threshing machinery, and traction engines. These underpinned the extraordinary growth of the industry, and in particular exports, after the early 1890s. Some notable successes were had with products such as cream separators and oil and petrol engines. Critically as it later proved, however, the British industry for the most part made a conscious decision to avoid the development of the motor tractor for direct field operations.

CHAPTER SEVEN

At the works in 1913

Expansion and the larger factory

The industry as a whole expanded rapidly after the early 1890s, buoyed up by some new markets at home, but mainly due to rising export markets. Few new firms entered the industry after about 1890, and the growth in demand was met by the expansion of existing works. Nowhere was this expansion more noticeable than in Lincoln, the most highly specialised city engaged in the agricultural machinery industry. In what could truly be called the capital of British farm machinery manufacture, some of the major firms had their principal factories, including Ruston, Proctor, Clayton & Shuttleworth, Foster, and Robey. Lincoln's excellent communications were one reason for this concentration. The River Witham, flowing in from Grantham in the south, turned east in the centre of Lincoln, at the ancient and enlarged dock basin of Brayford Pool, to exit on the coast of the Wash, at Boston. The Fossdyke Navigation stretched from Brayford Pool north-west, to join with the River Trent at Torksey only ten miles away, giving access to the river Humber some thirty miles to the north, and thus to the North Sea. The Fossdyke (originally a Roman canal) and the Witham met in the centre of the city, at Brayford Pool. Furthermore, the lines of four railway companies met at Lincoln – the Midland (1846), the Great Northern (1848), the Manchester, Sheffield & Lincoln (1848) and the Lancashire, Derbyshire & East Coast Railway (1896).[1]

In addition to this, there was a plentiful supply of labour. Lincolnshire had a large agricultural population, and few large towns or alternative avenues of employment. In the period 1871–1911, it showed considerable labour inertia, being one of the English counties whose rural population fell the least. One result of this was the holding down of farm wages. In 1907 the average weekly wage of an unskilled, 'ordinary' agricultural labourer in Lincolnshire was 15*s.* 4*d.*, almost as low as the poorest-paid ordinary labourers in England, those of Dorset, at 12*s.* 1*d.* Lincoln's factories were therefore attractive to the rural labourers, just as in the 1920s workers from rural Oxfordshire were to be attracted to William Morris's new motor car factory at Cowley in Oxford.[2]

The professional journal *The Engineer* summed up the reasons for engineering firms locating in Lincoln:

Iron and coal, if not mined on the spot, are obtainable from the neighbouring counties of Derbyshire, Nottinghamshire, and Yorkshire. The great industrial markets of the Midlands are close at hand. The immediate neighbourhood absorbs a considerable quantity of agricultural implements and machinery. Facilities for export are, as we have already indicated, excellent. Finally, while the town is in touch with all the principal home and foreign markets, its situation outside the exclusively industrial areas in its vicinity possesses advantages both from the employer's and the workman's point of view.*

One of the largest and most specialised firms in Lincoln was Clayton & Shuttleworth. Its main product lines were threshing machines, portable steam engines and traction engines, though its range also included chaff-cutters, maize-shellers, elevators, stackers and corn-mills. From its foundation in 1849 until 1906 it produced some 98,000 threshing

* *The Engineer*, 23 May 1913, 'A Visit to a Lincoln Engineering Works', p. 548.

Clayton & Shuttleworth's Stamp End Works, Lincoln, 1870. (*Engineering*, 12 August 1870)

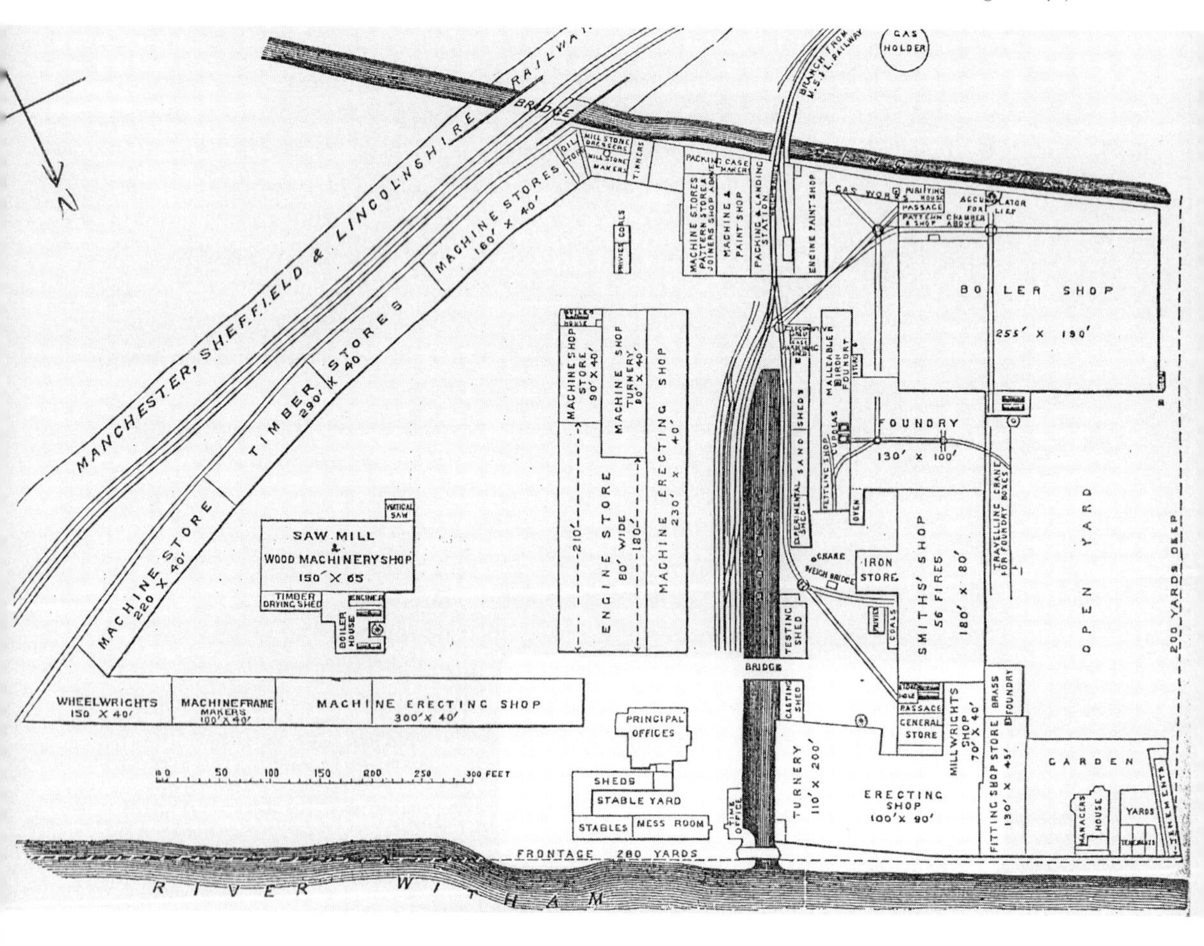

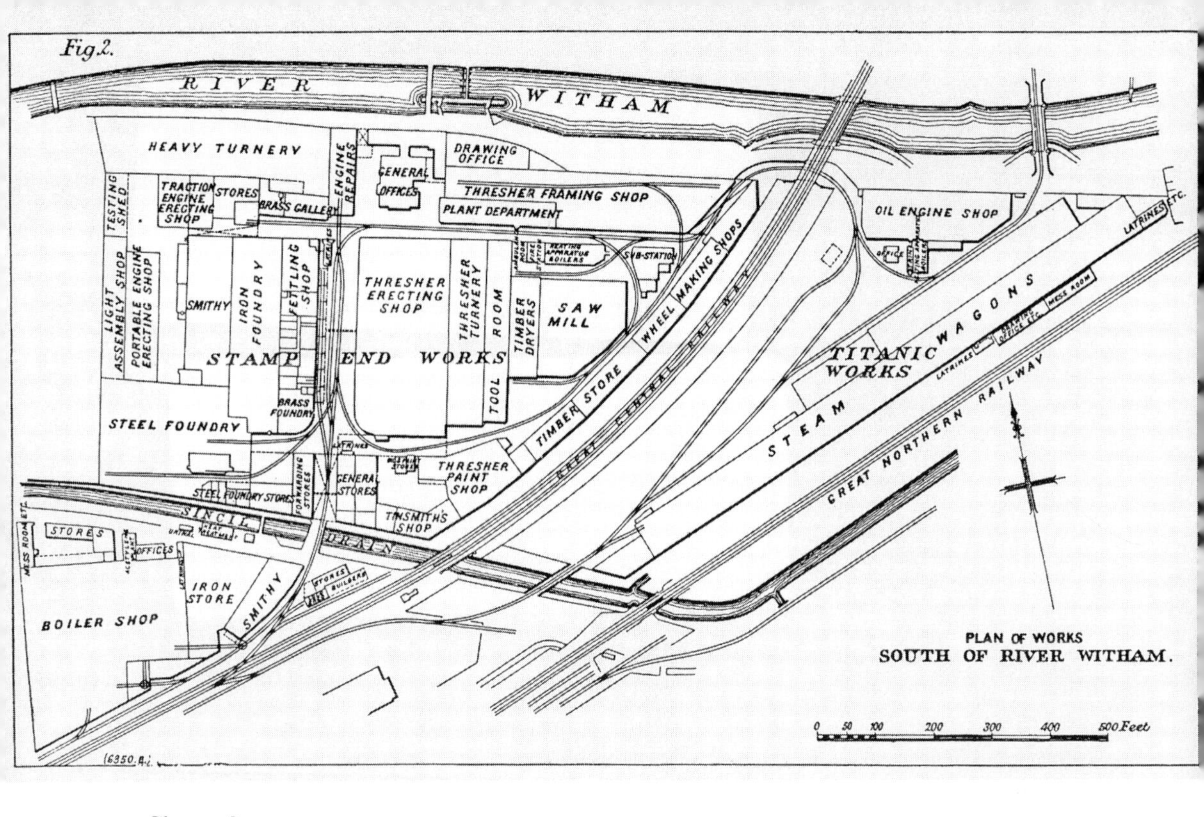

Clayton & Shuttleworth's Stamp End Works, 1920. (*Road Locomotive Society Journal*, 44/3, August 1991)

machines and portable engines. It employed 2,500 men in Lincoln, and several hundred at its branches in Vienna and elsewhere in Europe.[3]

In 1870, the Stamp End Works occupied 14 acres, on a site south of the River Witham, in the centre of the city. It lay immediately to the east of the eastern bridge crossing the river, which linked Broadgate and Melville Street, and was bounded on its south side by a drainage canal known as the Sincil Drain. Its south-eastern flank was bounded by the Manchester, Sheffield & Lincolnshire Railway. From the Witham, a dock extension ran due south into the centre of the works. It was met and abutted by sidings from the Manchester, Sheffield & Lincolnshire Railway running north. The works was thus uniquely located at the centre of the city, with excellent communications and transport links. The site was dominated by the large shops given over to boiler and 'machine' manufacture and assembly. The foundry and smith's shop are also prominent, as is the space given over to woodworking. It is notable that the manager still lived on the premises, occupying a large house and garden, and some workers also lived in tenements on the site. But there is also much unused space.

By the eve of the First World War, the site had expanded to 25 acres, and much building had been done, filling in the vacant spaces and converting some to new uses. To some extent the firm had diversified since 1870, and this is reflected in the names of the shops. The main products of the works were described in 1920 as: '... steam tractors, trailer wagons, traction engines, road rollers, portable engines, heavy oil engines, oil-driven chain-track tractors (a wartime innovation), thrashing machines, clover hullers, maize shellers, boilers and steel castings. Over 50,000 portable engines and approximately the same number of threshing machines have been turned out of these works.' Since 1870, the site had expanded to take in land south-west of the Sincil Drain, and this was now chiefly devoted to the making of boilers. The only post-1914 addition on the original Stamp End site was the building of the Steel Foundry during the war. Even more space than in 1870 is devoted to threshers. Traction engines have their own space. The manager's house and garden, and the tenements, have disappeared. The most notable addition was the building of a new works across the railway, the unfortunately named Titanic Works (built in the same year and named after the ship, as being of similar size and proportions), and devoted to the building of oil engines and steam and electric wagons for road use.[4]

The expansion of the Clayton & Shuttleworth works was merely a large example of a common trend in the years after 1900. Leading firms had recovered from the difficult times of the 1880s; home and export trades were good; profits were high. At the pre-war peak of 1906–12, the Lincoln firms and Hornsbys of Grantham together were exporting about 4,500 threshing sets a year. Continuous expansion was the rule. Sometimes events speeded change. In 1898 a disastrous fire destroyed the works of Fosters at Lincoln, and the board took the opportunity to buy a new six-acre site at Firth Road in the New Boultham district. This had the advantage of giving the firm their first rail connection. The new works was built and equipped to the highest standards of the time, and was something of a showpiece. The machines were now mostly driven by electricity, generated on the premises by large oil engines. The blast in the smiths' furnaces was also electrically driven, as were all the tools in the boiler shop, which used 4-ton hydraulic riveters, which must have been quieter than the old steam-driven ones. The move necessitated an increase in the company's nominal capital, from £90,000 to £125,000.[5]

Blackstones in Stamford expanded their works in 1904, on the back of the success of their oil and gas engines, first introduced in 1896. Ransomes of Ipswich, in the years 1900–14, built a new machine shop, machine fitting shop, plough works, and pattern shop, made extensions to the smiths' shop, boiler shop, and lawn mower works, and made 'considerable purchases of Cottages for pulling down for the utilization of their area'. Garretts of Leiston had occupied a site of ten acres in 1863. By 1907 this had risen to 'between 20 and 30 acres'. In 1913 there was a serious fire at the firm's now cramped village-centre site, and the opportunity was taken to build a completely new works close by, to the north-east, on land used by the company for farming – 'Leiston Works Farm adjacent to the Railway Station', as the 1862 catalogue had described it. The buildings on this site, which was rather larger than the original site, had not been finished on the outbreak of war, and the development was not quite completed even when the firm closed in 1932.[6]

The largest firm in Lincoln, and, along with Claytons and Marshalls of Gainsborough, one of the two or three largest in the industry, was Ruston, Proctor & Co. Established by Burton and Proctor in 1842, they were joined in 1857 by Joseph Ruston, whose dynamic leadership propelled it to new heights. In 1913 they boasted a turnover of £900,000 a year, although by then probably under half of the output was farm machinery. This had been joined by the making of gas, oil and steam engines, traction engines and road rollers, winding engines for mining, boilers, steam cranes and diggers, sugar cane mills and other items. In 1889 the works had covered 15 acres; in 1913 the total area of all the firm's Lincoln sites combined had risen to 52 acres. Again, the story had been one of continued physical expansion on cramped sites. The original works had been on the south side of the River Witham, immediately to the east of what became the Stamp End Works of Clayton & Shuttleworth. The Ruston, Proctor works, known as the Sheaf Iron Works, was supplemented by the Sheaf Wood Works, which specialised in agricultural machinery, and which lay to the west, across the High Street, in the area served by the River Witham and enclosed by Firth Road. These sites were becoming inadequate by 1900. In 1902–3, a new Boiler Works of 17,000 square yards was erected to the west, at Beevor Street, in the area known as New Boultham. This was to become the nucleus of the new Ruston and Hornsby works after 1918.[7]

Work in the factory *c.*1913

In many ways work in the Edwardian factory was little different from that in the mid-nineteenth-century factory. The pay was a little higher, and the working day a little shorter, but the work could still be messy, dirty, difficult and dangerous, although much of it could give scope for the exercise of considerable skill and craftsmanship. Unfortunately, there is no equivalent for the agricultural engineering business of the wonderful memoir by Alfred Williams, whose *Life in a Railway Factory*, originally published in 1915, went into considerable detail about the working and social life of the Great Western Railway Works at Swindon, where he worked as a 'hammerman', operating the steam hammer in the smithy/foundry. He was also a learned and published poet and was well able to paint an evocative picture of his working environment. A passage from it may help to give an idea of what went on in a smithy and forging shop:

> To stand in the midst of it and view the whole scene when everything is in operation is a wonderful experience, thrilling and impressive. You see the lines of furnaces and steam-hammers – there are fifteen altogether – with the monkeys travelling up and down continually, and beating on the metal one against the other in utter disorder and confusion, the blazing white light cast out from the furnace door or the duller glow of the half-finished forging, the flames leaping

> and shooting from the oil forges, the clouds of yellow cinders blown out from the smiths' fires, the whirling wheels of the shafting and machinery between the lines and the half-naked workmen, black and bareheaded, in every conceivable attitude, full of quick life and exertion, and all in a desperate hurry, as though they had but a few more minutes to live. And what a terrific din is maintained!
>
> You hear the loud explosion of the oil and water applied for removing the scale and excrescence from the iron, the ring of metal under the blows of the stampers or of the anvil under the sledge of the smiths, the simultaneous priming of the boilers, the horrible prolonged screeching of the steam-saw slowly cutting its way through the half-heated rail, the roaring blast, the bumping Ajax, the clanking cogwheels, the groaning shears, and a hundred other sounds and noises intermingled.*

While other branches of the agricultural machinery business, such as the paint shops, were quieter, there was still much sheer hard manual labour. At the allegedly up-to-date Stamp End Works of Clayton & Shuttleworth in 1920, 60 per cent of the castings turned out in the foundry were still hand-moulded, although the management intended to halve this proportion over time. Like other work in these factories, piecework was universal: the moulders were paid an agreed amount according to the size of each moulding box, regardless of the weight of the casting or castings. In Clayton's boiler shop, payment by results also was the rule, with the exception of the crane drivers. Even the men who came in to sweep out the shop and remove scrap were paid by a fixed price per barrow-load wheeled out. Work must have been a bit quieter in the woodworking shops, but the noise would have been of a different quality, since modern sawmilling machinery had been installed, and there were sanding, planning, morticing and spindling machines. These were all driven by electric motors, and provision was made for suction ducting to remove sawdust and shavings directly to the boiler house.[8]

Labour conditions and trade unions

As for most of British manufacturing industry, the period from the 1870s until the eve of the First World War marks the transition from a paternalistic era to one of larger, more impersonal firms, with a shift from individual bargaining to one in which trade unions were involved. Some things did not change very much. Hours of work had come down to between 54 and 57 in the period leading up to the early 1870s, and do not appear to have changed much thereafter until the end of the 1914–18 war. In 1872, Ransomes conceded a reduction in working hours from 58½ to 54, expressing the hope that the workforce would turn out as much

* Alfred Williams, *Life in a Railway Factory* (1915; repr. Alan Sutton, 1984), p. 152.

Notice to Ransomes' employees from the partners, concerning hours and conditions of work, 1872. Such publications give a sharp flavour of Victorian industrial labour relations. (*Museum of English Rural Life*)

TO THE WORKMEN

EMPLOYED AT THE

ORWELL WORKS, IPSWICH.

We have carefully considered the application made to us by a deputation appointed by you, to request us to revise the hours of labour and other regulations of the factory—to adopt 54 hours as a week's work, payable with the same wages that we now give for 58½ hours' work—and we have especially noted your views that such a change would be very beneficial to the workmen, and could in your opinion be accomplished with but little cost to ourselves, as you could do as much work in the 54 hours as you have hitherto done in 58½.

It has always been our wish that our men should have in the shape of wages a just proportion of the profit arising from our skill and capital and their labour.

The nature of our business and the occupation of our customers, who are chiefly agriculturists, prevents the fluctuation in our prices, which is usual in other trades according to the price of material and labour, and any change in our prices, as you suggest, is attended with many difficulties.

Competition of various kinds has reduced the prices of Agricultural Machinery, until it is only by the closest economy, and by the adoption of such improvements in plant and premises as you have seen us carrying out for some months past, that it has been possible for us to secure a moderate profit on the work which you have turned out.

We know therefore that our present prices will not enable us to afford the diminution of the hours of labour for which you ask, unless you turn out as much produce in the shortened hours as you do at present.

We believe that you can do this, and we have faith in your undertaking that you will try to do so. We are therefore prepared to try the experiment, and trust that you will take care that it is beneficial to you without being injurious to us.

We make this concession also with the understanding that you will continue as hitherto to work overtime when necessary in our judgment for the benefit of the business in which we are all engaged. We have adopted for this overtime the scale of payment which has been arranged by other houses in the same line of business in Lincolnshire, and in any exceptional case in which this scale appears to require it, the managers will modify it as seems to them fair and just.

Piece work prices must continue to be arranged from time to time between the workmen and managers in the same manner as they have been heretofore. We trust that the new regulations which we are prepared to adopt on January 1st, 1872, will be accepted by you in the same spirit in which they are made and will promote the good feeling and confidence which has always subsisted between us and yourselves.

RANSOMES, SIMS & HEAD.

work as before, and with the proviso that overtime would be worked when requested.

These conditions seem to have been common to the industry, whatever the size of the firm. Detail has survived for the small- to medium-sized north Wiltshire firm of R. & J. Reeves of Bratton, Devizes. While the precise size of the labour force is unknown at that time, the firm's Sick Club had 40 members in 1871, so perhaps a workforce, including apprentices, of around 60 may be a reasonable guess. The *Rules to be Observed by the Workmen and Apprentices of R. & J. Reeves & Son, Bratton Iron Works* were printed in 1871. They show a working week of 56 hours, consisting of five days of ten hours, with six hours on Saturdays. Everyone was hourly paid. In winter the working day was from 6.30 a.m. to 6 p.m. In summer it was from 6 a.m. to 5.30 p.m. Saturday work was from 6.30 a.m. to 1.15 p.m. all year round. Breakfast break was from 8.30 to 9.15 a.m., and dinner from 1.15 to 2 p.m. Late arrivers were fined half an hour's pay. There were fines for many misdemeanours, such as:

for entering the premises by any other way than by the main gate . . . . 3*d.*
for taking time off for Luncheon or Tea . . . . 6*d.*
Idling or Wasting his Time . . . . 6*d.*
for neglecting or absenting himself from work . . . . 6*d.*
for cursing, swearing or using obscene or profane language . . . 1*s.*
for throwing at another workman, interrupting or annoying him . . . . 3*d.*
for waisting [*sic*] materials . . . . 6*d.*
for smoking on the premises . . . . 6*d.*

The regime was tempered by the permission to bring in one pint of beer per workman at mealtimes, but anyone bringing in alcohol at any other time or being intoxicated was to be fined 1*s.* All the fines were to go into the Sick Fund. Finally, workmen leaving should give one week's notice, as should the Masters to those being discharged; but: 'Improper Conduct to the Masters or Officials' would result in instant dismissal.[9]

The paternalism of the later nineteenth century was gradually tempered by the rise of trade unions. The earliest of significance was the Amalgamated Society of Engineers, Machinists, Millwrights, Smiths and Pattern-Makers, established in 1851, and usually referred to as the ASE. Designed for the skilled man, it had obvious application to farm machinery industry. The foundry workers were catered for by the Friendly Society of Iron Founders, which developed from the original Friendly Iron Moulders' Society of 1809. Woodworkers were catered for by the Amalgamated Society of Carpenters and Joiners, founded in 1860. The Boilermakers' Society claimed a high degree of membership in the industry. In the twentieth century, national union membership rose rapidly; from 2,022,000 members in 1900 to 4,135,000 in 1913.[10] By 1913 the skilled unions were part of the normal process of consultation in the industry, although unions for the less skilled had not made much progress in agricultural engineering. Thus in March 1913 the medium- to large-sized firm of Harrison, McGregor, of Leigh in Lancashire, was faced with a list of demands by the ASE. It was requested that Mr Topping, who worked on a sliding lathe, and Humphreys (on a boring machine) should be paid the standard rate of wages; that Hilton be reinstated on the automatic lathes, and at the standard rate of wages; and that eight non-members of the ASE be compelled to join the union. Having considered these requests, the board agreed that Topping and Humphries should be paid the standard rate of wages; that Hilton should be found other work, and paid the union rate of wages, but declined the demand that the four automatic lathes should be in charge of a skilled man. Finally, the board had no objection to the eight men joining the union, but declined to compel them to do so.[11]

The rise of trade unions took place against a background of paternalism on the part of the employers. This was so whether the firms were large or not. At Garretts of Leiston, which had grown to employ 1,200 men by 1910, 'Frank Garrett the elder ruled like a benevolent despot', with the aid of his three very capable sons. But in 1911 the firm resisted an attempt to unionise the boiler shop, and locked the men out. The resulting dispute was bitter; it was said that the outer wall of the works was daubed with the whitewashed message: 'Mr Garrett – God doesn't pay his debts in money'.[12]

The year 1911 was one of general labour unrest, affecting Lincolnshire also. After several wage disputes had been settled, in July the Boilermakers' Society put in a claim for an increase of 2*s.* per week, and handed in their notices, all the larger firms in Lincoln being affected. In support of the Lincoln firms, Marshalls of Gainsborough dismissed their boilermakers. Labourers and machinemen became involved in the strike, which merged with the national railwaymen's strike, with much picketing and disorder. The trouble in Lincoln culminated in a riot lasting all night near the Great Northern Railway crossing. County police joined the city police. Troops were called (although no one knew who had called them), and the crowd, which was estimated at 4,000 strong, was told that if they did not disperse the street would be cleared by the troops. Order was not restored for some hours. Later, ten men were convicted, the judge describing their conduct as 'no better than untutored savages'. After prolonged discussions, exacerbated by Marshalls' action, the employers offered terms, which were accepted. The following year a long coal strike seriously affected production at Marshalls. In 1913 there was a long strike by the foundry moulders at Marshalls which lasted 23 weeks, and the firm was forced to raise its bank overdraft to £250,000.[13]

The growth of labour unrest before 1914 was a national phenomenon. In Gloucestershire, on 13 July 1914, 35 men in the engine-fitting department of Listers went on strike. This was deeply galling to the founder of the firm, Ashton Lister, who with his two sons owned and ran the firm, and was resolutely opposed to trade unions. The occasion for the dispute seems to have been a move by Listers to give some jobs to semi-skilled men which formerly had been entirely the province of the skilled men. The paternalistic nature of the management meant that communications between masters and men were poor, and the strike went ahead, receiving the support of the unions; in the end some 800 men were on strike. Listers countered with what was in effect a lock-out of the whole works, and the firm shut down completely, for the only time in its history. On 4 August war was declared on Germany; the works opened again for business on 10 August, as the management had promised, but the world was never to be same again.[14]

Masters and men

As throughout the rest of the economy, there was a huge difference between the incomes of masters and their men. Even by 1878 the two brothers who ran the Marshall business had achieved salaries of £2,000 a year, quite apart from any income from their shareholdings, on which the company declared an annual dividend of 10 per cent from 1888 to 1910. When Henry Marshall died in 1906 he left an estate valued at £443,523. This was a far cry from the wages of a foundry labourer at Harrison McGregor in 1913, which were 23*s.* a week.[15]

An estimate made by Josiah Stamp was that in 1914 the top 1 per cent of income receivers in Britain received about 30 per cent of the national income, and that the top 5.5 per cent received 45 per cent, leaving the remaining 55 per cent of the national income to be shared out amongst almost 95 per cent of the population. As a general rule, the point at which income tax became payable (£160 a year, or approximately £3 a week) was taken to be the dividing line between the middle and working classes. The overall tax burden was much lighter before the Great War: the government spent only about 10 per cent of the national income, compared with about 40 per cent after the Second World War. In such circumstances successful businessmen could more easily acquire and keep large fortunes.

For a working-class family to be reasonably free from the worry of poverty, a regular, year-round income of around 40*s.* a week was needed. Yet very many semi-skilled men in the agricultural engineering industry earned only around 23*s.*–25*s.* a week in *c.*1913; moreover, in common with other engineering businesses, labour was regularly laid off when times were hard, so that their families would be in a state of potential poverty much of the time. The title of a famous enquiry into urban poverty in 1913 by Mrs Pember Reeves, *Round About a Pound a Week*, is eloquent.[16]

CHAPTER EIGHT

Dynasties around 1914

Founding families

One of the earliest and most successful manufacturers, Richard Hornsby, was the son of a farmer. Having completed his apprenticeship as a wheelwright he was engaged in 1810, at the age of twenty, by Richard Seaman, whose business at that time must have been '... little more than a superior blacksmith's shop, capable of making farming implements; as well as the metalwork for farm carts and the like'. Hornsby's five-year apprenticeship to a wheelwright had given him a training in iron as well as wood, particularly in metal fabrication and forge work. In 1828, Seaman retired, and Hornsby traded on his own, later bringing his three sons into the business. His brothers, William and James, became his co-partners, but lacked the interest in the firm evinced by Richard and his other son, Richard Jnr. On the death of Richard Hornsby Snr in 1864, he was succeeded as sole proprietor of the firm by Richard Jnr. He, although presiding over the continuing expansion of the firm, died at the early age of 50, in 1877. His two sons being disinclined to take over the whole financial responsibility for running the firm, it was felt that it should be turned into a company, with limited liability. This was done in November 1879. The share capital was £300,000, in 30,000 shares of £10 each. All the shares were taken up within two days. Some 10,000 of the shares were taken up by Richard Jnr's brothers, James and William Hornsby, who became directors of the new company, which was styled as Richard Hornsby & Sons Ltd. The family received £235,000 for the sale of the greater part of their interest in the firm. James became Chairman, but the firm was increasingly influenced by professional engineers, most notably David Roberts, who joined the firm in 1895, and became managing director in 1915.[1]

In many ways the founder of Hornsbys was typical. He had humble origins, and was technically a craftsman; the family influence was diluted by the sequence of events following his death. None of these features applies to the firm which was eventually to merge with Hornsbys in 1918, Ruston & Proctor. The dynamic element in the history was Joseph Ruston, who was born in 1835, the son of a farmer, who was also a Wesleyan local preacher. His father died in 1850, leaving enough money to start Joseph in business. A committed Christian, Joseph at first intended to join the ministry, and was sent to Wesley College, Sheffield, in 1851. However, he then elected for a commercial career, and joined the Sheffield cutlery manufacturer, George Wostenholme, as an apprentice; there his training covered the technical aspects of the steel and cutlery business.

Once his apprenticeship had been completed, Ruston's ambition led him to use his father's legacy to buy a partnership in the Lincoln firm of Proctor & Burton, a millwright and agricultural machinery manufacturer with staff of 25. The firm seems to have been very good technically, but lacking commercial drive. References having been exchanged (the Mayor of Sheffield providing Ruston's reference), the new firm came into being on 1 January 1857. Ruston was to have the modest salary of £150, and take over the commercial side of the business entirely. He was to put in new capital – £500 in the first year, and then as required. The partnership was to be for ten years in the first instance, and the new firm styled Ruston, Burton & Proctor. However, Ruston's business ventures (moving heavily into steam engine manufacture, and even building portable steam engines ahead of orders for them) proved unacceptable to Burton, who was bought out of the business in July 1857, for £855 10*s.* 0*d.*, this being one-third of the firm's estimated value. Ruston was now free to develop the firm along his own lines, and did so with unremitting energy and single-mindedness. He recalled later that in the early days he

> ... had no hobbies. I had no shooting, no hunting, no racing, and none of the amusements which many people take to divert themselves from their business. I continued at the business early and late ... In my early days, twenty-five years ago, when striving to establish an agency abroad, I have travelled night and day, and been in bed only three nights a week, so that I might quickly be back in Lincoln and prevent the wheels from stopping.

Joseph Ruston's intuition let him to believe that there was a brighter future producing portable steam engines than in the more complex business of traction and ploughing engines, as Fowlers were doing. The technology of the portable engines and their associated threshing machines was relatively simple, and the volumes of production higher. By March 1897 Rustons had produced 20,500 steam engines (almost all portable), 19,700 steam boilers and 10,750 threshing machines.[2] But by that time Ruston had moved the firm away from agricultural engineering, and was also engaged in boilers, steam navvies (diggers) and oil engines. The success of the firm was staggering. In 1889 it was converted into a public limited liability company, and Ruston was paid £465,000 for it. At that time the firm had expanded to cover 15 acres, and employed 1,600 men.[3]

Ruston took a leading part in local and county government, becoming a city councillor in Lincoln in 1865. He became high sheriff of Lincolnshire in 1891, and served as MP for the town in 1884–86. His wife Jane bore two sons and six daughters. On his death he left an estate valued at £929,348 gross. One of his sons, Joseph Seward Ruston, had been a director of the firm before the death of his father (1897), and become chairman in 1901, remaining so until 1939. Although possessing outside interests (notably Colonel of the local Territorial battalion), he seems to have served the firm well, and it cannot be said that the entrepreneurial spark was perceptibly diminished.[4]

The success story of Hornsby and of Ruston could be paralleled in the case of the business of John Wallace & Sons, of Glasgow. The brothers John and James Wallace were blacksmiths, with a smithy at Hill of Haldane near Dumbarton. They began in the agricultural machinery business in 1857, and invented as well as sold machines and implements. Like other firms of the time, they took the trouble to exhibit at agricultural shows, and they won many medals. One of their best decisions was to acquire the sole British distributorship for the American Oliver Corporation's chilled iron plough. They produced also their own mowers and reapers, and prospered on a dealership for Massey-Harris binders from 1892. By this time they had moved their works to Glasgow, and just before the First World War moved their works to a new 2½-acre site, retaining their original showroom in Graham Square. They produced their own tractor, the 'Glasgow' in 1916. Although this was a failure, the firm, by this time run by the grandsons of the founders, survived into the post-Second-World War period. One of the founders, John Wallace, had died in 1904, leaving an estate worth £29,060, which included 2,000 shares of £5 each in the firm, worth £11,000; the value of his house was estimated at a further £1,400.[5]

The final example of the family firm is the rather smaller business of Wm Elder & Sons Ltd, of Berwick-upon-Tweed. This began in 1865. The founder of the firm, William, was followed into the business by his four sons, and the firm remained a family business. In their centenary year, of the seven board members, four were Elders (two grandsons and two great-grandsons of the founder). Like other small- to medium-sized firms, they relied on a mix of manufacturing and selling other makers' goods. The firm's main growth was in the early days. When incorporated as a limited company in 1914, the family was paid £7,000 for the assets of the firm. Annual profits before 1914 were never more than £750 (in 1914), and the inter-war peak was £3,509 (in 1935), on total sales of £44,493, of which more than half were agency sales. The agency side of the business had always been important. In December 1890 the firm had obtained the sole distributorship for the Canadian Harris (later to merge with Massey to form Massey-Harris) binders in Berwick, Northumberland and the Borders. They survived the First World War and prospered in and after the Second World War, retaining strong links with the locality.[6]

It is clear from many case histories that the founding families came from a wide variety of backgrounds. Some began as blacksmiths, and ended as captains of industry. Some were definitely from the middle classes. It all depended on the energies of one or two of the original partners as to whether the firm grew substantially and rapidly. It was also a matter of being at the right place at the right time, and producing the right product for the market. The potential for making and selling large numbers of portable steam engines and threshing machines was clearly greater in Lincoln, with its large agricultural hinterland, and many large tenant farmers, than (say) Berwick-upon-Tweed; the mania for steam on the farm could only carry a firm so far. But once established, the captains of the industry had placed themselves and their families firmly in the upper bourgeoisie of the mid- and later nineteenth century, although few aspired to join the aristocracy. Within their localities, they were often highly influential, as employers, politicians and philanthropists.

Local influence and social responsibilities

The degree of local influence exerted by the founders and their families was notable. It is remarkable how the leading men in the industry threw themselves into public life, even while their businesses were still relatively young; several, such as Ruston, Howard and Samuelson entered the House of Commons. Some of the founding fathers of the industry, like Charles Burrell, who ran the firm from 1836 until 1900, devoted most of his energies to the family business. The same was true of R. H. (Harry) Fowler, who took over the running of Fowlers of Leeds as managing director in 1888, following the death of his uncle, Robert. He took little part in public life, although he was a Conservative and freemason, as well as being a member of the executive committee of the Engineering Employers' Federation. But almost all the other notable leaders of the industry played an active part in local and sometimes national political life, and were prominent in local charity work.

Joseph Ruston had a particularly full public life. In 1865 he became a Lincoln city councillor. In 1869 he was mayor, and in 1874 an alderman. He became a JP in 1885 and in 1889 a deputy lieutenant of the county. He was high sheriff for 1891, and invited to be mayor in Victoria's Jubilee year, 1897, though his failing health prevented this. He had also been a Liberal MP for Lincoln in 1884–86. In addition to these high-profile functions, he was President of the Mechanics' Institute, the YMCA and of the NSPCC. He also showed much interest in the provision of Saturday evening entertainment, and his work in this sphere was continued by the Lincoln Temperance Society. In 1892 he was president of the New Permanent Library, and one of the trustees who managed all the city's charities. His beneficence helped build the Drill Hall in Lincoln, and a children's ward at the county hospital. He retained his interest in the cutlery trade, and became chairman of William Rodgers & Sons Ltd of Sheffield.[7]

Similar involvement with the local community is shown in the case of Nathaniel Clayton, who was Joseph Shuttleworth's partner in the firm which they founded in 1842 in Lincoln. He was mayor in 1857, and was one of the few cases in this industry of a manufacturer being 'received' into local county society, becoming a county JP in Lindsey. In 1875 he bought a substantial agricultural estate at Withcall in the Lincolnshire Wolds, but treated it as an investment and lived in some state in Lincoln, in Eastcliffe House. He was high sheriff in 1881. He used his wealth for various philanthropic purposes – better public sanitation, education and hospitals, most of which he pushed through as member of the relevant corporation committees. When he died in 1890 his fortune – £1,365,496 – was the largest of all the agricultural machinery makers in the pre-1914 period. While Clayton was the commercial drive of the firm; Joseph Shuttleworth was the technical inspiration. The latter also led an active public life, being at various times a town councillor and chairman of the magistrates' bench. He was chairman of the Lincoln Liberal Association, and had many external business interests, chiefly two directorships of major railway companies. He was a council member of the RASE, and vice-president of the Agricultural Engineers Association on its formation in 1875.

The Ransome family showed less outstanding involvement in public life, except for Robert Charles Ransome, the grandson of the founder. He effectively controlled the firm from the mid-1860s until his death in 1886. After the formation of Ransomes, Head & Jefferies in 1881, he held 40 per cent of the company's shares. Described as 'the first citizen of Ipswich', he devoted himself to public works, strongly motivated by his Quaker upbringing and his strong Liberalism. He was at various times mayor, alderman, town councillor, JP for the county and borough, and chairman of the Ipswich Liberal Association, although he declined an offer to stand for Parliament. He was a staunch free trader, in spite of the effect of overseas protectionism on his firm's sales. His other interests stemmed from his profession. He was a council member of the Smithfield Club and the RASE, and a founder member of the Agricultural Engineers Association.[8]

A sense of social responsibility, usually stimulated by Christian principles, was common among company founders and their successors. In a paternalistic age, before trade unions developed, and before the welfare state of the twentieth century, large local employers found themselves confronted directly with the social and economic questions of the day. Their response was bound up with the paternalism of their role as employers. In future years, the waning of their control and responsibility, as trade unions and the state took over functions formerly performed by employers, would lead to the waning of their civic involvement.

Gracious living for the third generation?

Historians have long debated the performance of British manufacturing and industry in the period 1870–1914. The economy grew slowly compared to those of the USA and Germany, and overseas markets were being lost to British exporters all over the world. While some of this may have been inevitable, it has also been argued that the quality of British entrepreneurship had deteriorated. The essence of the charge is that the second or third generation had been brought up far from the industrial centres in which the founders of the firms had been happy to live; often attending public school or the old universities, they learned to appreciate classical learning above money-making. Some decided to go into other, less industrial businesses, or even to retire from business and buy a country estate. The debate continues, although the recent work of F. M. L. Thompson has cast serious doubt on whether the purchase of landed property was much to the fore as a means of absorbing the wealth created by successful Victorian industrialists. He also considers that the acquisition of landed property did not necessarily lead to an decline in the entrepreneurial spirit. In his words:

> ... landed gentrification emerges as not necessarily a badge of entrepreneurial decadence, but equally and more frequently as a badge of entrepreneurial success and symbol of the penetration of the upper class by cultivated businessmen.*

In the case of agricultural engineering, three of the founding fathers of the industry did indeed purchase large country estates. James Howard (1821–89, who left £82,703 of personal wealth) purchased the Clapham park estate in Bedfordshire (1,000 acres). Joseph Shuttleworth (1819–83, who left £554,613) purchased the Hartsholme estate near Lincoln in 1861, and the Old Warden estate in Bedfordshire in 1871. And Nathaniel Clayton (1815–90), who seems to have been the only actual millionaire created in the industry, purchased Waddington, Lincolnshire (350 acres) and Withcall, Lincolnshire (*c.*2,600 acres), and instructed his heirs (his two grandsons) to purchase estates also. One of them, Nathaniel Clayton Cockburn, did so, at Harmston, Lincolnshire. But none of them took this as a cue to retire from business, and remained active in the running of their firms.[9]

To what extent, if any, can one discern these alleged failings within the farm machinery industry? Even if few from this sector became seduced by a life of county balls and fox-hunting, did success breed complacency? After the passing of the original founders, were their successors simply not up to the job? The enormous and rapid growth of the farm machinery

* F. M. L. Thompson, *Gentrification and the Enterprise Culture: Britain 1780–1980* (Oxford, 2000), p. 97.

industry in the nineteenth century certainly transformed the position of the leading families in the business, and provided a very adequate living for many of the families running medium-sized firms. Did the original business drive of the founding entrepreneurs become diluted by affluence in the second or third generation. There are no cases of 'clogs to clogs in three generations' before the First World War, although many of the leading firms did come to grief after 1914. But is it possible to discern signs of the corrupting influence of a gilded and protected upbringing? Some case studies may help to provide an answer. One major firm – Marshalls – and the medium firm of Bamfords will be examined here, and evidence also drawn from the history of some smaller firms.

The son of a shipwright/block and tackle manufacturer, William Marshall (1812–61) went as apprentice millwright to William Fairbairn's famous engineering works in Manchester. After finishing his articles, and after a spell as Fairbairn's representative in Russia, he set up his works in Gainsborough in 1835. His elder son James entered the business in 1849, and the younger son Henry followed in 1853. The working partnership between father and sons seems to have been very harmonious, and the firm prospered. It was incorporated in 1862, shortly after the death of its founder, and the subsequent decades were ones of rapid growth (see pages 79–80, 132). The firm was run by James and Henry Marshall. Henry married in 1862, and originally he and his bride lived adjacent to the works, but in 1865 moved to a 'house with a pleasant frontage on the River Trent', and to a larger house in 1868. Their final move was to a spacious house named Carr House near the Lea Road railway station. Both James and Henry had large families. By 1887 the brothers had seen their salaries rise from £200 to £2,000 a year. The next two decades saw outstanding growth, especially in exports. The company remained firmly in the hands of the family. Henry Marshall died in 1906 at Carr House, leaving an estate valued at £443,523. His energy and drive had been remarkable. His brother James remained as chairman. His eldest son Herbert, and Herman, Henry's youngest son, were made assistant managing directors. In 1911 the firm had purchased Carr House, intending to build a new factory on that site for the making of tea machinery. In the event, it was used in the First World War for the making of aeroplanes.[10]

What conclusions can we draw here? The founder and his sons were clearly men of great ability. The sons had a long reign, and in that time managed to steer the firm into profitable channels as conditions changed. Although still heavily involved in agricultural machinery, they moved at various times into building oil engines, tea-processing plant, and gold dredging machinery. By 1914 the board comprised James (chairman and managing director), Herbert and Herman (assistant joint managing directors), William Whiffen (a manufacturing chemist), James Hugh Marshall, and George Cook, the company secretary. In 1921 James Marshall died. Herbert Marshall became chairman, and Herman managing director. Thereafter, the history of the firm was caught up in the general problems facing the industry. Whether a different board could have done better is uncertain. It certainly does not seem to be the case that affluence reduced entrepreneurial spirit before 1914.[11]

The case of Bamfords of Uttoxeter, Staffordshire, is somewhat different. The firm began comparatively late, in 1871, and was unusually specialist. Its progress until after the First World War was based on a range of hay-making mowers and reapers, and feed grinding mills. Thus it was well placed to take advantage of the shift in agriculture from cereals to grass and dairying which took place after the early 1870s. The founder of the firm, Henry Bamford, began by running an ironmongery business purchased for him by his father-in-law in 1845. Henry and his son Samuel developed the business towards the agricultural engineering side, and the firm established its own ironworks in 1871. Beginning with six employees, the family fulfilled all the essential operations. By that time Samuel had been joined by his four brothers – John, Henry, Robert and Joseph. John managed the office; Henry did purchasing and advertising; Robert was on the sales side; and Joseph was an engineer who also pursued overseas orders very successfully. Their No. 5 mowing machine of 1882 made the company famous. By 1893 the works covered ten acres and employed 400 people. In 1896 the founder died, but the firm was joined by Samuel's sons Henry and Oswald. In 1912 John Bamford's son, Henry John, came of age. The celebrations began with the roasting of an ox in front of John Bamford's home, Oldfields Hall, Uttoxeter. A salute of 21 'bombs' was fired from the grounds, and in the evening 800 people were entertained with a meal in the new foundry completed in that year.

By 1914 the firm, which remained a partnership until becoming a private limited company in 1916, had had a remarkably successful record. It was its good fortune to have had excellent inventive and engineering skills, and to have concentrated on the areas of farming which were expanding. It was not until 1920 that anything other than farm machinery was made. In that year they began making small oil engines for barn work such as sheep shearing, driving cream separators etc. Samuel, the son of the founder, was chairman until his death in 1932, by which time a grandson of the founder, Captain H. John, and a nephew of Samuel, Cyril, had joined the board. If the third generation was suffering from a privileged upbringing, it does not seem to have harmed the firm, which continued in business until it was sold in 1981, being renamed Bamfords International, and being sold again in 1988 to Bensons of Knighton.[12]

Marshalls, and to a lesser extent Bamfords, was among the aristocrats of the industry. Most firms were much smaller. One such was E. & H. Roberts, of Deanshanger Iron Works, Stony Stratford, Buckinghamshire. This originated in the business founded by either Richard Roberts in 1821 or John Roberts in 1817, and operated from two cottages and a smithy. Ploughs seem to have been made from the start, and formed the basis from which other developments were made, although they continued to form the backbone of the business until its end in 1927. The son of the founder, John, took over in 1843, and in 1857 his grandson Edwin took over. By this time there was a foundry and ten employees. In 1875 Edwin's brother Henry joined him in the firm as partner. In 1890 the firm became incorporated as a limited company, with an authorised capital of £50,000, in 5,000 shares of £10 each. The brothers Edwin and Henry each got 1,200 shares; Albert Roberts (builders' merchant) 150; Thomas Roberts (engineer) 100; Lavinia Roberts [Henry's wife] 100; Caroline Kidman [farmer's wife – presumably née Roberts] 50, and Louisa Roberts (spinster) 50. This was clearly a family firm. By this time it had a branch works and depot at Buckingham, and implement depots at the cattle markets of Northampton and Towcester. The catalogue of 1 January 1900 lists a wide variety of field cultivation implements, beginning with no fewer than 22 basic models of plough, marketed generally under the name of 'Mephisto', presumably to indicate their magical properties; the firm claimed that Mephisto ploughs had been awarded 3,750 prizes. The firm made also wind and water pumps and elevators, and did a thriving export business. It also did more general engineering business, including large waterwheels.

Thus, even though by 1914 the firm had a long history, there is no evidence of lack of entrepreneurial vigour. The most serious incident before 1914 was a very large fire in 1912, which however allowed for a new and modernised works to be built with the insurance money. The fire made no impression on the company's balance sheet or its profits; even by 1907 the reserves stood at £4,000, and they reached a peak of £7,000 in 1916. But the high profits of the war years were dissipated in unwisely high dividend payments to the shareholders, and the firm was left ill-placed to cope with the slump of 1919–21. After 1919–20 the firm made losses in every year, and the demise of the firm came in 1927, following the General Strike. It is difficult to escape the conclusion that the firm was ill-equipped to cope with the radical alteration in its economic environment after 1918 (although it made efforts to produce ploughs suitable for tractors in the early 1920s), but that is not the same as saying that it had lost its entrepreneurial flair – or, at least, not before 1914. If Roberts evinced a failing before 1914, it would seem to be a certain lack of diversification. Even so, the firm was making good profits from arable machinery for most of the period from 1870 to 1914, although by 1911 the chairman was referring to fierce competition reducing profits.[13]

A more obvious example of inertia may be had from R. & J. Reeves of Bratton, Wiltshire. The earliest sales ledger of this firm shows largely repair work, supplying new ploughs etc. By 1870 repair and retail work still dominated, but the firm undertook a variety of construction work, notably the building of the local Baptist chapel. They were now manufacturing, making 'night soil carts for export'. By 1890 they had a catalogue of their own manufactures, mainly comprising ploughs, harrows and other cultivating machinery. They also were making an all-steel elevator in 1914. In 1912 they employed 66 people. A brief note of the history of the firm refers to its products being determined by 'the very extensive tract of arable and sheep-grazing land on the north side of Salisbury Plain.' Its products were sold 'all over the west and south country, and to a limited extent on the east side as well'. But the firm remained small; the partners' investment in the firm was only £7,500 in 1881; in 1923 the works were valued at only £3,500; and by 1930 net sales were only £20,000. By 1938 the firm was trying to get an overdraft of £5,000 from Lloyds Bank to enable it to 'help us carry over the stocks left on hand due to the failure of the hay crop this year'. The troubles of the firm continued until its final demise in 1970. It is difficult to resist the conclusion that the firm had not sufficiently diversified, and probably employed too many men. The same 1938 letter to the bank manager stated that: 'our manufacturing business has grown up as a means of providing work for these [skilled] men during the autumn and winter months.' Rather than a case of success breeding complacency, this is a tale of a firm which was was unwise to get overly involved in manufacturing in a market which was certainly not going to grow; it remained committed to arable agricultural products right up until 1914.[14]

Conclusion

The history of the industry gives little evidence to support the view that it was a case of 'entrepreneurial failure' before the First World War. Indeed, it was riding the crest of a wave of unparalleled success. Total sales of the industry were now about £6.5 million (in 1913), and slightly over half of this came from exports. There were concerns about rising costs and labour unrest, but these were not fundamental. The more prescient might have been troubled by the fact that the industry was too much oriented to steam technology, and insufficiently diversified in both home and export markets. The reliance on exports could of course prove a serious source of weakness in the event of a general European war. This had been prophesied many times since the turn of the century, but most people believed that the threat of war had receded by 1914. When it came, the timing was a surprise, and the devastation wholly unforeseen. From the point of view of the manufacturers of farm machinery, it was to prove disastrous.

CHAPTER NINE

War work, 1914–1918

The background

One consequence of trench warfare as it developed over the early years of the First World War was an enormous demand for munitions and military equipment, including shells, machine-guns and ammunition, barbed wire and sandbags. As the war progressed, new devices came into widespread use, notably the aeroplane, and the submarine menaced civilian shipping. The economies of the belligerents were hard pressed to supply the demands of the armed forces: already by April 1915, some 2 million shells had been sent to the British Expeditionary Force in France. At that point, the government decided to speed up the supply of ammunition, and set up the Ministry of Munitions, headed by its first Minister, the dynamic David Lloyd George. The Ministry gave out substantial contracts to private firms, and also set up its own shell factories. The number of shells sent to France had reached a total of 187 million by the end of the war. Along with other theatres of war the western front generated an unprecedented demand for munitions.

This enormous output meant that most productive effort became devoted to the war. At its peak in 1916–17, government spending accounted for 58 per cent of national income. Partly this was funded by taxation, but about 70 per cent of expenditure was financed by borrowing; the National Debt rose tenfold to about £8,000 million in 1918. This, coupled with the high cost of imports, led to inflation. Prices in almost every year of the war rose by about a quarter: to a generation accustomed to price changes of only a few per cent a year, this was deeply disturbing.[1]

Government armament contracts

Prior to the war, the arms industry comprised the three government arms factories, at Enfield, Waltham Cross and Woolwich, and a number of private enterprise suppliers such as Vickers. These would quickly prove incapable to meeting the demand for munitions, so the military authorities began issuing contracts, some of enormous value, to ordinary engineering and chemical firms. Firms scrambled to tender for War Office and Admiralty work.

Most engineering firm of any size were quite capable of doing at least some work of this nature, and munitions contracts flooded into the agricultural engineering sector as well as into general engineering. On 23 February 1915 the directors of Richard Hornsby & Sons of Grantham were informed that the value of contracts for war work so far undertaken by the firm was over £100,000. Two months later, on 27th April, they were informed that further orders had been received, including: '950 .303 Maxim Tripods value £22,000; 8,500 18 Pounder Bombs (Vickers) £7,400; 100 Sets of Control Gear (Vickers) £8,700'. They were by this time also acting as sub-contractors for large armament firms such as Vickers. By September a further 28 contracts reported to the board included: '48 Oil Engines for Pumping in Flanders [to relieve the waterlogged trenches]; 8 45 H.P. do. for Hospital Lighting in the Dardanelles; 60,000 18 pdr H.E. [high explosive] Shell value £60,000.' The same meeting was told that the managing director had joined a deputation to attend Christopher Addison, the Minister of Munitions, to see if the government could be persuaded to sanction an increase in agricultural machinery output. Addison told them that the government's first priority was munitions, followed by machinery exports, but, 'they were not sympathetically disposed towards the output of agricultural implements'. His advice was to put the unskilled and semi-skilled workers and women to the production of munitions, and to work night shifts. By early 1917, when the Ministry of Munitions took over responsibility for the supply of agricultural machinery, its official history records that agricultural engineering was producing munitions 'to the virtual exclusion of normal output.'[2]

The flow of contracts affected firms of all sizes. Thus the medium-sized business of Nalder & Nalder (sales of £17,779 in 1915) entered into contracts with the West of England Munitions Committee to make shells, although the quantities were not large – just 50 shells a week in 1915–16. Taskers of Andover, which normally made traction engines, carts and wagons, found that these routine products were now in demand by the War Office. Large contracts were also received for trailers, sleeping-vans and timber carriages for the use of haulage firms engaged on government contracts. The machine shop turned over to the production of 18-pounder shell cases. The rather larger firm of Robert Boby Ltd, of Bury St Edmunds, which normally made feed preparation machinery, had by this time become associated with Vickers, and thus received contracts to make shells and paravanes (used for mine-sweeping). At a special board meeting in July 1917, 'it was reported that the Works were very busy, being chiefly engaged on the manufacture of 4.5 Shells and others ...'[3]

War contracts committed firms to pay the rates of war bonus dictated by the Ministry of Munitions., which involved a certain loss of control by the management. Thus Harrison, McGregor, of Leigh, Lancashire, one of the larger firms, had had union members of the ASE before the war, but on 18 August 1915 was declared a 'Controlled Establishment' by the

Ministry. The result was that war bonuses were agreed with the unions at various times during the war, and in April 1917 the firm was ordered by the Ministry to give an increase in wages of 6*s.* a week, this being a total increase of 12*s.* a week over the pre-war rate. Firms without munitions contracts could get by with less intervention from outside: Elders of Berwick-upon-Tweed received a letter from the Ministry of Munitions dated 25 October 1917 which confirmed that since the firm was not doing any munitions work, but was entirely engaged on agricultural machinery, no wage regulations were applicable to it. This may have been a relief to the management. The next month they were faced with a demand from the Workers' Union for a rise in the wage of women employed in the smiths' shop, but this does not seem to have been conceded, the firm taking the initiative on the same day as the Ministry's letter was written to set up a consortium of local employers, to be named the Berwick and District Engineering Employers' Association.[4]

The large firms were more capable of manufacturing large and complex military items. Garretts (sales of £92,407 in 1913) saw its business transformed by the war. By 1917, when there had been some revival of agricultural machinery production, it was still dwarfed by munitions output, and the total sales of the firm had multiplied many times over, rising even more in 1918:

Table 9.1 *Garretts of Leiston, Suffolk: sales, 1917–18 (£)*

	1917	*1918*
Shells and shot	408,720	456,603
Machinery – home	125,300	155,861
Aeronautical	24,106	137,884
Engine parts/repairs – home		38,670
Exports – machinery	}	19,064
Scrap metal	} 45,489	15,816
Woodworking	}	5,032
Exports – parts etc.	}	1,994
Total	**603,615**	**830,824**

Source: Suffolk Record Office, Ipswich, Garrett & Sons archive, HC30/A3/10–11.

Garretts may have been in a particularly good position to get contracts, since a disastrous fire in 1913 had been followed by the rebuilding of the woodworking shop and the foundry on modern lines, and these facilities, known as the Station Works, were now turned over to the production of war goods, including entire aeroplanes, 'hundreds of thousands of shells, and the design and manufacture of shell-turning lathes.'[5]

The ability of the large firm to adopt the latest technology quickly on a large scale was seen in the war record of Ruston & Proctor. Apart from undertaking contracts to manufacture shells and other munitions, the firm specialised in making aircraft. Starting in 1915 with 22 for the War Office, they then produced, up to 25 January 1918, 200 BE2 aircraft, 350 Sopwith 110s, and 500 F.1 Scouts. The firms's one-thousandth plane was Sopwith F.1 No. B7380; to celebrate, it was by special permission 'painted with a fantastic design', of which unfortunately no record survives. The firm went on to make another 1,452 aircraft until the contract was cancelled, in November 1919.[6]

The tank was one striking technological innovation of the war, and in this a leading role was played by the agricultural machinery firm Wm Foster & Co. of Lincoln. Arising out of the work of the Landships Committee established by Winston Churchill, the First Lord of the Admiralty, the project had passed to the Ministry of Munitions when it asked Fosters to produce a prototype tank. Christened 'Little Willie', this was first demonstrated on 19 September 1915, just 39 days after construction had begun in Foster's Wellington Foundry at Lincoln. It had been designed by William Ashbee Tritton, the managing director of Fosters. A second prototype, jointly designed by Tritton and an army officer, Lieut. W. G. Wilson, quickly followed. This, known as 'Big Willie', was the form of the tank as used in the war. As a result of successful tests, a contract was placed with a Birmingham firm, the Metropolitan Carriage & Wagon Company, for 75 tanks. Since Fosters did not have enough capacity at this time, they were given a contract for 25. The first tanks went into action on 15 September 1916 on the Somme. In February 1917 William Tritton was knighted for his part in the development of the tank. On 6 March 1916 Fosters had become a Controlled Establishment under the Ministry of Munitions, and at the annual meeting on the last day of 1916, the directors reported that all the company's resources were engaged in producing: '... the now celebrated Fighting Tanks.' In the period to the Armistice Fosters produced the astonishing total of 2,969 tanks.[7]

All of the major firms undertook large contracts. For some this came

A Mark 1 Tank made by Fosters of Lincoln, undergoing trials. (*Tank Museum, Bovington*)

A Mark 1 Tank by Fosters being prepared for shipment to Russia. (*Tank Museum, Bovington*)

as a welcome relief after the loss of overseas markets at the beginning of the war. Fowlers of Leeds, which had been of all the large companies the most heavily export-oriented, made a variety of *matériel* for military use: shells, artillery wheels (especially types fitted with pads to avoid sinking in the mud at the front, including some Boydell-type wheel girdles), some tanks, and, most notably, a large order from Russia for oil-engined tractors – most of which were eventually lost in the Masurian swamps.[8] Ransomes, which had been more diversified than Fowlers before the war, and was less dependent on the export trade, also found compensation for the loss of exports by turning over to war work:

Table 9.2 *Ransomes, Sims & Jefferies' output, 1914–18 (£000)*

	1909–13 average	*1914*	*1915*	*1916*	*1917*	*1918*
Home	167	197*	487†	232	321	422
Export	391	340	105	115	61	49
War work	–	–	–	574	867	1,675
Total	**556**	**537**	**592**	**921**	**1,249**	**2,146**

* Includes £26,000 of war work.

† About £307,000 of this may have been war work.

Source: Rural History Centre, University of Reading, Ransomes, Sims & Jefferies archive, TR RAN AD 7/17.

Labour during the war

Labour was short and relatively expensive during the war. There was an enormous flow of men into the armed forces, eventually by conscription, and the pressure of military contracts fed demand for labour yet further. This led to a continuous pressure on employers to raise wages, exacerbated by the high rate of wartime inflation. At first pay rises came not, in a rise in the basic rate of payment, since it was assumed that wartime pay rises would be temporary, and employers did not wish to be left with permanently higher rates after the war, but by payment of a 'war bonus', which was continually being augmented, as inflation continued. In practice, however, the bonuses became subsumed in general rates of pay. During the war there was also a narrowing of pay differentials between skilled and unskilled workers, as the outflow of skilled workers into the forces was offset by the employment of less skilled workers, who were trained rapidly to do skilled, or at least semi-skilled work. A further wartime theme was the employment of women in manufacturing on a large scale; this was unprecedented, and provided a cultural shock to the management and workers alike.

The upward pressure on wages greatly strengthened the position of the trade unions, which doubled their membership during the war. In June 1916, Harrison McGregor was faced with a demand for a rise in wages by three unions, representing between them the spectrum of skills in the industry – the Amalgamated Society of Engineers, the national Union of Stove, Grate, Fender and General Light Metal Workers, and the National Union of Gasworkers and General Labourers. The application was rejected by the industry conciliation committee (the 'Committee on Production in Engineering and Shipbuilding Establishments engaged on government work'), headed by the civil servant G. R. Askwith. But in November the firm offered war bonuses to ASE and to NUSGFGLMW members. The pressure from the workforce continued. In February 1916 there was the threat of a strike by the general labourers, which was settled by the offer of an extra 2*s.* a week, so that the general labourers' rate was now 25*s.* a week In April 1917 the firm was ordered by the Ministry of Munitions (it being a controlled establishment) to pay an extra 6*s.* a week to all workers, this now being an increase of 12*s.* a week on the pre-war rate. However, this sparked trouble with the foremen, who now asked for, and got, an increase of 6*s.* a week to restore the pay differentials. Shortly after this, in May, the clerical workers also got an extra 6*s.* a week.[9]

The government's main priority was to keep up the flow of military material from the factories. To this end, agreement was reached with the trade unions that unskilled labour could be employed on work formerly reserved for the skilled men. The 'Treasury Agreement' in March 1915 and the subsequent Munitions of War Act enshrined this agreement. Thus the unions gave up their opposition to the use of unskilled labour, and also

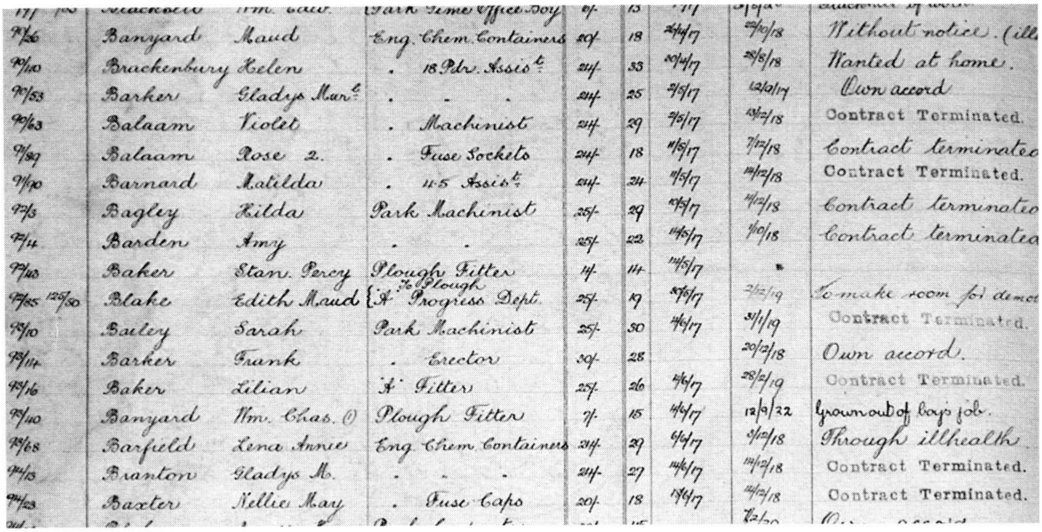

9/26	Banyard	Maud	Eng. Chem. Containers	20/-	18	24/4/17	27/10/18	Without notice. (ill
9/110	Brackenbury	Helen	. 18 Pdr. Assist.	24/-	33	20/4/17	28/8/18	Wanted at home.
90/53	Barker	Gladys Mar.	. - .	24/-	25	2/5/17	12/10/17	Own accord
90/63	Balaam	Violet	. Machinist	24/-	29	7/5/17	13/2/18	Contract Terminated.
9/89	Balaam	Rose 2.	. Fuse Sockets	24/-	18	11/5/17	7/12/18	Contract terminated
91/90	Barnard	Matilda	. 4.5 Assist.	24/-	24	7/5/17	14/12/18	Contract Terminated.
92/3	Bagley	Hilda	Park Machinist	25/-	29	10/5/17	14/12/18	Contract terminated
92/4	Barden	Amy	. .	25/-	22	11/5/17	7/10/18	Contract terminated
92/40	Baker	Stan. Percy	Plough Fitter	14/-	14	14/5/17	•	
92/65 125/50	Blake	Edith Maud	"A" To Plough Progress Dept.	25/-	19	8/5/17	2/12/19	To make room for demo
93/10	Bailey	Sarah	Park Machinist	25/-	30	14/5/17	3/1/19	Contract Terminated.
93/14	Barker	Frank	. Erector	30/-	28		20/2/18	Own accord.
93/16	Baker	Lilian	"A" Fitter	25/-	26	4/6/17	28/2/19	Contract Terminated.
93/40	Banyard	Wm. Chas. O	Plough Fitter	7/-	15	4/6/17	12/3/22	Grown out of boys job.
93/68	Barfield	Lena Annie	Eng. Chem. Containers	24/-	29	6/6/17	5/12/18	Through illhealth
94/13	Branton	Gladys M.	- - -	24/-	27	14/6/17	14/2/18	Contract Terminated.
94/23	Baxter	Nellie May	. Fuse Caps	20/-	18	14/6/17	14/2/18	Contract Terminated.

Ransomes' labour ledgers, 1918, showing names and details of employees. Note the number of temporary women workers, most of whom were dismissed after the Armistice. (*Museum of English Rural Life*)

the right to strike, on the understanding that normal working practices would be restored after the war; that the temporary wartime labour would be dismissed, and the skilled men away at the forces restored to their former jobs. The unskilled labour could, with the aid of automatic equipment, turn out work as good quality as that of the former skilled workers.

A large part of the unskilled wartime labour supply (officially referred to as 'dilutees') was female. Its employment in manufacturing was almost unknown before the war. Before the war, the only women employed at Ransomes were, 'the housekeeper, and sundry charwomen came in on Saturday afternoons to scrub the offices'. But the pressure of military work caused the firm to rethink. Having received a big order for 4½-inch shells, and having prepared for the move by installing ranges of automatic lathes, it began to employ women in February 1916, via the local Labour Exchange. In accordance with government regulations, they were supplied with working overalls and caps; supervisors were appointed; suitable lavatory provision was made; and canteens were provided. By the end of the war, the total number of women at work at the Orwell Works was over 2,000. They performed very well, the company history noting that, it was found possible to utilise their services in many directions, as most of them were very quick in absorbing instructions, and very clever in the use of their hands'.

At the signing of the Armistice on 11 November 1918, all existing war contracts were cancelled, and the women, and most of the male dilutees, were soon discharged. The only women retained were those on clerical work, and certain canteen attendants for the workmen's mess room.[10] The historian of the Garrett works was less complimentary to the women who had worked for the firm in wartime, writing that the labour force was erratic, turbulent, mainly temporary ad predominantly female. The women included some very hard cases, and any sign of weakness at the top would have quickly brought the situation to anarchy ... With the end of the shell contracts went the thousand or so women who had been employed on them, the turbulent multitude which had disturbed Leiston days and nights for nearly four years.[11]

Other large firms also employed women in large numbers. Marshalls lost about 2,000 men to the armed forces, but the total number of employees, male and female, peaked at over 6,000 in 1917, women having been employed since 1915.[12]

The loss of exports and overseas assets

The war was disastrous for the export trades. Apart from loss of markets due to actual fighting and enemy action, the military demands on the British and Allied shipping fleets led to a serious shortage of shipping. The result was an enormous rise in the cost of shipment, and the added dangers of sea trade in wartime led to a sharp rise in insurance costs. The loss of exports, and the associated losses of earnings, cash, and stock held overseas, were the most serious blows dealt to the industry by the war. The fall in exports was especially serious in comparison with sales for 1913, which had been the peak of the pre-war export boom:

Table 9.3 *British agricultural machinery exports, 1913–1918 (£)*

	1913	*1914*	*1915*	*1916*	*1917*	*1918*
Implements	745,883	529,436	377,059	549,670	502,651	494,331
Prime movers*	1,360,954	1,157,661	180,313	164,458	63,342	71,575
Other machinery	1,628,232	1,156,010	276,568	250,513	86,002	81,724
Totals	**3,735,069**	**2,843,107**	**833,940**	**964,650**	**651,995**	**647,630**

* Steam, oil and petrol engines.

Source: Annual Statement of Trade, 1913–18.

The decline in volume was even greater than that of value, since wartime inflation was so high. A price index of 1919 shows that the average unit price of exports had risen by 177 per cent since 1914. Applying this to the money value of exports of agricultural machinery in 1919 shows that the real volume of exports compared with 1913 had declined to just 21 per cent of its former level: exports were only one-fifth of their pre-war figure. In 1918 it would have been much less; by 1919 the current money value of exports had recovered to £2,177,856.[13]

As can be seen from the above table, the greatest loss was in 'prime movers', which were steam and oil (and presumably some petrol) engines.

The largest of these trades was the export of ploughing engines and steam portables to Russia, eastern Europe and South America. The European trade was severely affected by military action and transport problems from the start of the war. The South American trade was hit by the shortage of shipping, the rapid rise in freight costs, and the rise in the cost of insurance. As the war went on, these factors became more acute. The Russian Revolution in 1917 led to the complete loss of any assets remaining in Russia. 'Other machinery' held up better, but by the end of the war that trade had fallen almost as much as that of the engines. The implement trade held up better than either.

The losses of particular firms followed this general pattern, with variations dependent on their particular export exposure and administrative arrangements. Fowlers of Leeds suffered much loss in the trade in ploughing sets to Russia and eastern Europe. This was complicated by the existence of their overseas branch at Magdeburg. This made, as well as sold and serviced, Fowler products, and was the only manufacturing undertaken abroad by any of the big companies in 1914. The branch continued under Fowler's management until 1916, when it was compulsorily liquidated by the German government, and the assets sold to Fowler's main competitor, R. Wolf AG. At the time, the branch was owed debts in Germany to the value of £102,314. From then on, no income was derived from the branch. The subsequent post-war struggles with the Clearing Office for Enemy Debts, and the Reparation Department of the Mixed Arbitration Tribunal were not concluded until the final settlement in 1929, by which Fowlers was awarded £425,000. It is not known if any of this was in fact paid to the firm.[14]

The complexity of the losses sustained by the firms with overseas assets is shown in the case of the Russian losses sustained by Ruston & Proctor, of Lincoln. These were itemised after the war as follows:

Table 9.4 *Ruston & Proctor's Russian losses in the First World War (£)*

Bank balance in Russia	16,380
Debts	64,312
Office furniture	648
Trade debts	42,035
Bank balance in Odessa	111,289
Stock	3,146
Property in Moscow	11,315
Rouble bank deposits and bills*	70,788
Total	**319,913**

* Converted to sterling at the rate of 15.85 roubles/£, as at 1 January 1917.

Source: GEC-Alsthom, Firth Road, Lincoln; Ruston & Hornsby archive, 'The Odessa File'.

These losses may have been magnified by the defalcations of the firm's Odessa agent, R. A. Jahn. Before the war he was regarded as a distinct asset to the firm, and was even presented with a large photograph album, bound in snakeskin, with elaborately decorated pewter covers as a testimonial present. But the exposure of his fraud in 1916 must have come as a shock to the firm. To this day the file in the Ruston & Hornsby archive is known as 'The Odessa File'. It must be assumed that none of these losses was ever recovered.

The other large firms to suffer losses in this way were Ransomes (amount unknown) and Garretts. The latter put its losses due to the Russian revolution as £200,000 in unpaid customers' debts.[15]

Profits and taxes

There is no doubt that the war proved hugely profitable for the industry in the short run. In this, its experience was similar to that of large sections of British manufacturing. War profits quickly became a reality after August 1914, and the introduction by the government of taxes on 'excess' profits came late and was uneven and much avoided in its application.

The experience of high profits may be seen from the surviving financial records of some of these, mainly larger firms. The cases of Ruston & Proctor of Lincoln and Richard Hornsby of Grantham are set out below:

Table 9.5 *Profits, 1914–18: Ruston & Proctor and Richard Hornsby (£)*

	1913	*1914*	*1915*	*1916*	*1917*	*1918*
Ruston & Proctor: gross profits, 1913–18 (to 31 March)	117,630	119,118	110,405	158,518	159,333	169,744
Richard Hornsby & Co.: gross profits, 1913–18 (to 30 September)	51,578	35,949	53,653	82,299	*	*

* Directors unable to present balance sheet, owing chiefly to unresolved accounts on Government contracts.

Sources: GEC-Alsthom, Firth Road, Lincoln, Ruston & Hornsby archive, Ruston & Proctor Annual reports 1913–18; Richard Hornsby Directors' Minute Book, 1913–18.

The Ruston & Proctor accounts have a note which indicates that the gross profit figures for 1916–18 were arrived at after allowing for enough to pay for the 'estimated liabilities to the Government', implying Excess Profits Duty. Thus the wartime accounts, unlike the pre-war figures, are post-tax, and the real pre-tax level of profit was clearly much higher than pre-war. A similar story is given in the accounts of Richard Hornsby of Grantham, except that figures are not presented after 1916, since the firm was in dispute with the government over the level of Excess Profits Duty.

Similar rises were recorded at Marshalls, where gross profits (after tax) rose from £50,568 in 1913 to £82,236 in 1916. There was a decline in 1917 to £62,574, and no information is available for 1918. The profits of Fosters, a rather smaller firm, rose in 1914, but from 1915 onwards, with the introduction of taxes on profits (at first Munitions Levy, later replaced by Excess Profits Duty), the firm was unable to agree their accounts with the Finance Department of the Ministry of Munitions.[16]

While the larger firms seem to have made good profits, the smaller firms had mixed experiences. Unless engaged on munitions, they were unable to protect their workforce against military recruitment. Before the introduction of conscription in 1916, their men would have been subject to the prevailing public pressure to enlist, and, after conscription began in January 1916, to compulsory recruitment. While engagement on munitions would lead to the firm becoming a Controlled Establishment under the Ministry of Munitions, and thus give the labour force protection from military service, this would have to occur early on if the workforce were to be retained. The history of Nalder & Nalder, of Wantage, shows that not all firms made a lot of money from the war. Their net profits in 1913 and 1914 had averaged £290, and although they rose to £429 in 1915, they fell to £44 in 1916 and to a large loss, of £358, in 1917. In 1918 there was a small profit, of £44. The problems facing the firm for the years after 1915 were described in the 1916 annual report:

> During the year there was a considerable drop in the output, entirely due to the shortness [*sic*] of men. There was also a considerable rise in price of raw material and wages, which was not compensated for by a corresponding rise in price of the Company's Manufactures. The Works are well employed, but the shortage of men still continues, and orders have to be refused. Female labour is being employed as much as possible to augment the output. Forty-seven of the employees have left to serve in HM Forces. The Works were declared a Government Controlled Establishment on 31 January 1916.

Nalders had not been especially profitable before the war, and its failure to make more money from its munitions business is a little puzzling, The directors' report has a somewhat self-exculpatory air, and the complaint about raw material prices was not a new one in the firm's annual reports.[17]

On the other hand, the comparatively small firm of Elders of Berwick-upon-Tweed (worth only £7,000 at its incorporation in 1904), which had total sales of £26,909 in 1913, saw total sales (excluding sales of other manufacturers' goods) rise to £31,678 in 1915, £38,851 in 1916, £50,130 in 1917, and £64,662 in 1918. This was achieved entirely by making agricultural machinery, since the firm did not have any munitions contracts. This had the advantage of leaving the firm free to set its own wages, and it took the trouble to check with the Ministry that this was the case when, in the autumn of 1917, the Workers' Union complained to the firm that it was paying low wages.[18]

While the profits of the larger firms were impressive, they were less so when inflation is taken into account. On the other hand, the surviving accounts do not give any indication of what firms may have done with their profits to reduce their liability to Excess Profits Duty. Much of the subsequent history of the major firms was determined by how far the war was used as an opportunity to strengthen the underlying financial situation of the firm. Harrison, McGregor followed a conservative financial policy, continuing the pre-war practice of keeping dividends modest, and either put some cash into the reserves each year, or carried most of the current year's profits forward into the next year, to strengthen the capital side of the balance sheet:

Table 9.6 *Harrison, McGregor & Co. Ltd: trading profits and reserves, 1913–18 (£, year ending 30 September)*

	1913	*1914*	*1915*	*1916*	*1917*	*1918*
Sales	210,096	205,492	136,637	151,586	184,256	289,124
Trading profit	16,772	12,342	3,949	8,469	9,348	9,781
Brought over from previous year	7,460	10,404	16,862	16,398	18,982	19,503
Total available for distribution	24,232	22,747	20,811	24,867	28,330	29,284
Distributed as:						
to reserve account	5,000	–	–	–	–	–
dividend*	8,827	5,885	4,413	5,885	8,827	8,827
Balance carried forward	10,404	16,862	16,398	18,982	19,503	20,456

* Unlike most firms, the share capital seems to have been entirely in Ordinary shares, having no Preference or Debenture shares.

This modest policy meant that the firm finished the war with slightly higher reserves (£50,000, rather than £45,000), slightly higher property and financial assets (£140,259, as against £136,115), but no more share capital (static at £117,704). However, along the way it had bought 400,000 francs' worth of French government Defence Bonds (on 13 November 1916), and, that apart, the total assets had risen from £393,451 to £422,186.[19]

A similarly cautious policy was followed by Ruston, Proctor & Co. (see Table 9.7). Paying modest dividends and raising the proportion of profits put into reserves was far-sighted. In 1913 the Reserve and Equalization of Dividend account had stood at £126,679. After the 1917 accounts had been agreed, it stood at £295,000. At the merger with Hornsbys in 1918,

Table 9.7 *Ruston, Proctor & Co.: financial record, 1913–17* (£)*

	1913	*1914*	*1915*	*1916*	*1917*
Gross profits	117,630	119,118	110,405	158,518	159,333
Balance brought forward	14,930	17,049	16,820	13,353	15,626
Less: depreciation	17,890	17,930	17,322	20,996	20,109
Total available for distribution	114,670	118,237	109,903	150,875	154,850
Distributed as:					
to reserve account	28,320	25,000	25,000	45,000	45,000
interest and dividends	77,500	72,625	69,500	87,500	87,500
Carried forward	17,049	16,820	13,353	16,326	19,599
Total	**122,869**	**114,445**	**107,853**	**148,826**	**152,099**

* In 1918, the company merged with Richard Hornsby & Sons, of Grantham.

Source: GEC-Alsthom, Firth Road, Lincoln, Ruston & Proctor annual reports.

Ruston Proctor would in any case have had a dominant role, since it was the bigger of the two, but its strong reserves position gave it an even more powerful influence in setting the terms of the merger.

While it might seem that the industry had profited from the war, this was more apparent than real. The nominal value of profits had gone up, but they had probably not done much more than keep pace with inflation. The more cautious companies had followed a conservative policy of increasing the cash reserves rather than increasing dividends. This would help to cushion post-war adjustments, but on the other hand the war had seen the physical capital of the industry get four years older. Even in cases where the demands of munitions-making had led to the installation of new machinery, this would have to earn its keep after the war, or be in effect worthless. And hanging over all, there was the anxiety of how to restore exports of peacetime goods.

The agricultural market and the food production programme

For the first two years of the war, farming was not directly controlled by the government. In that time, it was subject to various pressures, which affected its demand for agricultural machinery. There was a shortage of labour, since recruitment to the forces, although voluntary until conscription in January 1916, was a large movement. By the end of 1914, over a million men from all industries had enlisted in the armed forces. In all, by the Armistice on November 11 1918, a total of 5.67 million had served in the forces.[20] The extent to which recruits were drawn from agriculture is uncertain, and estimates of the numbers who left agriculture for the forces or for other civilian employments vary. One estimate is that, up to the end of the war, 273,000 men had left agriculture, out of the 760,000 male employees in agriculture recorded at the last pre-war population census, in 1911. A loss of this magnitude would be a very severe one. However, other evidence suggests that the loss of labour was less than this, and that the effect on farm output was not large. In addition, after the government adopted a policy of raising food production, in December 1916, farming was given substitute labour, in the form mainly of soldiers, but also women (from both the newly formed Women's Land Army and the local population), prisoners of war, and schoolchildren. Thus amended, the labour position does not look too serious. An estimate of the total labour supply in agriculture in England and Wales suggests that the supply of labour fell by about 7 per cent by 1916, and that the use of substitute labour reduced this loss to about 3 per cent by 1918.[21]

This labour shortage was relatively minor, and presumably did not give rise to any increase in demand for labour-saving machinery. A greater problem, which did alter the agricultural demand for the products of the agricultural engineering industry, was the shortage of horses. These were required in enormous numbers on the western front, for the transport of supplies and the movement of guns and ammunition. Initially, the army made up the supply by impressing horses from the British countryside, and from farming in particular. In 1914 the number of horses on farms at the annual farm census day (4 June) was 926,820, declining to 858,032 in June 1915. Thereafter, the army relied chiefly on imports and on non-farming sources, and the number of farm horses had almost recovered in 1916, to 906,233. The shortage of farm horses was not severe enough to curtail the tilled area, which remained almost unchanged in 1914–16, but farmers had recourse to expedients to eke out their supply of horsepower, such as using breeding mares for field work, which in ordinary times they tried to avoid. Finally, the supply of power for farming was affected by the shortage of working steam ploughing sets. By early 1917, it was thought that about half of the 600 pre-war sets in the UK were inoperative, due to shortage of skilled operators, spare parts or coal. Even earlier, the lack of sets had aroused official concern, and in March 1916 the drivers, attendants, and mechanics of steam ploughs and threshing machines had been given exemption from military service.[22]

Thus, while the machinery industry was turning over to munitions contracts, it does not seem that the conditions in the first two years of the war were such as to generate any particularly large change in the demand for agricultural machinery. The preoccupation of the manufacturers with war work could also be offset by farmers to some extent by using machinery for longer. Most farm equipment had a fairly long life. This was particularly true of barn machinery, although cultivation implements such as ploughshares and harrows wore quite quickly, and the complex machines such as binders and threshers needed a continual supply of spare parts. Such evidence as exists suggests that farmers were keeping their stock of implements and machines in good condition. The only survey of farming equipment on any scale is that for Essex in July 1917 (there was no national survey until 1942). The greatest degree of disrepair was in binders, where 3,999 were reported to be in good repair, with 330 being 'capable of being repaired'. For ploughs the numbers were 13,373 and 464; for threshing machines, 420 and 22.[23]

The only case in which it is clear that the war changed the demand for machines in a marked fashion was the motor tractor. In January 1916 the *Implement and Machinery Review* commented on the machinery section at the annual Smithfield Show:

> Motor ploughs and milking machines were two outstanding classes of mechanism which received the closest attention, whilst agricultural motors were also keenly examined in view of the growing scarcity of horses.*

The Board of Agriculture (the precursor of the Ministry of Agriculture, and now DEFRA) helped to publicise the new motor tractors and ploughs. In November 1915 the *Journal of the Board of Agriculture* carried a report of recent demonstrations of motor ploughs and tractors in the eastern counties. Four types of motor plough and ten tractors were demonstrated. Most of the tractors were petrol driven, although two were steam; three of the petrol tractors were from the USA. The main types of tractor in British farming before 1917 were: from the USA, the Overtime (there known as the Waterloo Boy), the International Harvester Titan and Mogul, and the more minor types such as the Interstate, and the Avery. The Parrett probably occurred in smaller numbers, and probably even fewer of an overweight oddity, the Bates Steel Mule. The main British makes were the Saunderson and Ivel, with a few Clydesdales (from Wallace of Glasgow). There were also two British motor ploughs, the Wyles and the Crawley Agrimotor. The number of each make working on farms is unknown. The government estimated that there were 3,500 private tractors in the UK by the end of 1917. Most of these would have been imported from the USA. The total number of Saundersons produced is not known, although the total number of Ivels produced between 1904 and the demise of the company in 1919 is estimated at 900. The dominance of US models on the British market was assured by the middle of the war.[24]

By the end of 1916, manufacturers had moved over to war work. Whether this left spare capacity for increasing deliveries to the home market is doubtful. It was not a matter of farmers not having the money to buy machinery, since they made high profits during the war. It was rather that the demand does not seem to have been significantly different from pre-war. Analysis of the production of the main firms suggest that war work neatly replaced the lost exports, and home output was much

* *Implement and Machinery Review*, 1 January 1916, p. 1041.

C.6646
I.H.C. MOTOR TRACTOR
R.Y.L.T.
FURROW

Before 1917, the tractor market was dominated by US tractors, of which the most popular were the International Harvester Mogul and Titan, and the Waterloo Boy, which was marketed in Britain as the Overtime.

Left: International Harvester Mogul

Bottom left: International Harvester Titan

Right: The Overtime. (*Museum of English Rural Life*)

the same (after allowing for inflation) as before the war, but certainly had not increased.[25]

This was now to change. In December 1916 there had been a change of government; the slow-moving Asquith administration giving way to that of Lloyd George. One of his main new policies was to increase home food production, in order to offset the losses of imported food due to the German submarine attacks, both present and anticipated. To this end, he changed the President of the Board of Agriculture. R. E. Prothero (later Lord Ernle) assumed the office, and established a sub-department of the Board, charged with the mission of raising food production. The Food Production Department (FPD) was headed by the energetic Sir Arthur Lee. The aim of the government was to have a 'plough policy', which, by growing cereals and potatoes, would supply more food per acre than if the land had been left to grassland and animal raising. The object was to plough up the 3 million acres which had changed from cereals to grass since the early 1870s. The policy of the FPD would be implemented through committees established in each county. These, the County Agricultural Executive Committees, would survey each farm and give each farmer a ploughing target. Farmers were given guaranteed prices for wheat and oats (but not barley, in deference to the temperance movement), and assured that the supplies of labour would be protected from further recruitment. Farm labourers were given a guaranteed minimum wage. Farmers were also promised an enhanced supply of machinery. One of the last acts of the Asquith government had been to make the Ministry of Munitions responsible for the manufacture of agricultural machinery, via a newly established Agricultural Machinery Branch. On 2 January 1917, S. F. Edge, one of the pioneer pre-war motorists, and a close associate of the late Dan Albone, was made the first Director of the AMB.[26]

The functions of the AMB were to place contracts, control manufacture, and meet the requirements of the Food Production Department. Manufacture of agricultural machinery was prohibited without a licence from the Ministry of Munitions. When the licence had been given, priority for materials was granted, and manufacture could begin. Home production for private customers continued, but on a reduced scale. Private imports were much reduced. The bulk of supply, whether home or imported, was now in the hands of the Ministry of Munitions. In 1917 and 1918, the MoM placed orders for machinery to the value of £4.7 million:

Table 9.8 *Ministry of Munitions machinery orders*

From the USA, 1917–18:
6,000 Fordson tractors
3,750 Oliver tractor ploughs (for use with the Fordsons)
2,632 International Harvester Corp. 'Titan' tractors
5,000 binders
6,000 two-furrow Fordson tractor ploughs
From home manufacturers:
500 caterpillar tractors (Clayton & Shuttleworth)
400 Saunderson tractors
65 sets steam ploughing tackle (Fowler & Co.)
6,500 harrows
393 threshing machines

Source: Ministry of Munitions, *History of the Ministry of Munitions*, vol. XII, pt VI, pp. 8–9.

As can be seen, the FPD orders were for cultivating machinery of the usual types, with the novel addition of a large order for motor tractors. But the most striking feature of the list was the strong dependence on imports from the USA. This was not surprising in the case of the motor tractor, since Henry Ford had finally, after years of experimenting, brought out his tractor, the Fordson (though still in prototype) at the end of 1916. This was so clearly ahead of any other type of tractor that the British government really had no alternative but to opt for it. Apart from lightness (only 1.5 tons) and simplicity in operation, the Fordson had the virtue of being cheap. Henry Ford agreed to offer the machine to the government at the amazingly low price of £150, plus carriage. The British farm machinery industry might have benefited from the original plan, which was to manufacture the Fordson in England. But the government was so worried by the air raids on London in June 1917 that they decided to make the resources over to aircraft production instead. The plan was altered so that the Fordson was now to be built at Detroit, and shipped over to England, to be assembled and tested at Trafford Park, Manchester, which was Ford's UK base.

But the dependence on the USA did not stop with the Fordson order. There were also Titan tractors from International Harvester, the Oliver ploughs to accompany the US tractors, special ploughs for the Fordsons (partly to replace the Oliver ploughs, which proved unsuitable for British conditions), and a substantial order for 5,000 binders. While some British firms profited from the revival of cultivation and the FPD orders in 1917–18, some did not. Hornsby, which had been a large producer of binders before the war (producing an average £37,400-worth in 1909–13), was down to £24,900 in 1916, only rising briefly to £29,40 in 1917, falling again, to £20,600-worth in 1918. But Ransomes benefited a lot from the revival of home trade. In 1909–13 it had sold £95,000 each year on average of ploughs and parts on the home market, and in 1918 this rose to £342,000. Harrison McGregor was in July 1917 awarded a large order (1,000) for binders by the Ministry of Munitions.[27]

The growing dependence in the second half of the war on imports can be shown by the trade figures. In 1908–13, retained imports of agricultural machinery had been worth £527,000 a year. In 1917 they rose to £3.56 million, and in 1918 to £5.69 million. Thus by 1918 they were worth almost the same value in money terms as the whole of the output of the British industry before the war – although since prices had risen, the rise was less in real terms. A rough estimate would be that before the war retained imports were equivalent to about 10 per cent of the output of the home industry, and that in 1918 they would be equivalent to about half its output.

There was a growing dependence on the USA. Of the imports of 'prime movers' (tractors and engines) in 1916–18, which averaged £1,254,000 *per annum*, 97 per cent came from America; some 60 per cent of 'other machinery' also came from the USA. American trade statistics shed some light on the type of machinery exported: of the total of $6.79 million of US machinery exports to Britain in 1917 and 1918, 'other' machinery (chiefly tractors and their spares) amounted to $3.29 million, and $2.01 million was accounted for by ploughs and cultivators. When inflation is taken into account, it becomes apparent that US exports fell off early in the war, and had not quite recovered their pre-war level by 1917, although there was an enormous increase in 1918. At 1909–13 prices, US exports of agricultural machinery to the UK were worth $1,182,000 in 1914, but fell to $757,000 in 1915, recovering slightly to $944,000 in 1916 and $993,000 in 1917, and rising sharply, to £3,197,000, in 1918.[28]

The rise in exports in 1918 was due to the programme encountering delays. In particular, the Fordson delivery programme was behind schedule. The first Fordsons arrived in England in October 1917, but not in any quantity until the beginning of 1918. In most cases this was too late for the ploughing season of 1917–18, although they could be put to work on ploughing for the harvest of 1919. However, in practice, the 1919 programme was abandoned by the government, due mainly to a defeat in the House of Lords on the enabling legislation. The FPD began to sell off its tractor stock, beginning on 1 June 1918, when 1,000 tractors were released from the county agricultural executive committees (CAECs, or ECs for short) for sale to the public. The ECs took the opportunity to rationalise the number of tractor types they used, reducing the types to six: Titan, Overtime, Clayton & Shuttleworth (caterpillar), Saunderson, 25 hp Mogul, and the Fordson.

Fordson Model F tractor, 1917

The first Fordson tractor, the Model F, was a truly revolutionary design. Ford's mass production techniques, helped to make the Fordson the cheapest and best value tractor in the world when it was launched in 1917. Henry Ford wanted a small tractor which could do most farm tasks, and still be within the financial reach of the small farmer. The Fordson weighed only 2,700 lbs (1.7 tons), stood less than 5 feet high, had a wheelbase of only 63 inches, and could turn in a 21-foot circle. In a typically Ford move, the tractor was made short enough to fit sideways on a rail wagon, to minimise transport costs.

Previous tractors had been usually built with the engine and transmission resting on large longitudinal steel girders or frames. This made them heavy and unwieldy. In addition, much of the transmission and drive mechanism was exposed, leading to severe wear and tear. The Fordson had a unit construction, in which the engine, gearbox, transmission and rear wheel differential were enclosed in hollow cast-iron units, which were bolted together to form a rigid structure. This gave the tractor lightness and great bodily strength. Use of heat-treated chrome carbon steel (thought superior to the vanadium steel of the Model T car) also made for strength, and an engine design which minimised the number of parts meant that maintenance costs were low. The unit design meant that the three main components – front wheels, engine/transmission, and the rear wheels – could be brought together in a simple final production technique similar to that of the Model T car assembly line, thus cutting production costs. The four-cylinder engine was not large (3.9 litres), and only gave 20 hp, but it was sufficient for the light weight of the tractor. There were three forward gears, all enclosed in an oil bath for protection. The Fordson fulfilled Ford's idea of making a simple, affordable tractor for the 6 million US farmers. It was the agricultural counterpart to the Model T car.

The Fordson did have disadvantages. It had poor fuel efficiency, and its light weight and short wheelbase led to a tendency to tip it backwards. The starting system, which entailed starting on petrol until the engine was warmed up, and then switching over to kerosene (paraffin) was temperamental. The final drive to the rear wheels was via a worm gear, which overheated (and overheated the driver's seat), and reduced the fuel efficiency. But it was cheap, simple and inexpensive to run and maintain, and was the model for every subsequent tractor. Ford's mass-production techniques meant that it could undercut every other tractor on the market. On its launch it cost $795, when most other tractors cost over $2,000, and when the tractor market collapsed after 1920, Ford cut the price relentlessly. At one time in the early 1920s the farmer could buy a new Fordson for $230, a staggeringly low price.

The late delivery of Fordsons was the main check to the success of the import programme. The other problem was the failure of the Oliver plough. This was designed for horse work in US conditions, and its breast board was short and stubby, designed to break up the light soils of North America. It was unsuitable for tractor work, since tractors went faster than horses, scattering the soil widely and breaking up the furrow slice. Attempts to modify it were made, but they do not seem to have been on a large scale, and eventually an order was placed for 6,000 special ploughs

A 1917 Fordson Model F photographed at Lamport Hall Agricultural Museum, April 2005. *(Peter Dewey)*

The first Fordson Model F arriving at Trafford Park, Manchester, 1917. (*Museum of English Rural Life*)

for the Fordsons. The recipients of this order are unknown; it may account for the sharp rise in Ransomes' output, but the company accounts are silent on this point.[29]

The revival of cultivation in 1917–18 had mixed results for the industry in Britain. There were some clear beneficiaries – Saunderson and Fowler received orders, and Ransomes also did well. But the reliance on imported tractors and other equipment reduced its impact on the industry, which in most cases remained locked into munitions contracts. Apart from tractors and their equipment, steam ploughing sets and binders, the FPD ordered only a relatively modest amount of equipment – about £1.9 million in value. The plough programme of 1917–18 came too late to affect the fortunes of the industry except in a superficial and short-term way.

Fears for the future

The boom in the years immediately before the First World War was the best ever for the industry, with its highest output level yet seen. Profits were also respectable. The post-war enquiry into the industry by the Agricultural Engineers Association analysed the 1913 financial results of 27 companies. The report considered that the firms in the survey accounted for 90 per cent of the industry, in terms of capital invested and manufacturing capacity. The value of output (of 26 of the companies) was £5,849,140. Deducting costs of £5,249,503 left a profit of £599,637. This was enough on average to pay the preference shareholders just over 5%, and ordinary shareholders just over 7.5%. Of 26 companies, ten paid dividends of over 8%, eight paid between 5% and 8%, and eight paid nothing – so not all firms were making (or at least not distributing) high profits. But overall it was a profitable industry, at a time when inflation was about 3 per cent a year (average of 1909–13), and the Bank of England's bank rate was varying between 3 and 5 per cent (in 1909–13).[30]

Yet there was cause for concern. It cannot have escaped the notice of the leaders of the industry that their prosperity was based very largely on the export trade. Nor would they have been immune from the increasing talk of war in the years before 1914. This was common currency in the press and in imaginative literature – perhaps most famously enshrined in the gripping novel of German invasion preparations by Erskine Childers, *The Riddle of the Sands* (1903). In the event of a major war, British trade would certainly suffer badly. Also, a large part of the exports were to the Russian empire, which was chronically unstable. There had been a full-scale revolution in 1905, which had led to a degree of parliamentary democracy being conceded by the ruling family. The agricultural system had been reformed, so that former peasants could create their own discrete farms, apart from the communally held village lands. But this had not laid revolutionary agitation to rest, nor solved the problem of intermittent famines, which threatened public order.

Finally, by 1913 it was becoming clear that the industry's products had become seriously outdated. The rapidity of the growth of the motor vehicle industry had taken British industry by surprise. The Ford Motor Company had been formed in 1903; in 1906 it produced only 8,729 cars, but that was enough to make it the largest producer in the fledgling US industry. After several early models, the Model T was produced in 1908, the car being displayed at the British Motor Show at Olympia in November of that year. By 1913 output was a staggering 168,220 Model Ts. In this context, and given the growing interest in motor tractors, the industry must have felt increasingly vulnerable. Although the only major firm positively to set its face against the internal combustion engine was Fowlers, and some had by this time diversified into civil engineering, railways, oil engines or food processing, the remainder were dependent on a market with limited potential for growth, and considerable risk of trading conditions deteriorating at short notice.[31]

Considerations such as these must have been in the backs of the minds of the leaders of the firms who contemplated mergers before 1914. Merger proposals were in the air for some years before 1914, and these were revived when war broke out in 1914. Concerned at the loss of exports at the outbreak of war, Marshalls tried to merge jointly with Clayton & Shuttleworth and with Ruston, Proctor, but this was rejected by the Marshall board after protracted negotiations.[32] The war then took people's minds off mergers. But by 1918 they were again in the air. In many industries, boards of directors saw merger as a way of meeting the expected resumption of international competition after the war. In addition, there was the prospect of making a capital gain for themselves and their shareholders, by capitalising the high profits of wartime. Thus capital reconstructions, sometimes stopping short of mergers, occurred in large industries such as cotton, shipbuilding and steel making, and actual mergers took place on a large scale in railways and banking. In the case of the agricultural engineering industry, there was probably a third factor: a realisation that the future of the industry was particularly bleak. The fear that exports could not be revived, and in particular the fear of competition from the USA, both in third markets and in Britain itself, probably also played a part.

Before the war ended, this movement had produced only one merger, but it was of two of the largest firms, and pointed the way to others. The firms were both from Lincolnshire: Ruston. Proctor & Co., of Lincoln, and Richard Hornsby & Sons, of Grantham. The details of their amalgamation throw light on the financial position of the industry, and the motives for the merger. Both were making good profits. Hornsby was paying a dividend of 6¼% on the Ordinary shares, and in 1916 added a bonus of a further 3¾%, making ten per cent in all. But even after paying all dividends, only half the disposable profits had been disbursed, and there remained some £44,000 which was carried forward. The question of merger with Ruston, Proctor was first reported to the Hornsby board on 19 June 1917, although the reasons were not recorded in the minutes. After rejecting at first a scheme for the merger prepared by the accountancy firm headed by Sir William Peat, an alternative scheme prepared by Mr Roberts, the managing director, was adopted. The basis of it was that, 'any union of the two companies should be on the parity of the ordinary share capital'. At the time, Hornsby's capital structure was: £100,000 of 6% preference stock; £355,450 of £10 ordinary shares; £150,000 of 4½% debentures, and £70,000 of 5% debentures. There was also a Special Reserve Fund, which was partly cash and partly trustee securities. Hornsby's position, arrived at after talks with Colonel Ruston, was that, if the Special Reserve Fund could not be handed over to Rustons without risk, then each paid-up £10 Hornsby ordinary share should be exchanged for nine Ruston £1 ordinary shares. If the Special Reserve Fund could be handed over completely and without risk, the exchange should be ten Rustons £1 ordinary shares for one £10 Hornsby ordinary share. The Debentures should be ranked equally with Rustons debentures, with some minor qualifications. The preference shares should rank equally with Rusons preference shares in the event of a winding-up, but Rustons preference shares should have priority as regards dividend.[33]

In the event, the final agreement with Rustons followed the proposals for the fixed-interest shares, but the ordinary shares of Hornsbys were to be exchanged for Ruston ordinary shares in the basis of 17*s*. for £1. The question of the Special Reserve Fund would be taken to court. If it were decided that the fund was the property of the company or of the shareholders, it would be transferred to Rustons as part of the assets of the new company, but if the courts held otherwise, Hornsbys should not be under an obligation to do so. Thus the final agreement was a significant loss for Hornsby shareholders. It is difficult to escape the conclusion that this was less a merger than a takeover of Hornsbys by Ruston, Proctor. It was the bigger firm; in 1912, its capital value was £1,381,193, and Hornsbys £918,254. By 1917, its issued share capital was £1,200,000 (£600,000 as ordinary shares; £350,000 as preference shares; £250,000 as debentures). In addition, the Hornsby Special Reserve Fund does not necessarily seem to have been treated as part of the company's assets, being regulated by a trust fund. But by the time of the merger Rustons' cash reserves stood at £290,000, and Hornsby's Special Reserve Fund at £175,000. It was thus a case of a cash-rich firm taking over a smaller one with a poorer cash flow. Finally, since Rustons had moved further from agricultural machinery than had Hornsbys in the pre-war years, they had more post-war potential. The reasons of the merger were explained to the Hornsby shareholders at a meting on 19 June 1918, by the chairman:

> Your Directors have been giving very careful thought how best to meet the difficult conditions which will prevail after the cessation of hostilities in meeting foreign competition, and they have come to the conclusion it is of the utmost importance for businesses to unite forces in order that low costs of production may be obtained by greater efficiency, specialisation and concentration ...[34]

The merger came just in time for both firms. The Armistice was followed by the boom of 1918–19, but this proved short-lived and was certainly not the precursor to a long-term revival of the export trade. In particular, the Russian trade was lost forever. For Hornsbys, although the courts eventually decided that their Special Reserve Fund should remain the property of the Hornsby family, this was cold comfort in the depressed world of the 1920s. Demand was low, and short-time working endemic. But the merger may have prolonged the firm's business. It was generally agreed that by 1914 the firm had lost its way technically. Their basic type of steam engine had remained unchanged between 1892 and 1918; the 'hot-bulb' oil engine system which Hornsbys had pioneered and clung to had been superseded by cold-starting engines produced by other makers. Also, their engines, beautifully engineered and long-lasting, kept going for a long time, and replacement demand was low. Finally, the depressed state of the farm machinery market would have boded ill for Hornsbys in the absence of the merger. The merger gave the firm some resources to continue in business, but the new firm increasingly made over its agricultural work to Ransomes, after Ruston & Hornsby bought into Ransomes in 1919. After 1932, the Grantham factory of what had been Hornsbys was moved entirely into engines and other non-agricultural work. In the same year, the formation of Aveling-Barford benefited Grantham, since Aveling-Barford relieved Hornsby's of some 30 acres of their redundant premises.[35]

For Rustons, the merger provided a chance to use its cash (although not the Hornsby Special Reserve) to buy into Ransomes in 1919. In effect, Ransomes became a subsidiary of Ruston & Hornsby. R & H then passed over its agricultural work to Ransomes, ridding itself of a potential liability in the market conditions of the 1920s. It could then concentrate on engines, earth-moving machinery and cranes. In 1929 it was to conclude an agreement with the US Bucyrus Corporation, which was effectively an agreement to share world markets; this gave the firm renewed vigour in the next depression in world trade, in 1929–32.[36]

The merger of Ruston, Proctor & Co. and Richard Hornsby & Sons, which took effect on 1 April 1918, was the only case of a merger in the industry before the Armistice. From then on, firms would have to steer their own course and find their own salvation in the post-war world. Few would have predicted an easy future, but fewer still would have been prepared for the roller-coaster ride of 1918–23.

CHAPTER TEN

A new world, 1919–1939

THE HISTORY of the industry between the the two world wars was closely linked to the ups and downs of the British economy as a whole. There was a brief burst of prosperity in the 18 months after the war, when exports revived to some extent, although they were in volume terms much below the pre-war level. Prices rose even more rapidly than in the war, and although costs of materials and labour rose, there was enough surplus to provide good profits for manufacturers. It was not really a true boom, since output did not rise very much. It was rather that there was a high demand for replacing machines and implements which had been scarce in wartime, and the boom was one of prices and profits rather than of output. But the prosperity came to a sharp end with the collapse of exports and production in 1920–21. This was the sharpest fall in production, and rise in unemployment, in British economic history, and was accompanied by falling prices, falling wages and costs, and rising labour unrest. When the slump bottomed out in 1922–23, it was apparent that, although home production was slowly recovering, exports had not. This was as true of agricultural engineering as of the other great pre-war exporting industries such as cotton, coal and shipbuilding. The rest of the 1920s saw a continuing effort to develop lines of production which would either have a good future on the home market, or provide new export markets. But these efforts, temporarily thrown off course by the General Strike of 1926, were undermined in 1929, when a fresh slump began, which did not reach its trough until 1932. Recovery after that was slow and difficult, with foreign sales hard to find. But insofar as new home markets developed, aided by the technical changes of the period, centring on the motor and electrical industries, then firms had a brighter future. The growing perception after 1933 that another war was coming, led to efforts to assist the government in the coming conflict.

The boom of 1918–19 and the slump of 1920–23

After the war manufacturers and consumers naturally looked forward to the resumption of normal trading. While the sections of industry which were protected from foreign competition, or did not rely much on foreign markets, could profit from the resumption of pre-war spending patterns, this was not so for the agricultural machinery industry, which had problems from the start. This was in spite of the fact that farmers had made high profits in wartime, and had little on which to spend them. In fact many former tenant farmers used their savings to buy their farms in the great sales of landed estates which resumed after the war, rather than to buy machinery. In addition, the sudden cessation of government contracts left the manufacturers high and dry. Finally, the political and economic chaos of the post-war world prevented the resumption of exports. This was especially true of trade with Russia and eastern Europe, where the Bolshevik revolution, the repudiation by the new Russian government of its debts, and the freezing of all assets formerly held by non-Russian companies, severely hurt many of the larger firms which were particularly exposed to the Russian trade in 1914.

The common experience of the larger agricultural engineering firms was a large drop in sales immediately on the resumption of peace, as munitions contracts came to an end, and continuing slump thereafter, in the absence of anything like a full recovery in export markets. In the case of Ransomes, sales were:

Table 10.1 *Ransomes, Sims & Jefferies: sales 1913 and 1918–20 (£)*

	1913	*1918*	*1919*	*1920*
Home	187,905	422,412	505,075	662,535
Export	429,090	49,535	202,933	298,273
War work	–	1,674,882	378,568	378,480
Totals				
(at current values)	616,995	2,146,829	1,086,576	1,339,288
(deflated by the Board of Trade wholesale price index)	616,995	932,998	426,945	423,022
(volume of sales as an index, based on 1913 = 100)	100	151	69	69

Source: Rural History Centre, University of Reading, TR RAN AD 7/17.

These totals may be compared with the total Ransomes sales in 1913, which totalled £616,995 (home £187,905; export £429,090). One may deflate these figures by the official Board of Trade wholesale price index, which stood at 116.5 in 1913 (taking 1900 as 100), and was 368.8 in 1920. Thus the price level in 1920 was about three times the level of 1913. On deflating post-war sales by this index, it turns out that Ransomes home sales after the war had been more or less restored to their real pre-war level by 1920/21, but exports were much reduced, and so the total sales were, in real terms, about a third below their pre-war level. This decline occurred in 1919, while the rest of British industry was still enjoying its post-war boom, which was to go on until late 1920.[1]

The failure to restore sales after the end of the war had an adverse impact on profits. In the case of the amalgamated Ruston & Hornsby, gross profit in the first year of trading (1918–19) was £222,341. This was a return of about 10 per cent on shareholders' funds, and was enough to pay a dividend for the year of 8 per cent on the ordinary shares. But it was a cause for worry that the company had issued 316,000 new ordinary £1 shares early in 1919, in order to make their investment in Ransomes shares, so that the capital needing to be serviced had risen by that amount, and was now £2,128,238. The following year the gross profit was down to £165,397, and in 1920–21 it was £108,682, and the company recommended that no ordinary share dividend be paid. The smaller firm of Fosters of Lincoln made a gross profit of £37,994 in 1919–20, but this fell off in 1920–21 to £7,614. The 1921 directors' report noted that: 'trade was satisfactory until the late summer of last year, when a rapid decline set in'. Smaller firms also saw only a brief period of post-war prosperity. Elders of Berwick-upon-Tweed had seen their sales rise from £28,143 in 1914 to £64,662 in 1918, and to £68,225 in 1919 but thereafter decline was continuous. In the circumstances, it was odd that the board, having followed a policy until then of reinvesting profits, capitalised the £7,761 in the cash reserve, and issued it as fully paid-up shares to the existing shareholders. This seems to be a case of rewarding the shareholders at the expense of the longer-term prosperity.[2]

The brief period of prosperity in 1918–19 evaporated sharply towards the end of 1920, as the whole British economy (and that of western Europe and the USA) slid suddenly into recession. Exports fell precipitously and profits evaporated. The national rate of unemployment rose from the abnormally low level of 3.9 per cent in 1920 to 16.9 per cent in 1921. Inflation came to a sudden end, and prices began to fall. By December 1922 the Board of Trade wholesale price index had fallen 51 per cent below its peak 1920 level. This literal deflation of prices was aided by the Bank of England setting a policy of high interest rates. From April 1920, the bank's rate rose to 7 per cent, and was kept at that level (a near-panic level by contemporary standards) for twelve months. Wages fell along with prices, aided by an employers' offensive against trade union organisation. The upshot was that by 1922–23, much of British industry, including the agricultural machinery industry, was in a state of deep recession, with low sales, profits and employment.[3]

The recession may be traced in the fortunes of several firms. For example, in 1921–22 Ransomes' sales fell precipitously, by almost 60 per cent, from £1,362,002 the previous year to £554,212 (the firm changed their financial reporting year after 1920, 31 December to 31 March]. In 1922–23, sales were almost static, at £522,373. Only in 1923–24 did they begin to rise again. The loss had been mainly caused by the collapse of exports in 1920–21, although home trade had also fallen.[4] Other firms also slid into recession, and it was not long before losses were being recorded.

This is so in the case of Harrison, McGregor, although the period of loss was relatively short:

Table 10.2 *Harrison, McGregor & Co. Ltd, profits and losses, 1918–23*

	1918	*1919*	*1920*	*1921*	*1922*	*1923*
Sales (£)	289,124	281,591	368,173	not stated	not stated	not stated
Net profit (loss) (£)*	9,781	13,661	18,711	(9,968)	(2,601)	12,889
Dividend (%)	7½	7½	7½	5	3¾	5

* after providing for all interest, depreciation and other charges.

Source: Wigan Archives Service, Harrison, McGregor archive, Reports of Directors, 1918–23.

Ruston & Hornsby made a profit in 1920, but after that they were only able to keep paying a dividend by drawing on their reserves. This policy had to come to an end in 1923, when the balance of profit, after allowing for depreciation and debenture interest, was only £12,043, and the directors recommended that no dividend be paid. Dividends on the ordinary shares were not resumed until 1926, with the modest payment of a dividend of 2½%.[5]

In the difficult years of 1920–23, it was of course true that the industry was affected by external problems not of its own making. The recession of 1920–21 depressed export markets, which were further undermined by the collapse of currencies in Europe, thus making trade almost impossible. Many countries had succumbed to galloping domestic inflation. In Russia, the process had been topped off by civil war, and the currency was effectively worthless. As Colonel J. S. Ruston, the Chairman of Ruston & Hornsby, remarked in his 1922 report to the shareholders:

> On the 31st March last the German Mark had fallen to one-fifth of its value a year previously, the Greek Drachma to one-half, the Roumanian Leu is now merely nominal, the Polish Mark is about one-ninth, the Serbian Dinar one-third, the Turkish Piastre about one-half, the Austrian Krone was quoted at about one-twentieth, and has now collapsed altogether, while for the Russian Rouble there is no price at which it can be exchanged.*

There were also domestic problems, including severely worsening labour relations. In 1920 there was a 14-week national strike of woodworkers and a 17-week strike of iron moulders. In 1921 there had been a national coal strike and a railway strike. In addition, in common with manufacturing in general, the hours of work had been reduced from about 56 to 47 at the end of the war, entailing higher costs for all firms.

---

* Ruston & Hornsby, Thirty-Third Annual Ordinary General Meeting, 27 June 1922.

The collapse of export markets

Once the wartime munitions contracts had expired, the main immediate problem of the industry was the glaring hole in sales left by the disappearance of exports, particularly to Russia and eastern and central Europe. Since there were sharp changes in the prices of exports, thus making comparison with pre-war sales difficult, the current value of post-war exports is listed below, with the Board of Trade index of export prices. When the current values are deflated by the price index, one gets a view of the 'real' level of exports, after price changes have been taken into account:

Table 10.3 *British agricultural machinery exports, 1919–38*

	Column 1	*Column 2*	*Column 3*	*Column 4*
	Current value of exports	*Price index for all UK exports*	*Real value of exports (1913 prices)*	*Real export value as percentage of real 1913 value*
	£	(1938 = 100)	£	%
1913	3,735,814	68	3,735,814	100.0
1919	2,177,856	189	783,567	21.0
1920	1,711,955	245	480,453	12.9
1921	949,719	184	353,222	9.5
1922	948,808	136	474,404	12.7
1923	1,281,885	130	670,524	17.9
1924	1,612,998	129	850,263	22.8
1925	1,876,956	126	1,012,960	27.1
1926	1,682,577	118	969,621	26.0
1927	1,567,506	112	951,700	25.5
1928	1,817,067	110	1,123,278	30.1
1929	1,977,496	108	1,245,090	33.3
1930	1,718,663	103	1,134,651	30.4
1931	966,542	92	714,401	19.1
1932	620,135	86	490,339	13.1
1933	767,570	86	606,916	16.2
1934	795,673	87	621,905	16.6
1935	1,044,738	88	807,297	21.6
1936	1,151,335	90	869,897	23.3
1937	1,492,120	98	1,035,349	27.7
1938	1,365,598	100	928,607	24.9

Sources: P. Dewey, 'The British Agricultural Machinery Industry, 1914–1939: Boom, Crisis, and Response', *Agricultural History* (1995), p. 301: B. R. Mitchell, *British Historical Statistics* (Cambridge, 1988, p. 527.

The failure of exports is glaringly apparent over the whole inter-war period. Looking at exports in real terms (column 4 above) enables the full measure of the collapse to be appreciated. The most severe period was the 1920–21 slump, when the real value of exports fell to about one-tenth of the pre-war level. This collapse was catastrophic: it was much greater than the general collapse of British exports at this time, and much greater than that in the great exporting industries of coal, cotton and shipbuilding, which attracted so much contemporary public attention. While the agricultural machinery industry slowly rebuilt its exports in the late 1920s (with a blip in 1926–27 due to the General Strike), even at the decadal peak in 1929 they were, in volume terms, only one-third of the 1913 level. The fresh world economic crisis of 1929–32 saw exports fall sharply once again, to a low of about one-eight of the 1913 level. Thereafter there was only partial recovery, and exports by the late 1930s were lower than in the late 1920s.

Perhaps the greatest problem left by the war was the collapse of certain markets, on which the industry had particularly relied before 1914. This can be seen by analysing the export markets in the years 1909–13 (Table 5.4 above). The dominance of the Russian, German, and South American markets is clearly indicated, with important contributions from British India, other European countries, and from Australia and New Zealand. But the problems of the post-war markets were not confined to the Russian *débâcle*. Germany was in a state of political chaos until 1924. The currency collapse which affected chiefly the Russian and German currencies was also seen in Austria and Hungary. Both of these currencies had collapsed by 1924, when they were restored with the aid of the League of Nations. The French franc continued on a more gentle, but inexorably downward course, until *de facto* stabilisation in 1926. The only reverse to this story of currency depreciation was seen in Italy, when Mussolini ordered the revaluation of the lira in 1927, for reasons which remain obscure, and are likely to have been just the product of muddled thinking. This must have improved the chances of British exporters in Italy, but on the Continent generally the post-war era was one when governments adopted much higher levels of tariff protection. This was true of France, Germany and Italy. High protective tariffs were also resorted to by the new 'successor states' of eastern Europe (Poland, Czechoslovakia, Yugoslavia), which were desperate to earn foreign hard currency, and had only agricultural exports to offer the world. These were in excess supply in the 1920s, and all these countries fought hard to overcome balance of payments deficits, by the use of tariff protection, and the not always unwelcome depreciation of their currencies.

In many ways, the 1930s were a re-run of the foreign trade problems of the 1920s. The world export collapse of 1929–32 was succeeded by only a partial recovery, as excess supply of primary products (especially cereals) led to the dumping of produce on foreign markets. The financial collapse of world farming led to lowered demand for manufactured goods. The reductions in national income which ensued from the slump in farming and industry persuaded governments to resort to even higher levels of tariff protection. This process began in the USA with the Smoot-Hawley tariff of 1930. By 1932 even the traditionally free-trade British government joined the rush to protection: with the Import Duties Act of 1932, a minimum of 10 per cent by value was placed as tax on imports of manufactured goods. The rest of the decade was a story of higher protection, and competitive currency devaluations. The upshot was that world trade did not recover, and nor did British exports. In spite of the British attempt in 1932 to buck the depression in exports by creating a free-trade system based on the British Empire (called, in a low-key way, 'Imperial Preference'), the real level of British exports in 1938 was only about 80 per cent of that of 1929. By that time, rearmament was well under way, and normal trade even more difficult.[6]

But these events were, in the immediate post-war years, things yet to come. The farm machinery industry was immediately concerned with restoring old markets and finding new ones as soon as possible. The loss of pre-war markets had plunged it into a crisis much more severe than that faced by coal, cotton or shipbuilding. How would the industry leaders respond?

The first, and obvious course of action was to keep export markets continually under review, and to consider which were capable of being revived. If there was little to be done in this respect, then attention should be given to finding new markets elsewhere. This was much on the minds of the directors of Ransomes, Sims & Jefferies in the post-war slump of 1920–23. At their meeting on 22 December 1922, they heard a lengthy report on the Russian market. It concluded that pre-war traders would not be welcomed back in Russia; that it would be risky to make any financial investment there; that any transactions should be on a cash-only basis; that the Russian government's system of distributing imports was chaotic; finally, that business could only be done with difficulty in Russia, due to 'the heavy commissions extorted by Russian Government Officials.' (The phrase 'heavy commissions' should undoubtedly be read as 'large bribes'.) At the next meeting, on 24 January 1923, the export trade report noted that:

> Since the last Board Meeting the economic condition practically all over the continent has become much worse, and we are now further off regaining our footing on the continent that we were six weeks ago. While still keeping a watchful eye on any chance of doing trade in Europe, I feel we shall have to concentrate on doing an increased trade with the Colonies and Dominions.

The board considered whether the prospects of trade with South Africa and Australia would be helped by the appointment of a permanent representative in both those markets, to liaise with the agencies selling Ransomes' goods. While mulling over this proposal, which would be expensive to implement, the board was informed of an unforeseen reason for the slowness of the Argentine market to pick up. The farmers before the war used to scrap machines after two or three years, but during the war two large Argentine repair firms had greatly expanded and improved their operations, so that farmers were now patching up their machines rather than purchasing new ones. Further consideration was given to trading with the Soviet government, whose Arcos agency had requested trade and credit facilities. However, caution was again expressed at the idea of giving any significant amount of credit to the Russian government or its agencies; the securing of 'illicit commissions' was again referred to. At the meeting of 23 May 1923, the Russian situation was summed up succinctly: 'The whole thing is that no one knows whether the present Russian government will be there tomorrow.'[7]

Hopes for the revival of the Russian market were not realised, and as time passed, Ransomes concentrated on imperial and dominion markets. The size of the readjustment required may be seen from the comparison of its 1913 trade with that of 1923–24:

Table 10.4 *Ransomes, Sims & Jefferies: export markets 1913, 1923–24 and 1933–34*

	1913		*1923–24*		*1933–34*	
	£	%	£	%	£	%
Europe	199,401	47	35,208	14	56,486	32
Russia	165,258	39	96		negligible	
Asia	26,039	6	38,574	15	14,956	8
Africa	131,097	31	148,333	57	72,523	43
Union of South Africa	84,929	20	87,111	34	50,770	29
South America	57,521	13	11,271	4	16,424	9
Argentina	34,344	8	4,093	2	12,707	7
Australasia	12,667	3	26,246	10	10,792	6
Totals	**426,725**	**100**	**259,632**	**100**	**176,608***	**98†**

* Includes £2,183 to 'other countries'.

† Rounding error.

Source: Rural History Centre, University of Reading, Ransomes, Sims & Jefferies archive, TR RAN AD7/18–20.

Although Ransomes did make efforts to replace the lost European (especially Russian) exports of 1913, they were unsuccessful. The Russian market itself never revived. In 1923 it posted sales of £96. In 1933–34 it was not even recorded as a market. There was some revival in Argentina, but the South American market was much smaller than before the war. The South African market, although smaller than pre-war, held up better than the average. But the African figures were held down by losses of the Egyptian and Sudanese markets. Ransomes' experience was similar to that of the industry as a whole: in spite of vigorous sales efforts, the losses of foreign trade were not made good, and even by the early 1920s the leaders of the industry were contemplating organisational changes as a way out of their difficulties.

Attempts at restructuring the industry

The position of the industry by the mid-1920s was desperate. The gravity of the situation was realised by the trade body, the Agricultural Engineers Association, which commissioned a report into the industry. This comprised the collection of accounts from the leading firms (29 in all), accounting for about 90% of the industry's output, before and after the war. These were then analysed by the accountancy firm of Price, Waterhouse. This showed that the industry had shrunk drastically between 1913 and 1923. A comparison of the 1923 sales of 27 of these firms, taken at 1913 values, showed a decrease of 12.4% in home sales, and of 67.4% in export sales. Not surprisingly, employment had fallen in those ten years, from 27,411 in 1913 to 15,169 in 1923. Sales of these firms in 1913 had been £5,849,140, and a profit made of £599,637. Sales in 1923 (at current prices) were £4,544,608, and a loss of £423,467 was recorded. Of the 29 firms supplying details of output and costs, 15 were making losses. These firms employed the bulk of the labour force in the industry – 12,439 out of the 15,169. Action of some sort was clearly needed, whether of finding new products, or of bringing firms together in some sort of association, or even amalgamation, with a view to finding economies and stimulating efficiency.[8]

Even before 1914, some leading firms in Lincoln had contemplated amalgamation. The only one to occur had been that of Ruston, Proctor, and Hornsby in 1918. A less tight association, which was still designed to strengthen the firms concerned, was the agreement between Ruston & Hornsby and Ransomes on 31 March 1919 for the former to make a 'trade investment' in Ransomes. This took the form of Rustons acquiring 96 per cent of Ransomes ordinary share capital, in exchange for an equivalent number of Ruston shares, to the extent of £500,000-worth of shares, being the whole of Ransomes' ordinary share capital. In effect, Rustons acquired a controlling interest in Ransomes, although it did not propose to exercise full control. There was an exchange of directors: three Ransomes directors joining the Ruston board, and one Ruston director joining the Ransomes board). Ransomes welcomed the injection of cash, since they were in serious trouble due to the decline of home and foreign trade. The *quid pro quo* was that Ransomes would take over the manufacture of the

Rustons agricultural machinery lines (chiefly threshers, but also binders and mowers), since Rustons intended to move entirely into industrial machinery. Ransomes was strengthened by inheriting Ruston's agricultural business. Finally, Rustons benefited from the dividends on the shares (5 per cent at the outset of the agreement). It was a neat and mutually beneficial response to the altered circumstances of post-1918. By the end of the 1920s Rustons had almost eliminated their agricultural interests, producing only a few threshing machines. The last machine to be exported was a clover huller sent to Riga on 8 September 1931.[9]

Rustons apart, the only major realignment in the industry was the reorientation of Fosters of Lincoln. Their main business before the war had been steam engines, threshing machines, chaff cutters and elevators. However, they had made a start in the oil engine business, and had been instrumental in the evolution of the tank during the war (see chapter 9). But by the early 1920s it was apparent that the traditional lines were under threat, and by 1922 the firm had begun to make losses, which persisted until 1928. There was a feeling on the board that other lines of business should be sought. The solution eventually adopted in 1927 was to use the spare cash accumulated in the war, which had been carefully husbanded, to buy a business unconnected with agriculture – Gwynne's Engineering, of Hammersmith. This firm had got into trouble through building motor cars, and there was a personal connection, since Sir William Tritton, the managing director of Fosters, was a former Gwynnes employee. The buy-out cost about £19,000, but the transaction still left Fosters with about £40,000 in cash reserves. This was the turning point. The new firm began making profits, albeit modest, the following year. While Foster continued to trade under its own name, it ran its traditional lines down considerably. Although a new firm was registered, as Gwynne's Pumps, and traded separately, the whole of its share capital was held by Fosters, which itself moved largely into the pump business. At the Foster AGM in 1940, the chairman (still Sir William Tritton) noted that sales were at a record high, due almost entirely to the pump business, 'the Foster business affording only 5% of our turnover'.[10]

Fosters had the financial resources, and knowledge of available opportunities, to restructure successfully in time to ride the next depression. These advantages would have been less available to the smaller firms, and it was now just a matter of time before the industry began to witness business failure. The first was that of E. & H. Roberts, of Deanshanger, Stony Stratford, Buckinghamshire. A medium-sized company with a share issue of £34,000 and employing about 160 men in 1910, Roberts specialised in ploughs, and had tried to move into tractor ploughs after the war, but with what success is unknown. It had been making losses continuously since 1921, and by 1927 the unsecured creditors were owed £9,694. The General Strike may have been the last straw for an inland company which had only canal and road links to the nearest railway line.

It went into voluntary liquidation on 22 February 1927. At the dispersal sale the following year, its patterns were eagerly acquired by the rest of the trade.[11]

But amalgamation was still in the air. At first a scheme was canvassed for an East Anglian amalgamation, which might have included Clayton & Shuttleworth, Ruston & Hornsby, Ransomes and Garretts. However, this did not get very far, and foundered when the Garrett brothers, Frank and Alfred, concluded that it would entail the closing of their Leiston works. A more fundamental reorganisation, in which a large part of the industry was involved, and which was almost entirely disastrous for it, occurred shortly afterwards. This was the *débâcle* of Agricultural & General Engineers Ltd. This originated, according to R. A. Whitehead, the historian of Garretts, as a 'brain child of the ebullient Thomas Aveling (of Aveling & Porter of Rochester) and his friend and neighbour Archibald W. Maconochie, who owed his fame, if not, as the unjust suggested, wholly his fortune, to "tinned rations" or "plum and apple jam"'. Maconochie had made a lot of money in the war with his Maconochie Tinned Rations, and was still in the business of producing a famous bottled condiment, Pan Yan Pickle.[12] He was 'a man of standing and a director of the Great Eastern Railway Company'.[13] The proposal was that the leading agricultural engineering companies should amalgamate, with the intention of becoming more efficient and thus more able to ward off foreign competition. The proposal found favour with the directors of the leading firms, as 'the scent of amalgamation was heavy on the air'. Within a short time, the first five firms signed up: Aveling & Porter Ltd (Rochester); E. & H. Bentall (Heybridge); Blackstone & Co. (Stamford); Richard Garrett & Co. (Leiston); James & Fredk. Howard Ltd (Bedford). These each contributed one director to the new company. The other directors were Maconochie (chairman) and Sir Thomas Robinson (deputy chairman), who had been Agent-General for Queensland. The company was formed on 4 June 1919. Shares in the new company would be offered to the shareholders of the constituent companies, in exchange for the shares in the constituent companies. The shareholdings of the constituent companies would thus be extinguished, and in effect the companies would be owned by Agricultural & General Engineers. The enterprise was given a glowing send-off in an interview with Mr Maconochie in the main trade journal, *The Implement and Machinery Review*, in a masterpiece of puffery and empty bombast:

> 'This is the beginning of great things for the agricultural machinery business of this country,' he [Maconochie] asserted. 'Do you think so,' was the query, but barely had it been put when promptly came the reply: 'I *know*.' [original italics]

Those words give a thumbnail sketch of the man. Shrewd, sagacious, alert and full of confidence, Mr Maconochie is determined to achieve his set purpose, and that is to make this new undertaking pre-eminent.

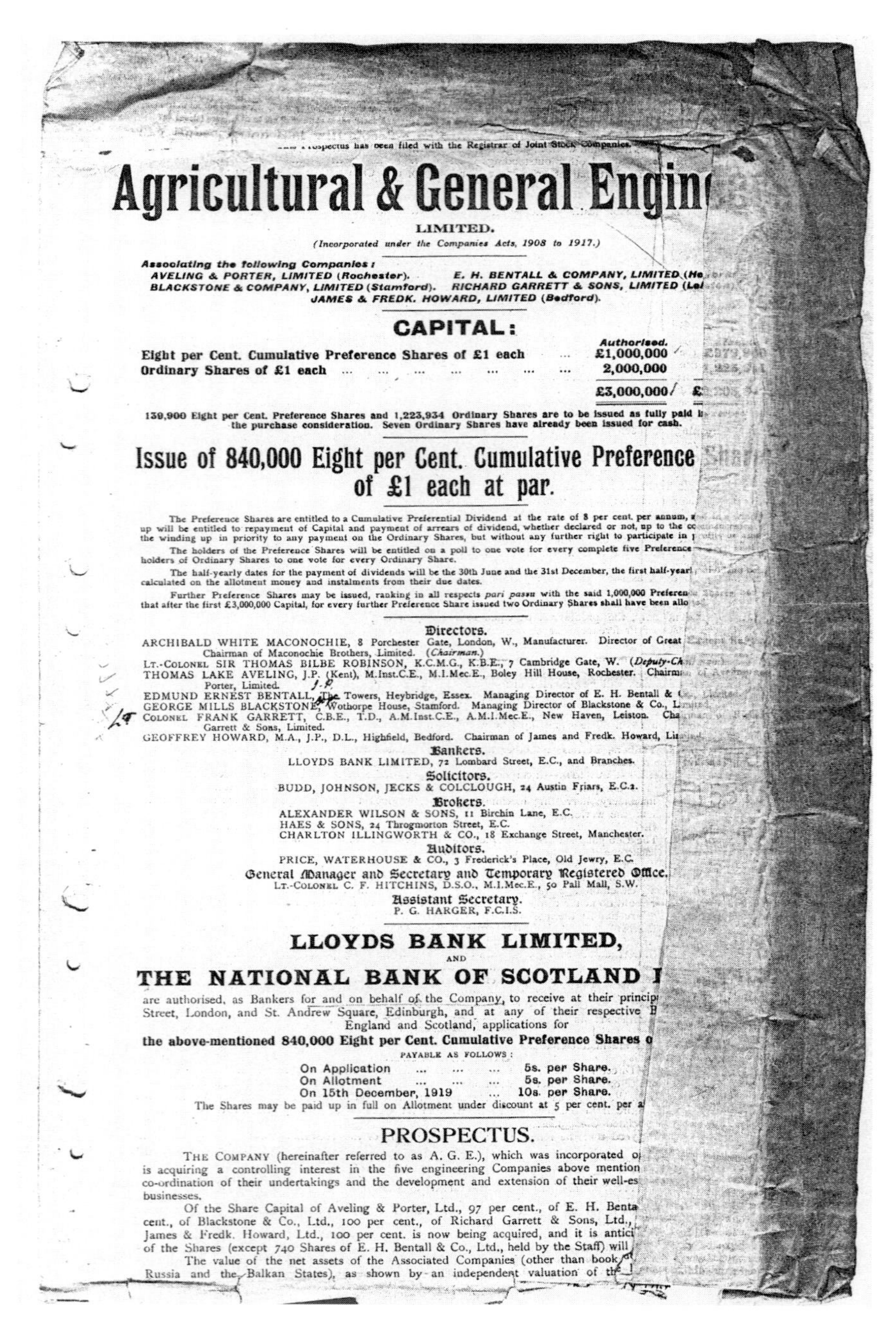

... Prospectus has been filed with the Registrar of Joint Stock Companies.

Agricultural & General Engin

LIMITED.

(Incorporated under the Companies Acts, 1908 to 1917.)

Associating the following Companies:
AVELING & PORTER, LIMITED (Rochester). ***E. H. BENTALL & COMPANY, LIMITED (He***
BLACKSTONE & COMPANY, LIMITED (Stamford). ***RICHARD GARRETT & SONS, LIMITED (Lei***
JAMES & FREDK. HOWARD, LIMITED (Bedford).

CAPITAL:

	Authorised.
Eight per Cent. Cumulative Preference Shares of £1 each ...	£1,000,000
Ordinary Shares of £1 each	2,000,000
	£3,000,000

139,900 Eight per Cent. Preference Shares and 1,223,934 Ordinary Shares are to be issued as fully paid i the purchase consideration. Seven Ordinary Shares have already been issued for cash.

Issue of 840,000 Eight per Cent. Cumulative Preference of £1 each at par.

The Preference Shares are entitled to a Cumulative Preferential Dividend at the rate of 8 per cent. per annum, up will be entitled to repayment of Capital and payment of arrears of dividend, whether declared or not, up to the co the winding up in priority to any payment on the Ordinary Shares, but without any further right to participate in

The holders of the Preference Shares will be entitled on a poll to one vote for every complete five Preference holders of Ordinary Shares to one vote for every Ordinary Share.

The half-yearly dates for the payment of dividends will be the 30th June and the 31st December, the first half-yearl calculated on the allotment money and instalments from their due dates.

Further Preference Shares may be issued, ranking in all respects *pari passu* with the said 1,000,000 Preferen that after the first £3,000,000 Capital, for every further Preference Share issued two Ordinary Shares shall have been allo

Directors.

ARCHIBALD WHITE MACONOCHIE, 8 Porchester Gate, London, W., Manufacturer. Director of Great Chairman of Maconochie Brothers, Limited. (*Chairman.*)
LT.-COLONEL SIR THOMAS BILBE ROBINSON, K.C.M.G., K.B.E., 7 Cambridge Gate, W. (*Deputy-Ch*
THOMAS LAKE AVELING, J.P. (Kent), M.Inst.C.E., M.I.Mec.E., Boley Hill House, Rochester. Chairm Porter, Limited.
EDMUND ERNEST BENTALL, The Towers, Heybridge, Essex. Managing Director of E. H. Bentall &
GEORGE MILLS BLACKSTONE, Wothorpe House, Stamford. Managing Director of Blackstone & Co., L
COLONEL FRANK GARRETT, C.B.E., T.D., A.M.Inst.C.E., A.M.I.Mec.E., New Haven, Leiston. Cha Garrett & Sons, Limited.
GEOFFREY HOWARD, M.A., J.P., D.L., Highfield, Bedford. Chairman of James and Fredk. Howard, Li

Bankers.

LLOYDS BANK LIMITED, 72 Lombard Street, E.C., and Branches.

Solicitors.

BUDD, JOHNSON, JECKS & COLCLOUGH, 24 Austin Friars, E.C.2.

Brokers.

ALEXANDER WILSON & SONS, 11 Birchin Lane, E.C.
HAES & SONS, 24 Throgmorton Street, E.C.
CHARLTON ILLINGWORTH & CO., 18 Exchange Street, Manchester.

Auditors.

PRICE, WATERHOUSE & CO., 3 Frederick's Place, Old Jewry, E.C.

General Manager and Secretary and Temporary Registered Office.

LT.-COLONEL C. F. HITCHINS, D.S.O., M.I.Mec.E., 50 Pall Mall, S.W.

Assistant Secretary.

P. G. HARGER, F.C.I.S.

LLOYDS BANK LIMITED,

AND

THE NATIONAL BANK OF SCOTLAND

are authorised, as Bankers for and on behalf of the Company, to receive at their princip Street, London, and St. Andrew Square, Edinburgh, and at any of their respective England and Scotland, applications for

the above-mentioned 840,000 Eight per Cent. Cumulative Preference Shares o

PAYABLE AS FOLLOWS:

On Application	5s. per Share.
On Allotment	5s. per Share.
On 15th December, 1919 ...	10s. per Share.

The Shares may be paid up in full on Allotment under discount at 5 per cent. per a

PROSPECTUS.

THE COMPANY (hereinafter referred to as A. G. E.), which was incorporated o is acquiring a controlling interest in the five engineering Companies above mention co-ordination of their undertakings and the development and extension of their well-es businesses.

Of the Share Capital of Aveling & Porter, Ltd., 97 per cent., of E. H. Benta cent., of Blackstone & Co., Ltd., 100 per cent., of Richard Garrett & Sons, Ltd., James & Fredk. Howard, Ltd., 100 per cent. is now being acquired, and it is antici of the Shares (except 740 Shares of E. H. Bentall & Co., Ltd., held by the Staff) will

The value of the net assets of the Associated Companies (other than book Russia and the Balkan States), as shown by an independent valuation of th

The company prospectus for their cumulative preference shares issued by Agricultural and General Engineers in October 1919. Optimism for the future and fear of renewed foreign competition combined to drive many of the industry's famous firms into this ill-starred holding company. Aveling, Blackstone, Bentall, Garrett and Howards had already joined. (*National Archives*)

This country alone is not our market, nor is the Continent of Europe. The *world* is our market. If I can get such 'overheads' in my particular business that are actually lower than those of our great American competitors, don't you think I can do the same in this combination? To prove that, I will tell you that we have immediately arranged for the production of one line alone at the rate of 5,000,000 per year. And what competition can stand against that? Works are being extended, other schemes are afoot, and we are going out for the best. You can't have more than that.*

The prospectus of the company drew attention to the advantages expected to accrue from it. They were: mass production; increased employment; increased efficiency in production and selling; economies due to the reduction of overlapping lines of business and from the establishment of central selling and purchasing organisations; simplified financial arrangements; higher exports. For the sake of these hypothetical benefits, the shareholders would cease to be interested in their particular companies, and would hold instead shares of the so far untried AGE.

In view of the sorry later history of AGE, interest has centred on the original participants, and their motives. The cynical view is that they had a financial interest in doing what they did. Maconochie received a commission of 3 per cent in AGE shares for his services in arranging the merger, as well as his salary as chairman.[14] For the directors drawn from the family firms, the incentive cannot have been their fees as directors, which were modest. They were paid £200 a year, the chairman got £500, and the deputy chairman £1,000. Edward Barford, who became a central figure in the story, alleged that they were paid commission on the deal, but there is no evidence of this. Only Maconochie seems to have received a commission, and that was in shares rather than cash. Their financial incentive, if there was one, was that they swapped their shares in family firms with uncertain futures for shares in a much larger entity with what must have seemed a brighter future, or at least better than might have been expected from remaining outside the combine. A further consideration was that, although loath to relinquish their paternalistic control, they hoped that they would thereby safeguard local employment and the future of their home towns. As Table 10.5 shows, the directors in fact swapped large numbers of shares, and thus diluted their control over their family businesses. Thomas Aveling, Edmund Bentall, George Blackstone and Geoffrey Howard (and possibly Frank Garrett) were giving up a dominant stake in their firms for the uncharted waters of shares in the new AGE. In 1919 it was possible to be optimistic, but by the end of 1920 the post-war slump had hit the industry, and the advantages anticipated to flow from the merger proved to be a chimera. But by then other firms had been persuaded to join, and the original five firms had grown to fourteen.† For the most part, AGE owned 100 per cent, or nearly so, of these, although

* *The Implement and Machinery Review*, 1 November 1919, p. 988.

† The other firms to join were: Barford & Perkins (Peterborough); Clarke's Crank and Forge (Lincoln); L. R. Knapp (Clanfield, Oxon.); E. R. and F. Turner (Ipswich); Bull Motors (Stowmarket); Charles Burrell (Thetford); Burrell's Hiring Co. (Thetford); Davey Paxman (Colchester); and Peter Brotherhood (Peterborough).

Table 10.5 *Financial structure of Agricultural & General Engineers Ltd, 1919*

No. of shares in AGE issued in respect of:				
Aveling & Porter	*Bentall*	*Blackstone*	*Garrett*	*Howard*
225,091	216,324	419,921	272,598 Ordinary, 54,000 Preference	90,000 Ordinary, 85,900 Preference
Total no. of shares in constituent companies exchanged for AGE shares:				
Aveling	*Bentall*	*Blackstone*	*Garrett*	*Howard*
202,801	145,480	10,200*	10,000 Ordinary*	90,000 Ordinary
Of which: no. of shares exchanged by:				
Thomas Aveling	*Edmund Bentall*	*George Blackstone*	*Frank Garrett*	*Geoffrey Howard*
110,060 + 566 joint	107,930 + 26,500 joint	721* + 965 joint‡	† Ordinary, 1,257, * Preference	† Ordinary, 27,600 Preference

* £10 shares; all the rest were £1 shares

† Number unknown: the entry in the prospectus has been covered over by subsequent binding at PRO.

‡ 'joint' means that the shares were held jointly with another person.

Source: Public Record Office, *Prospectus for Cumulative Preference Shares of Agricultural & General Engineers* Ltd, 30 October 1919. Board of Trade BT 31/32289/155725.

only 70 per cent of Peter Brotherhood. This expansion increased the authorised share capital of AGE substantially, from the initial £3,000,000 to a final £8,000,000 (£1 million in preference, £100,000 in second preference, and £6,900,000 in ordinary shares) in May 1920. Of this, £1,649,392 in ordinary, £999,900 in preference and £28,320 in second preference shares was issued. This was a total of only £2,677,612. The fact that the issued capital was much below the estimated value of the capital assets of the constituent firms, as well as of the authorised capital of AGE, was made much of in the AGE promotional literature, but could not disguise the fact that the fortunes of AGE were a product of the fortunes of the constituent firms. Perhaps people were hypnotised by the sheer size of the sums involved, which would make the proposed combine one of the biggest companies in Britain.

The optimistic tone of the launching prospectus – 'the future offers great scope for extension of the business carried on by the Associated Companies, and a largely increased output may confidently be expected' – could not disguise the fact that the associated firms were a mixed bag, and in fact had little in common.[15] The only really specialist agricultural engineering firms of any size were Garretts, Howards and Bentalls.

Knapps was a small firm, whose presence is difficult to account for. The rest had moved away from agricultural work, and were oil and steam engine builders, locomotive builders, steam roller makers or in the business of showmen's engines (Burrells). There was thus no common interest to build on apart from the fact that they were all engaged in mechanical engineering of some kind, and it was difficult to see how their operations and factories could be integrated in order to save costs and engage in mass production. The latter was little more than a dream, given the propensity of the industry to work on a bespoke basis, with great variation of product types to suit individual and often small markets. Also, the agricultural machine makers were well aware that their products were outmoded, and that they were in need of a lot of capital to make good the depreciation of the past, write off the old wartime equipment, and tool up for new lines of production. Finally, most of the leading firms in the agricultural machinery business proper stood aside – Ransomes, Clayton & Shuttleworth, Fowlers, and Marshalls, so that an industry-wide reorganisation was clearly out of the question.

However, the centralisation of buying and selling was proceeded with. The intention was to centralise all purchasing at head offices, down to the last nut and bolt. This was admirable in principle, but was not really workable for a combine with such scattered manufacturing centres, and with large differences in product types. Victor Garrett, who had had experience of such a system when serving at the Ministry of Munitions, was put in charge of it, but he soon concluded that the system was so inflexible as to be unworkable, and on his recommendation it was abandoned in July 1920. Central selling survived a little longer, until October 1922, but from then on the companies reverted to their own direct selling organisations. Sales promotion, however, remained a central function, and that was in the hands of Frank Garrett as Sales Director – 'a not unsuitable candidate for the position, for he was a clever and imaginative, if not always a wise man', and under his direction sales offices were set up in South Africa, India, Chile, Australia and New Zealand. It becomes difficult to avoid the conclusion that as far as employment was concerned Garretts benefited more than others from the new company structure, since none of the other families seems to have supplied senior executives to AGE.[16]

From the start AGE had a millstone around its neck: early in 1920 it leased a site at Aldwych in central London from London County Council. Here AGE built a new headquarters building. Completed in 1923, it cost some £400,000. A large part was occupied by AGE itself, so the building was only partly producing revenue by being let out to other firms; a large part of its construction cost and upkeep therefore became a charge on the group itself – in effect, on the constituent firms. A further company, called Aldwych House Estates, was formed to control the property, but did not lease parts of the building to other firms until much later in the 1920s, when the financial problems had become insurmountable.[17]

The high hopes which many must have pinned on AGE evaporated after the industry was hit by the slump which began so abruptly at the end of 1920. From then on the troubles of the constituent firms became the troubles of AGE, and were reflected in its share price. The only dividend ever paid on the ordinary shares was 4.86 per cent free of tax for the year ended 31 March 1920. In 1920 the price range of £1 preference shares had been from 20*s.* 6*d.* to 18*s.* 6*d.* By 1929 it was from 9*s.* 6*d.* to 5*s.* 6*d.* The £1 ordinary shares had dropped to a low of 2*s.* 6*d.* in 1929.

Failing to reap the expected benefits from centralising purchasing or selling, the only other way in which AGE could contribute to a solution of the problems of the industry after it fell into serious difficulties after 1920 would have been to rationalise it – in effect, to close down the most serious loss-makers, and hope that business would revive in time. There was some action here. The first casualty was Burrells of Thetford, which had been acquired totally by AGE between 1920 and 1924. It had been making losses continually since 1920, and in 1928 the decision was taken to close the works at Thetford, and to transfer production to Garretts at Leiston. C. J. Lines refers to one other instance of rationalisation, Bentall's giving up the making of petrol engines in exchange for the 'right' to make coffee plantation machinery, which was of use to the firm later on.[18]

These attempts at rationalisation came too late and did not touch the core of the industry's problems: lost markets, technological backwardness and outdated capital equipment. But AGE continued in business. Maconochie left in 1921, and was replaced by as chairman by Gwilym E. Rowland, a successful tax consultant with considerable personal contacts in the City. He had a powerful and manipulative personality, and soon dominated his other directors, for whom he did not bother to disguise his contempt. He was also good at psychological warfare, and could sense an office plot in the making. These attributes stood him in good stead, and ensured that he remained chairman to the end, but did not help him in addressing the problems of the industry. However, there is no doubt that he was sincerely trying to do his best for AGE. He was already a rich man, and did not need the money, but thought that it would be a national shame for AGE to collapse, and that his strong management could rectify the situation. He was closely observed at work by the young Edward Barford, son of James Barford, joint owner of Barford & Perkins. In 1923 he took a job with AGE in export sales, and at some time later Rowland offered him the post of his personal assistant, at the large salary of £2,250. In early 1930 he became privy to the information that Rowland was exploiting the affairs of Aldwych House Estates, and the system of export credit guarantees recently established by the government, via the Export Credit Guarantees Department, in order to disguise the failing financial position of AGE. On confronting Rowland with these facts, he was sacked. Subsequently, with the support of the city editors of the

Daily Express, *Daily Mail* and *The Times*, the young Barford, a comparative newcomer to the City and its ways, convened a shareholders' meeting, at which his resolution was carried. 'Directly after the meeting Rowland fled from Aldwych House, leaving no address, pausing only in his flight to scoop up all the petty cash.' He was never heard of again.[19]

In February 1932, on the petition of Barclays Bank, which was owed about £400,000 by AGE, a receiver was appointed; a liquidator was appointed in April 1932. He decided that the constituent firms should be sold off when possible, to pay the debts of AGE. The two casualties, both in 1932, were Garretts and Howards. Garretts was bought by Beyer, Peacock & Co., of Gorton, and a new firm known as Richard Garrett Engineering Works Ltd formed to run the works. But it was the end of Garretts' agricultural work. It was also the complete end of Howards. Their works in Bedford was closed immediately by the receiver, being purchased shortly afterwards by one Edward Le Bas on behalf of a new company – the Britannia Iron & Steel Works Ltd. This leased the works to a Swiss company, George Fischer Steel & Iron Works Ltd, in 1933. In 1936 the Britannia Iron & Steel Works took over the works, and was in business on the Howard site until the 1970s. The beneficiary from the demise of Howards was Ransomes, which bought out Howards' agricultural interests. After the closure of the Howard works in 1932, no more agricultural machinery was produced there. The new firm produced steel tubing. In both cases it was a serious blow for the employees and the townspeople; the more so for Leiston, which was a much smaller town than Bedford, and which had depended much more on the now-defunct business. Finally, the families themselves lost their shareholdings. It seems that they had promised Rowland that they would not sell their shares, and they seem to have stuck to their promise. As the Bedford local newspaper commented: 'This means, therefore, that the members of the Howard family share the misfortune that has befallen their workpeople.'

There were no more large failures. The remaining firms which had been part of AGE unscrambled themselves painfully from the wreckage. Davey, Paxman and E. R. and F. Turner were reconstructed, but never returned to agricultural work. Edward Barford was instrumental in arranging a merger between Aveling & Porter and Barford & Perkins, and became the owner and managing director of the new firm, Aveling-Barford. This was with the help of Ruston & Hornsby, which wanted to dispose of a surplus factory at Grantham, and of Listers of Dursley. The new firm then moved from Rochester to Grantham. Edmund Bentall, who had lost his shareholding in AGE, managed to borrow some capital and purchase Bentalls back from the receiver in 1933, and by 1938 it was once more in profit. Knapp, whose history is rather obscure, survived until the 1960s. Blackstones was purchased at first by the Blackstone brothers, then by Barclays Bank, and finally by Listers of Dursley in 1936, thus acquiring a range of large oil engines to add to the small ones it already produced.

This was in spite of an earlier agreement in 1931 with Rustons that they would confine themselves to the small engine sector.[20]

It is difficult to decide whether there would have been more, or earlier failures without the existence of AGE. Perhaps it papered over the cracks for a few years more. It is certain that, the case of Burrells apart, it did nothing of any significance to reorganise the industry. On the other hand, it is almost equally certain that the families who still ran the firms would have clung to power and influence until the bitter end, and, in the absence of a receiver to pick up the pieces, more firms might have collapsed after the Great Depression struck in 1929–32. In all, with its lack of realism, excessive London expenses, and lack of original rationale, it may be said to represent the triumph of hope over reality; a rose-coloured view of the most desperate period in the history of British agricultural engineering.

Changing patterns of demand at home

The collapse of the export trade was the largest blow to the industry after 1920, but the home trade was also much reduced. For a while, the higher cereal acreage due to the 'plough policy' of the government was more or less maintained. The tillage area in Britain, which had been in the region of 10.4 million acres since 1913, had risen to 12.4 million acres in 1918. The Corn Production Act of 1917 had offered farmers guaranteed prices for wheat and oats until 1922. Post-war policy for a time favoured the continuation of the cereal guarantees, and the Agriculture Act of 1920 continued them in a modified form, with the promise that Parliament would be given four years' notice of any government intention to abolish them. Even in 1920, although the cereal acreage had reduced slightly, as the world was beginning to find that it had a surplus of wheat on its hands due to the enormous wartime increases in production in the USA, Canada and the Argentine, the British tillage area was still 11.5 million acres. Meanwhile, farmers had plenty of money from wartime profits.[21] By 1919–20 many landowners were beginning to take advantage of the high prices of farm land at the end of the war, and were fearful of higher taxation in the future. They were looking to sell, and farming tenants (especially those in land-hungry small farm areas, such as the uplands of Wales) were looking to buy. The subsequent flood of estate sales has been described by F. M. L. Thompson in these words:

> Precisely one quarter of England and Wales had passed … from being tenanted into the possession of its farmers in the thirteen years since 1914. Such an enormous and rapid transfer of land had not been seen since the dissolution of the monasteries in the sixteenth century, perhaps not even since the Norman Conquest.*

* F. M. L. Thompson, *English Landed Society in the Nineteenth Century* (Routledge & Kegan Paul, 1963), p. 332.

While only about 11 per cent of farm holdings in England and Wales had been owner-occupied in 1914, the proportion had risen to 36 per cent by 1927. This had profound consequences for the financial position of farmers, and thus their ability to purchase farm machinery.[22]

The expansion of owner-occupancy coincided with a sharp turnaround in the fortunes of arable, *vis-à-vis* grassland farming. In 1920 the world surplus of cereals which had been growing since the middle of the war peaked. Since the British market was still open to imports free of tax (as it had been since the Repeal of the Corn Laws in 1846), cheap imports flooded in, depressing prices severely. The price of wheat and oats, respectively, had peaked in 1920 at 18*s.* 10*d.* and 20*s.* 5*d.* They then fell continuously until they reached lows in 1923 of 9*s.* 10*d.* and 9*s.* 7*d.* This sudden fall took the government by surprise. Fearful of having to pay out enormous sums by way of compensation, as the prices of wheat and oats fell below the guaranteed levels, it quickly repealed the 1920 Agriculture Act without notice, and settled the matter with a one-off compensation payment to cereal growers of £20 million. This 'Great Betrayal' as farmers' leaders came to describe it, was merely the precursor to several years of sharply falling prices, incomes and profits for cereal farmers. To a lesser extent, grassland farmers suffered the same problems, since prices of all agricultural produce were falling. It was a time of general deflation, and all producers suffered to some extent. The agricultural depression bottomed out in 1923, and produce prices thereafter recovered somewhat, but they never regained the 1920 levels.[23]

The fluctuations in the value of British farmers' output have been calculated recently by Dr Brassley:

Table 10.6 *Value of gross output of British farming, 1911–39*

	£ million	*Index (1911–13 = 100)*
1911–13	188.05	100
1914–19	334.29	178
1920–22	409.35	218
1923–29	283.67	151
1930–34	245.47	131
1935–39	293.00	156

Source: P. Brassley, 'Output and technical change in twentieth-century British agriculture', *Agricultural History Review*, vol. 48/I (2000), p. 84, Table A4.

The result of all these financial changes was that farmers drew back from cereals, and moved instead over to meat, milk and other dairy products, for which the market was more favourable. Throughout the inter-war period the area of permanent grassland rose, and the tillage area fell. In 1920 there had been 11.5 million acres of tillage and 15.8 million acres of permanent grassland in Britain. By 1939 tillage was down to 8.3 million acres, and permanent grass had increased to 17.3 million acres.[24]

The effect on the farm machinery manufacturers was that the home market, especially for arable equipment, after a brief spurt in 1919–20, remained weak for most of the next twenty years. For most of the 1920s, farmers, especially the ones who had bought their farms at the high land prices of 1919–20, were financially embarrassed, often lacked the capital to make the transition to dairying or cattle (the need for altered farm buildings, and a piped water supply in the buildings and fields being prime requisites, apart from the actual cost of buying animals), and were often kept going by bank overdrafts. This situation was described by A. G. Street, who was himself a large cereal tenant farmer (in south Wiltshire) struggling in the 1920s to find a way out of his financial difficulties:

One man said to me in conversation the other day, when we were discussing this matter:

> 'Grievance? I reckon I've got a just grievance. On the strength of the Corn Production Act I bought my farm and paid cash for it. I've been let down.' I murmured that he was fortunate in that he was able to pay cash, and asked what of the man who had purchased on a mortgage. 'God help him,' he replied. 'Look at old So-and-so. He bought his place for sixteen thousand pounds. 'Twas either that or turn out. He put down half of it, and the bank advanced the rest. Now it's been revalued at eight thousand pounds, and the bank are still willing to advance half of it as before, but they want him to find a matter of about four thousand pounds to put things straight.'
>
> 'What'll happen?' I asked.
>
> 'What can happen?' he retorted. 'Nothing! If they insist they'll bust him, and that'll do 'em no good. They'll wait for something to turn up, and meantime keep on worrying him.'*

The weakness of the home market for arable equipment can be seen from the experience of Ransomes, the largest maker of ploughs and field implements. It was not until 1937 that the value of plough and implement sales exceeded the levels of 1919–20 (see Table 10.7). Ransomes, with other irons in the fire, were in a relatively fortunate position. For smaller firms, particularly if they could not replace lost export markets, the future was bleak in the mid-1920s. This was so in the case of Smyth of Peasenhall, whose sole product was its seed drill. Honed to perfection by 1913, the product depended on the export market. Home arable farming could not replace exports in the 1920s. Before the war, about half of Smyth's output had been exported. After 1920 exports never took up more than 12–15 per cent of total output, although the firm continued to sell to Holland and France. There was a continuous decline in production. By 1925 it

---

* A. G. Street, *Farmer's Glory* (Faber, 1932), pp. 230–1.

Table 10.7 *Ransomes' home sales of ploughs and field implements, 1919–29 (£)*

1919	249,779	1929–30	141,637
1920	308,400	1930–31	117,126
1920–21	273,943	1931–32	132,572
1921–22	167,613	1932–33	134,332
1922–23	123,441	1933–34	178,652
1923–24	133,129	1934–35	200,661
1924–25	141,056	1935–36	245,152
1925–26	155,089	1936–37	228,564
1926–27	145,593	1937	368,347
1927–28	136,226	1938	321,443
1928–29	149,286	1939	491,131

Source: Rural History Centre, University of Reading, Ransomes, Sims & Jefferies archive, TR RAN AD 7/17.

was down to about 170 drills a year. Final near-collapse came shortly afterwards, and in 1926 and 1927 output averaged 92 a year. By 1927 Smyths had been making losses for two years, and was described in an accountant's report as 'moribund'. It staggered on into the 1930s, and for the next several decades survived on a shoestring, probably by doing a lot of repair work. It finally closed in 1968.[25]

As tillage areas shrank in the 1920s, so the 'permanent' grass acreage (i.e. that not forming part of the crop rotation cycle, but intended to be used in the long term) rose. This was not just a matter of farmers reacting to the decline in cereal prices. The balance of consumer demand for food was moving away from what agricultural economists decribe as 'inferior' foods (cereals and potatoes) and towards 'superior' foods (meat, milk, dairy produce, vegetables, fruit). This shift underpinned the improvement in the prices of superior foods, *viv-à-vis* inferior foods. In certain regions such as the uplands of Wales, the west of England and Scotland, local soil, topography and climate already supported a predominance of grass and animal husbandry. In the fenlands of east Anglia, cereals and potatoes predominated. Here farmers had their capital committed to their existing farming systems, and change was slow. However, the shift from tillage to grass continued apace, and presented opportunities for the manufacturers of agricultural machinery. Thus the Wilder firm at Crowmarsh, Oxfordshire, formerly engaged in field machinery, contract threshing and structural building work, diversified further. Two Wilder inventions which saw the firm through these two difficult decades were the 'pitch-pole' self-cleaning harrow, and a grass-cutter-elevator, useful for supplying the grass dryers which were coming into vogue at the time, and which was marketed as the Wilder Cutlift.[26]

The major product change was the growth in milk production.

IT SAVES HOURS IN A DAY

AND POUNDS IN A WEEK.

A TRIO OF TRUCKS THAT PLAY NO TRICKS!

Here we see three Auto-Trucks in the capable control of feminine drivers, at the works of Messrs. Sunbeam Motors Ltd. Here is ample proof of the ease with which Lister Auto-Trucks are operated.

As a result of four years' experience, many new models have been added to the range of Lister Auto-Trucks, but the original model P.N., described within, retains the basic design from which all other models have been evolved.

Improvements have from time to time been made, but the carefully thought out design of the original model, in spite of modifications, remains intact.

FURTHER PHOTOGRAPHIC PROOF OF THEIR POPULARITY.

This shows a consignment of 19 Trucks ready for despatch to fulfil one order received from a prominent Railway Company.

CANADA: *Toronto, Winnipeg, Regina, Montreal, Edmonton, Vancouver.*

'A trio of trucks that play no tricks!' A page from Lister's brochure advertising their 'Autotrucks', 1931. *(Peter Dewey)*

Demand was stimulated by a growing body of medical and nutritional opinion extolling its health-giving virtues; there had also been much more attention to providing clean milk, uncontaminated by mud and dung; and pasteurisation and sterilisation were being adopted to reduce the incidence of TB bacillus in milk. All these helped to expand the market. In 1906–10 the average number of cows in milk in Britain had been 2,330,000. By 1931–35 it had grown to 2,933,000.[27] One notable beneficiary of this shift was Listers of Dursley. Its cream separators sold well, and its petrol engine range, which was used to drive them was enhanced in 1929 by the introduction of a small cold-start diesel. These 'prime movers' were a growing feature of the British farm, and they also were used for electricity generation on farms and in country houses, in a rural Britain still many years away from having mains electricity. Lister diesels also found marine uses. The firm also moved into the materials handling business, producing from 1926 their 'Autotruck', a small self-propelled wagon for use around the factory; this was a great success, selling 10,000 by 1938. The firm opened a new cream separator works in 1933. It was still a major manufacturer of sheep shearing sets, and churns and other dairy utensils. By 1929 it was making a profit of £74,000 after depreciation. Between 1919 and 1936 the numbers of employees rose threefold, and in 1932 it made its shares available to the public.[28]

There was a growing market also for milking machines. Attempts to produce reliable machines went back to the 1880s, and 237 patents were filed between 1860 and 1915, but the early models were crude, the rubber connections deteriorated rapidly, and the pulsation was erratic. It was only in 1918 that the De Laval company put on the market the first milker with pulsation control for uniform milking. However, farmers were slow to take up the device. For farmers, it was not just a matter of buying the milking machine: a reliable and smooth-acting power source was needed, and though petrol engines would do, electricity was best, and this was rare on farms before the 1940s. In addition, the cowhouses had to be adapted, and supplies of running water laid on for the animals and for washing and sterilising the equipment. It is thus perhaps not surprising, that the take-up of milking machines was slow in the inter-war period. One view is that there were fewer than 10,000 installations in use by 1936. Even by 1942, when many more had been installed, the first official count showed only 23,860.[29] In the 1930s, the two largest firms in the milking machine business were Alfa-Laval and Gascoignes (of Reading). If they had depended solely on milking machines, they would probably have failed, and they were supported by other activities; Alfa-Laval in effect subsidised its milking business with the making of brewery equipment and industrial separators and evaporators.[30]

The use of milking machines meant that farmers had to bring feed to, and remove dung from, a herd of cows gathered together in the cowshed. One novel and creative solution was the design in 1922, by A. J. Hosier, a north Wiltshire farmer, of a system which brought the machine to the cows in the field. This, his 'milking bail', was a linked group of portable cow stalls, the whole unit being moved on wheels. To it was attached the milking machines, and a petrol engine as a power source. The cows were herded into the bail, the milking performed, the cows released, and the milk taken to the farmyard. The bail could be then moved to a new part of the field, which thus received the dung from the animals. This saved time and labour, produced cleaner milk, did not require the mucking out of cowsheds, and fertilised the land. Hosier went into the business of manufacturing and selling the bail, and similar designs were bought out by other manufacturers such as Alfa-Laval (in 1928). The bail was especially suitable for the high, smooth, downland fields that Hosier farmed, but less suitable for the small farms of the highlands and the narrow roads of the lowlands. In spite of its technical excellence, it was not a runaway commercial success. A survey of 1932 showed 102 Hosier sets in operation. At the outbreak of the Second World War 90 per cent of herds were still milked by hand.[31]

The search for new products

Much of the technical change after 1918 was based on, or stimulated by, the new farm tractors. From the USA came the Overtime (there known as the Waterloo Boy) and the International Harvester Titan and Mogul. The main British makes were the Saunderson and Ivel, followed by several motor ploughs – the Clydesdale (from Wallace of Glasgow), the Wyles (and Fowler-Wyles) and the Crawley Agrimotor. By the end of the war, they had been joined by the Fordson, which was the largest single make at work. This was a period of rapid development in tractor design, on both sides of the Atlantic. In the euphoria of peace, and with concerns about the future of the industry, many firms had high hopes of leading the nascent British tractor industry. The government estimated that there were 3,500 private tractors in the UK by the end of 1917. To these may be added the 6,000 Fordsons, 500 Claytons and 400 Saundersons ordered by the government, and later released for private purchase, and whatever private imports there had been in the war. In relation to the 250,000 farms in Britain, this was not large, and the future for a British-built tractor must have seemed bright.[32]

At the end of the war, several firms made attempts to get into the tractor market. One of the most publicised was the 'Glasgow' tractor, here illustrated from a publicity poster (opposite).

The Glasgow had been produced since 1916 by the firm of J. Wallace, of Glasgow. The company had ambitious plans for its development, and in 1920 took over a redundant armaments factory at Cardonald. A new company was formed to produce the whole Wallace range, including the tractor, and articles appeared in the farming press extolling the virtues

Advertising for Wallace's 'Glasgow' tractor, 1919. (*Glasgow City Archives*; *Museum of English Rural Life*)

The 'GLASGOW' FARM TRACTOR

At the Lincoln Trials the 'Glasgow' Tractor strikingly demonstrated its superiority in the matter of wheel-grip over all other tractors of medium weight. The draw-bar pull was given officially as 3,550 lbs.—a very much higher figure than that of any competitor in its class—but even at that strain the slipping-point of its driving wheels was not reached. The 'Glasgow' Tractor excels in the matter of wheel-grip owing to its positive drive on all three wheels. This has been proved under the most exacting conditions—in snow, wet soil, frosty ground, etc.

of the tractor and the works. One described the aim of the firm as being to produce the tractor at the rate of '£7,000,000' [*sic*] a year for the next five years. The tractor, which had three wheels, all driven by the engine, performed well at the 1919 Tractor Trials at Lincoln, but the 1920–21 recession was just around the corner, and most farmers were not prepared to invest in a tractor which was much more expensive than a Fordson; by this time second-hand Fordsons could be had even cheaper. Production ceased in 1924, and the total output is unknown, although a guess of 200 has been made; the highest serial number of the tractor recorded was 234.[33]

Many of the other smaller firms which had produced tractors or motor ploughs before 1918 either failed or carried on at a low level. The pioneer firm, Ivel Motors, had lost momentum after the death of Dan Albone in 1906, and went into liquidation in 1921. The Saunderson carried on, and made a later appearance under licence to the French manufacturer SCEMIA in the early 1920s, but thereafter faded from view. Similar obscurity overtook many of the motor ploughs. The number of tractors at work probably declined in the 1920s. In 1920–21, UK International Harvester sales had exceeded 2,750, while sales from the newly opened Fordson plant in Cork amounted to about 7,700 between 1919 and early 1923. This was Ford's first British manufacturing facility. He closed it in 1923, due to losses and operating problems. The Fordson continued to be made in the USA until 1928, when Ford, judging the competition now too keen, closed down the plant. But in the same year he took the decision to produce in England, having already bought 66 acres of land at Dagenham for the purpose. Cork production was restarted, and ran from 1929 to 1932. In that year Dagenham came into production, and Cork was finally closed. The stage was set for a new era in British tractor history. But for the moment the market was weak. From a peak of about 25,000 in early 1921 the national inventory steadily declined as sales dried up and older machines wore out. The 1925 census of agricultural production returned 16,731 tractors (14,565 field and 2,166 stationary). Sales began to revive from the late 1920s. There were a variety of firms still interested in making tractors in the late 1920s and early 1930s. However, only one (Austin) had anything approaching the mass-production facilities employed at Ford's Dearborn plant in the USA, and later Dagenham.[34]

The variety of models on the market was illustrated at the 'World Tractor Trials' of 1930 organised by RASE, and held at Wallingford. Eight different British/Irish tractors were represented, as well as several motor ploughs. The British makes were Austin, Aveling & Porter, Blackstone, Clayton & Shuttleworth, Fordson, Marshall, McLaren, Peterboro, AEC, and Vickers. Strictly speaking, the Austin and Clayton should have been classed as non-British, since neither was made in the UK. The Austin had begun being made in England, but the firm took the view that the market was unpromising, and had transferred production to their French factory. The Clayton was actually made by the Hungarian firm with which Claytons had an agreement, and the tractor was officially known as a CSHS (Clayton Shuttleworth Hofherr-Schrantz) (This connection dated from 1912, when Clayton sold its Vienna factory to Hungarian interests, thus avoiding the war losses suffered by some of its competitors). The Fordson was a Cork model.

One of the interesting features of the line-up was the attempt to make a diesel tractor. The Aveling & Porter (entered by Agricultural & General Engineers) was a diesel, as was the Blackstone, the CSHS, the Marshall, and McLaren. The Austin derived its main technical features from the Fordson, but less successfully, and more expensively. The Vickers, which did not take part in the trials, was a 23 hp model, starting on petrol and running on paraffin. The AEC entry (from a firm which made lorries and buses), known as the Rushton, was, like the Austin, indebted to the Fordson for many of its features; it came in wheeled and tracked versions (the latter by Roadless). None of these machines, apart from the Fordson apart and to some extent the Austin, was in anything like mass-production; and they all cost more than a Fordson, which was then retailing at £170. Its cheapest competitor, the Austin, was £210, and the others in the range cost between £300 and £500.[35]

None of these tractors provided the solution to the problems of the industry. The search for new products took up a lot of time and energy among the older firms with a limited product range. This was particularly apparent with Fowlers, which had been so heavily skewed to traction and ploughing engine production before 1914. In 1909–13 it had produced 4,616 steam vehicles, of which 55 per cent had been ploughing engines. But in the whole decade 1921–30, only 179 ploughing engines were produced. In the 1920s, the firm kept going by making road rollers, which were in demand as the national road network was being improved, but these could not make up for the loss of the ploughing engines. In 1921–30, only 916 rollers were made. A shift to road rollers was one of the changes keeping afloat another old-established major firm with readjustment difficulties – Marshalls of Gainsborough. The Fowler board toyed with many ideas, but these were mainly variations of old products , such as the flirtation with diesel or diesel-electric cable ploughing systems. Much effort went into making steam railway locomotives; this had some success overseas, but little in the home market. By 1929 a diesel-powered tractor was on the market, and in 1931 a diesel lorry, but these proved disappointing technically, and engendered few sales (only ten lorries were delivered to buyers). In retrospect, it may have been unwise to venture into the uncharted seas of diesel engine development. But the firm persisted, and produced two variants of a diesel crawler tractor from 1932. Again, these were not without technical problems, and only 117 were built.

Fowler's most spectacular attempt at diversification was the evolution, from 1927, of the 'Gyrotiller'. Harking back to the nineteenth-century 'diggers' of Darby and others, this was an enormous ploughing and cultivating machine, originally designed for cultivating land for sugar-cane. It was a diesel tractor, the first version of which weighed twenty-three tons. It had at the rear two revolving horizontal digging wheels, which ploughed, pulverised, aerated and ridged the soil in one pass, leaving it instantly ready for sowing, at any depth up to about 12 inches. This monster was introduced in 1932, initially priced at £8,000, although by 1934 the smallest of the four versions offered cost only £1,150. Not surprisingly, in view of the renewed depression of 1929–32, it failed to sell, and sales of only about 80 were recorded. The net result was that by 1932 the firm was effectively moribund. Even as early as 1925 a director had proposed that it should be wound up. In 1930 a partial liquidation and capital reconstruction took place, but the firm staggered on until 1941, when it was compulsorily acquired by the government.[36]

The failure of Fowlers to diversify more effectively owed something to pre-war specialisation, a certain lack of realism, and a lingering commitment to steam. Some of these features were present in Clayton & Shuttleworth, which also toyed with the internal combustion engine in the 1920s, although somewhat half-heartedly. Its main market, in threshing machines and steam engines, was never recovered in the 1920s, and although it produced a diesel crawler, it was not the commercial success the company expected. In 1928 the firm went into receivership. The steam engine, oil engine and boiler side of the firm was sold off to Babcock

The Fowler Gyrotiller. This one has the usual rear sets of rotating tines, and two plough ridging bodies. An enormous machine, it could prepare a seed bed in one pass, but its cost (£1,150–£8,000) denied it a market, and it was a commercial failure. (*Museum of English Rural Life*)

and Wilcox; the moderately successful wagon business being formed into a separate company. The threshing machine, elevator and chaff cutter business, together with the newly developed Clayton combine harvester, was offered to Marshalls of Gainsborough. This proved to be a mistake for Marshalls. In particular, the combine harvester, which was the first on the British market, proved disappointing. The machine was technically very good, but the launch of an expensive piece of arable equipment in 1930, just as cereal prices had begun another sharp slide downwards, was doomed to failure. In its first year on the market, the combine harvester made losses of £24,628, and in 1932 it was discontinued. The new company, Clayton & Shuttleworth Ltd, in which Marshalls had a controlling interest, was thus a factor in the passing of Marshalls into receivership in 1934.[37]

A number of other firms, regardless of size or experience, thought that it would be worthwhile producing a tractor during or soon after the war. These included such short-lived failures as the Garner, the Pick three-wheeler, and the Alldays and Onions 'General' tractors. Allured by the bright prospects of the motor industry, some thought that it would be a good idea to develop a motor car. On the whole, these were expensive failures. One early pre-war saga was that of the Bentall car, which had been a commercial failure. Post-war, many small motor car companies failed, as competition increased and prices fell. The Pick Motor Co. of Stamford, which experimented unsuccessfully with tractors in 1919–20, was one such. Founded in 1903, it went into liquidation in 1925. But the largest example was Ruston & Hornsby, which had three models in production by 1923. The cars were of a very high quality; an engineer's dream of perfection. All parts were said to be made on the premises, and it was assembled by craftsmen. The 1923 brochure stated: 'Every detail in the construction of motor cars at the extensive Lincoln Motor Works is literally a piece of craftsmanship such as is understood the world over to be peculiarly British'. That it was aimed at the upper end of the market is attested by the wording of the brochure, such as:

> Aristocratic Appearance. That the moderate price [the cheapest was £475] of the Ruston-Hornsby Car is no criterion of its real value is established by the list of those who have made it their choice, for among these can be named many of the highest rank to whom only a car of perfect refinement would possibly appeal.

But this was also a failure commercially. Sales were described by the historian of the firm as 'meagre', although a recent estimate is that the company sold a total of 1,500 cars after the war. Some comfort for R-H was to be had immediately post-war by making furniture, using the resources of the woodworking shop, which had expanded greatly during the war.[38]

Diversification saves some firms

Some firms did manage to diversify successfully. Perhaps the most successful story among the old-established firms was that of Ransomes, which happened to be the oldest, and had the biggest commitment to the agricultural market, having more men employed on agricultural work than any other firm. To some extent, the ground had been laid before the war, in that a large trade had developed in lawn mowers, chiefly for the home market. In 1913 these had accounted for about 10 per cent of the firm's sales. This rose steadily after 1918, reaching 21 per cent by 1930. This was a considerable achievement in view of the low entry costs to the mower industry and the keen competition. As well as light mowers for households, gang mowers for institutional use were also made. The expansion of golf course acreage between the wars was considerable, and playing fields were also increasing, and mowers alone probably kept the firm afloat through the 1920s.

Ransomes' other diversifications were many and varied, and often took them away from agricultural work and into newer technologies. Unlike Fowlers, which had closed down its electrical engineering work before the war, Ransomes entered the electrical side, producing motors, generators and vehicles. A notable and successful initiative was the provision of electric trolley buses for Ipswich Corporation,. This was a logical extension of a relationship going back well into the previous century, when it had supplied gas to the town from its own generating plant. In 1927 the firm entered the fairground business, making electrically driven fairground cars known as 'Wobblums'. They also made precision tools, and were not too proud to make the relatively simple product of steel aircraft hangars for the RAF. By the end of the 1920s the output structure was much more diversified:

Table 10.8 *Ransomes, Sims & Jefferies' output structure, 1919 and 1929–30*

	1919		*1929–30*	
	£	%	£	%
Ploughs & implements	191,524	31	290,224	33
Threshing machines	138,353	22	118,565	14
Corn mills	901	negligible	1,097	negligible
Lawn mowers	59,686	10	181,484	21
Engines and boilers	213,667	35	104,480	12
Vehicles and trucks	0	0	151,668	17
Other products	13,134	2	23,113	3
Totals	**616,995**	**100**	**870,631**	**100**

Source: Rural History Centre, University of Reading, Ransomes archive, TR RAN AD 7/17.

A sort of reverse diversification was evident in the case of Harrison, McGregor, of Leigh, Lancashire. Best known before the war for its harvesting machinery, it had also been prominent in barn machinery for animal feed preparation. In the 1920s, the development of prepared animal feeds undercut the latter market, and the firm moved back into the field machinery market, where its reputation for quality seems to have been the key to success; it made horse rakes, corn drills and swathe turners. By the end of the 1920s it was developing implements for use with motor tractors, such as power-driven mowers and binders, and special land-wheel driven mowers for use with tractors which did not have their own power take-off point. Their principal export market had always been France, and this proved a source of strength in the longer run. Opportunity was also taken to make some products under licence. Unlike Fowlers, their main technology was not out of date, and in the 1930s they became the only British manufacturer of binders. Again, unlike Fowlers, their products had high depreciation rates; many Fowler engines were still working after half a century or more (one Fowler double-engine ploughing set, built in 1918, and subsequently rebuilt, was still being employed by an agricultural contractor in Sussex in 1983 for such tasks as dredging ponds). But field machinery was reckoned to have a life of only about five years.[39]

Small firms which regarded themselves as part of the agricultural engineering industry had many paths towards diversification and financial survival. Many of these might be essentially blacksmiths/repair shops, or threshing or ploughing contractors, and had long been accustomed to move into and out of manufacturing as occasion presented itself. Some acted as agents for the sale of the machines of the larger manufacturers. By the 1920s, there was also the possibility of operating a motor agency and/or running a garage. Finally, those with an iron foundry could exercise some choice as to whether they did agricultural or non-agricultural work. In the medium term, the choices made by such small firms often took them away from agricultural work.

Examples are many and varied. Elders of Berwick-upon-Tweed reduced its own manufacturing business, and expanded its agency sales. In 1913 goods for resale had been about half the value of total sales, but by 1929 they were about two-thirds. By 1931, the firm also styled itself 'motor engineers', and in December 1938 acquired the local Fordson agency. In Norfolk, the firm of F. Randell & Co., of North Walsham, which certainly regarded itself as in the agricultural engineering business, and had established a foundry in 1867, had not filed an agricultural machinery patent since 1890, and was also going into the motor business, purchasing a motor engineering firm at Reepham in 1931. By this time it seem to have quit the business of making farm machinery. The firm of Smithdale, although on the list of privileged firms entitled to an allocation of steel in the Second World War, probably got its place on the list due to its repair work, rather than to manufacturing. If it had engaged in manufacturing previously, this has left no trace in its archive. The position of Frederick Springall, of Horsham St Faith's, seems similar; the firm is described after 1918 as an: 'agricultural engineer, blacksmith, carpenter and cycle agent'. Although it did employ three moulders, and clearly had a foundry, it is not known whether it actually made farm machinery or merely repaired it.[40]

A move away from manufacturing is also evident in the record of J. Gibbs, of Bedfont, Middlesex. By the 1930s it had moved into the motor and garage business, although continuing to make some agricultural products. The move away from manufacturing came even earlier for the Bridge End Foundry, Cardigan, since by the time the firm's archive commences (1922), it seems already to have become an agency for the sale and repair of motor cars. The same defensive reaction to hard times for manufacturers may also be traced in the history of J. W. Titt, of Warminster, Wiltshire, who in the later nineteenth century had described themselves as 'engineers and implement makers'; by 1931 the description had altered to 'windmill manufacturers, agricultural engineers, mechanical engineers, well borers and sinkers, and pump manufacturers'. The nearby firm of T. H. White, of Devizes, seems to have had only a brief period of manufacturing, and had turned to agency sales and service before the First World War. Doubtless the firm was grateful for this in the troubled times of the 1920s and 1930s, when it gained a leading position in the agricultural machinery selling business, which it retains to this day.[41]

Fordson, Ferguson and the revival of the market

The world depression heralded by the Wall Street Crash of 1929 was the most severe on record, and the world economy shrank severely. The British national income contracted by 11 per cent between the start of the slump in 1929 and its trough in 1932. Mostly this was accounted for by falling prices, the volume of national output falling by just 4.5%. The volume of industrial output was worse affected, declining by 11 per cent, while the decline in volume of the engineering industries (mechanical and other) was much greater, at 24 per cent.[42]

This depression, following so soon after the earlier one of 1920–21, from which many of the agricultural engineering firms had not recovered, was very serious for the industry. By the early 1930s, three notable firms had disappeared or abandoned agricultural engineering: Burrells, Howards and Garretts). Claytons had been taken over; Bentalls was in receivership; and Fowlers and Marshalls were in a serious condition. However, paradoxically, the later 1930s was a period of partial recovery for the industry, as home agriculture revived under the stimulus of changed government policy, and, later, rearmament. This can be seen from the censuses of production, with account taken as far as possible of alterations in the relevant price index between censuses:

Table 10.9 *Output of the agricultural machinery industry, 1913–35*

	1913	*1924*	*1930*	*1934*	*1935*
Total value of output (£000, current values)	6,500	3,363	4,044	3,507	4,395
Board of Trade wholesale price index for iron and steel (1913 = 100)	100	142.9	112.7	109.6	111.6
Volume of production (1913 = 100)*	100	36	55	49	60

* Derived by deflating the current value of production by the wholesale price index for iron and steel.

Source: Board of Trade, *Census of Production*: Prices from Mitchell, *British Historical Statistics* (Cambridge, 1988), pp. 729–30.

In spite of the loss of exports after the war, the home market had continued to develop, and did so more rapidly in the 1930s. Some of the new products and firms entering the market in this period have been noted above. These were joined by other new products and firms. The largest single new product was the Fordson, and the largest new firm the one which made it, the Ford Motor Co. of Great Britain. Fordson tractors made in Cork in 1919–23 and 1929–32 had been the largest single British/Irish make. But the fragmentation of the rest of the tractor producers, and the competitive power of other American makers besides Fordsons, had meant that a large part of the British tractor supply had been imported:

Table 10.10 *Numbers of tractors imported into the UK, 1920–32*

	From USA	*From Irish Free State*	*Total*
1920	1,896	–	2,175
1921	1,209	–	1,328
1922	60	–	114
1923	180	–	253
1924	1,609	–	1,777
1925			
1926	931	–	1,011
1927	2,429	3	2,449
1928	1,414	5	1,671
1929	752	854	1,812
1930	867	1,197	2,230
1931	1,396	316	2,657
1932	379	1,781	2,566

Note: Irish Free State trade statistics were not enumerated separately from those of the UK until 1923.

Source: *Annual Statement of Trade*.

In the period 1920–32, approximately 17,000 tractors had been imported, plus whatever Cork-made Fordsons found their way onto the British mainland between 1919 and 1923, when Cork production totalled about 7,700. The British tractor stock, which may have been about 25,000 in 1921, had declined to 16,731 at the 1925 census of production, as older models were scrapped, and British manufacturers' sales dried up. It then rose somewhat, to reach 18,654 at the 1930 census of agricultural production (only 14,565 in field use; the rest were used for stationary work, presumably mainly threshing). It is likely that the great majority of tractors in Britain in 1932 were either Fordsons (some wartime imports; some post-war Fordson imports from the USA; some Cork-made) or imported other US makes; there was simply no British manufacturer of any significance apart from Fordson at Cork until 1932. British sales by International Harvester alone in 1926–31 totalled 3,468 (mostly 10/20 hp Titans), and this by itself must have exceeded all sales by British manufacturers in this period.[43]

In the four years in which Fordson production resumed at Cork, a total of 31,461 Fordsons were produced. Production was not spread evenly over the period, since about half of these (15,196) were produced in the single year 1930. As can be seen from the above table, relatively few of these came to Britain. But the British tractor scene was about to be transformed, as Dagenham Fordson production commenced. The factory, which occupied 66 acres on the bank of the Thames, had a deep-water jetty for the import of iron and coke, and its own blast furnace and foundry. The entire process, from raw material to finished vehicle, was performed on the site, and the products shipped out by river for export and by rail for home sales. It was a very modern concept which left the British vehicle producers years behind. It was, in fact the same system pioneered by Henry Ford at the River Rouge plant in the 1920s, transplanted to British conditions. Production rose rapidly:

Table 10.11 *Dagenham Fordson tractor production, 1933–38*

1933	1934	1935	1936	1937	1938
2,778	3,582	9,141	12,675	18,698	10,647

Source: M. Williams, *Ford and Fordson Tractors* (Ipswich, Farming Press, 1985), Appendix 2.

By 1937 annual output was virtually the same as the total number of tractors which had been recorded in the whole of British agriculture in 1930. This impressive performance dipped in the following year, as Ford suffered along with the rest of the engineering industry the brief national economic recession of 1937–38. However, a new era in the British tractor story had been opened up.

The saga of the failure to establish a British tractor industry in the

1920s, and of the disengagement and then re-engagement of Ford in British tractor production, may obscure the fact that tractor design in the inter-war period was evolving continuously, even if the Fordson remained comparatively unaltered from 1917 to 1939. One major problem concerned wheels and adhesion. Early tractors had metal wheels, with wheel slip in the fields prevented by protruding metal strakes. These were not ideal for their purpose, and were illegal on roads. Some experiments were made with caterpillar tracks, but British firms had little success with them, and most crawlers at work in Britain were imports from the USA. The most successful innovation in tractor design came with the use of rubber pneumatic tyres. These were first introduced by Allis-Chalmers in the USA in 1929. These solved the problems of adhesion, land compacting, and road work, and they were adopted by Fordson in 1933 and later by Ferguson. Their only drawback was the high cost. Perhaps for that reason, Ferguson did not specify them for his first batch of tractors in 1936, although he soon changed his mind.

The contribution from other British manufacturers was slight. The greatest success was had by Marshalls, which had evolved a diesel tractor, based on the German Lanz 'Bulldog'. The first version, the 15/30, was in production from 1930 to 1932, and its successor, the 18/30 from 1932 to 1938. The final form of this was the Model M, in production from 1938 to 1945. Like the Lanz this had the virtue of running on low-grade fuel oil. Rated at 18/30 hp, the Marshall was also more powerful than the Fordson, which was 11/20 hp. This, combined with the greater torque given by its heavy flywheel meant that it could be used for heavy tasks, of which threshing was the most important. But numbers produced were small. The combined sales of the 'M', and its immediate predecessor, the 12/20, were probably no more than 450. Fowlers produced a diesel crawler tractor in 1936–37, but the financial problems of the company led to its abandonment.

The only other tractors of any note on the market were some tracked conversions (mostly of Fordsons) by Roadless, and a small single-cylinder crawler for market gardening by Ransomes. There was a similar one by the Bristol company, but sales were very small. Only Ford had anything like a mass-production technique. All the other firms relied on hand-made, batch production.[44]

Ford quit the US market in 1928, and in 1929 the foundation of the Dagenham factory, which had been in the planning since the land was acquired in 1923, was laid. The impact of the Fordson in the UK in the 1930s was similar to that in the US a decade earlier. The Fordson was still technically superior in the UK (if not in the USA), and only Ferguson was to provide any real challenge, from 1936 onwards. But the Fordson was by this time an ageing design. It was to be given a new lease of life by wartime farming after 1939, and its engine was to survive in the revamped Fordson Major in 1945. But even in the 1930s, it was beginning to show its age. Its main advantage had been its unitary construction (see page 176), in which were located the gearbox and rear wheel drive. This made the machine much lighter (1½ tons) and more rigid, and also provided protection for the moving parts, including the engine. Other tractor makes had open, exposed workings, with consequent high wear and failure. Although there were various detailed modifications to the Fordson before 1939, they did not alter the essential nature of the machine.

The Fordson had a worrying tendency to rear up when its implement hit an obstruction, and it was alleged that as many as 136 drivers had been killed by 1922 in this type of accident. All tractors were potentially prone to this, but the Fordson's light weight contributed to the problem. The company changed the design of the rear mudguards: from 1924 until 1932 these were extended downwards to the rear, and one of them incorporated a toolbox.[45] The other main problem of the Fordson was its drive system. This was via a 'worm drive' wheel in the rear axle, which suffered from excessive friction. This in turn produced overheating and led to relatively high petrol consumption. Nothing was done to cure this problem, which remained with the Fordson until the advent of the Fordson Major in 1945. The only notable changes by 1929, when Cork production started, were the addition of a power take-off point, and the uprating of the engine to 27 hp. The only technical change of note in the 1930s was the replacement of the water air-washer (which tended to freeze in the winter) with an oil bath. There were various cosmetic changes, the most obvious of which was a change from grey to blue paint in 1932, and to orange in 1937.[46]

Fordson sales were buoyant in the 1930s, despite its ageing design. One reason was that the economic background of farming was changing. Although the crash in the prices of farm produce, and in the value of farm land, had been severe in 1929–31, there was a recovery thereafter. In 1932 the Wheat Act provided bounties for the growing of wheat. In 1934, dairy farmers' incomes were boosted by the establishment of the Milk Marketing Board, which in effect cartelised the industry, and was the salvation of many a small farmer in the upland regions. Producer marketing boards were also established for potatoes, bacon, hops and pigs. The subsidy on planting sugar beet, first introduced in 1925, was made permanent in 1935. These measures, plus the continuing shift of demand towards 'superior' foods revived farmers' incomes. Dr Brassley's estimates of British farm income show it as £245 million in 1930–34, rising to £293 million in 1935–39. Thus farmers had more disposable income. In addition, the machinery industry benefited from the imposition of tariffs. In 1932 the Import Duties Act set a basic minimum tariff on manufactured goods of 10 per cent by value. Following the Act, the Import Duties Advisory Committee adjusted the rates upwards in many cases. An extra 5 per cent was added to the rate for tractors and farm machinery, and an extra 15% on agricultural crawler tractors. Finally, the Fordson remained cheap. At the 1930 World Tractor Trials it was priced at

£170, compared with £220 for the nearest competitors, the International Harvester 10/20 and the Massey-Harris. The Austin was priced at £210, but it was hardly a competitor, and soon went out of production. The Marshall was £315. This differential remained for the rest of the decade, and a farmer could buy one on the active second-hand Fordson market even more cheaply.[47]

The expansion of the British tractor population in the 1930s also spawned a growing trade in tractor implements. The market for tractor ploughs rose, while the value of horse-ploughs sold declined. In 1930 the two combined had been worth £460,00. By 1935 this had risen to £593,000.[48]

To some extent, the market was supplied increasingly through imports. In the early years of the depression after 1929, the value of imports was surprisingly well maintained. The value of imports of tractors and parts had been £249,000 in 1929, and rose during the depression until 1932, when they were valued at £320,000. A brief decline in 1933 to £267,000 was followed by recovery the next year and continuous growth thereafter until 1937, when they were valued at £717,000. The numbers of tractors imported rose from 1,812 in 1929 to 2,566 in 1932, falling briefly to 990 in 1933. Continuous growth thereafter took the number to 3,245 in 1937, with a slight drop the following year to 2,860. Although imports rose in the later 1930s, the home market was still largely in the hands of British firms. On average in 1937–39, the annual supply of wheeled tractors in Britain was 10,489, of which 2,951 were imported; a ratio of 28 per cent.[49]

The impact of the Fordson was also felt in the export market. The rise of tractor exports in the 1930s was little short of remarkable. Although not negligible in 1929, and declining in the early 1930s, by 1937 they provided about half of all exports of the whole agricultural machinery industry. Exports ranged far and wide, although there were a few substantial markets in their own right. It was notable that the biggest single market was the USA. This was facilitated by the development of a row-crop tractor version of the Fordson, the 'All-Around' in 1936, specifically for the US market. In 1937 the biggest market for British tractors was the USA, at £205,000, followed by Sweden at £158,000, and Australia at £139,000:[50]

Table 10.12 *British tractor exports, 1929–38 (£000)*

	1929	*1930*	*1931*	*1932*	*1933*	*1934*	*1935*	*1936*	*1937*	*1938*
Tractors and parts	232	88	126	45	156	144	416	602	1,036	621
Total exports	1,977	1,385	1,015	727	904	957	1,205	1,505	2,039	1,522

Source: Annual Statement of Trade.

The final chapter in the British tractor story of the 1930s belongs to Harry Ferguson. His entry into the industry was sparked by his interest in improving tractors to deliver more power and improved safety. In charge of government tractor ploughing in Ulster during the war, and agent for sales of the 'Eros' ploughing attachment for the Ford Model T motor car, he had become aware of the deficiencies of early tractor work, which merely hitched horse-drawn implements to the tractor. This was inefficient and dangerous, as tractors could rear up disastrously. He evolved a system whereby the implements were carried on the back of the tractor, and raised by hydraulic power. The linkage between tractor and implement was a three-point hitch. The lower two points carried the implement, and the top one adjusted the depth at which the implement

The Ferguson-Brown tractor, 1936–39

Harry Ferguson had been involved with the development of tractors and their ploughs for over twenty years before he was able in 1936 to produce his own tractor, incorporating his own system for the control and better utilisation of the plough and other implements. This tractor was officially called the Model A, although better known as the Ferguson-Brown. Its importance for the future development of the world tractor industry lay in its hydraulic implement control system. This consisted in a three-point hitch at the rear of the tractor. The two lower arms were used to raise and lower the implement, by the use of hydraulic pressure. The top hitching point controlled the depth of operation of the implement. In order to fit the implement to the tractor the driver had merely to back onto the implement, attach it to the three links, and operate a lever for the implement to be raised automatically. The tractor with implement could then be driven to the scene of operations. The third (top) link was via a small hydraulic piston, with a spring between it and the implement. When the implement was lowered into the working position, the spring compressed as the implement moved through the soil. This regulated the hydraulic pressure in the piston, which in turn moved in or out to maintain the implement at a constant working depth. Thus a plough could be set to maintain a constant depth in work, regardless of variations in the contours of the soil. The final refinement was that, if the plough met an obstruction, the hydraulic pressure in the piston fell abruptly, so that the plough automatically raised itself out of the ground. This prevented the tractor tipping up, a much-needed safety improvement.

With the Ferguson System, tractors could now attach, raise, lower and operate their own implements, whereas formerly they had been trailed behind the tractor. This has been dangerous (in case of hidden obstructions), had often meant that field operations needed two men (one to sit on the implement), and had meant that the tractor and implement together had a very large turning circle, so that it was impossible to plough the headlands at the start and finish of the tractor's run. Now the whole field could be cultivated; Harry Ferguson enjoyed demonstrating how his tractor ploughing in a roped-off enclosure measuring 6 × 4 yards, showing that it could work in limited spaces and plough up to hedges and walls.

The Ferguson System was not the only feature of interest in the Ferguson-Brown. Use of aluminium components in the hydraulic pump and the transmission housing enabled the weight to be kept down to 16½ cwt. This was much less than the 34 cwt of the Fordson Model F, the nearest competitor. The lower weight of the Ferguson gave it a much higher power to weight ratio, and the fact that the implement was fixed to the tractor meant that the tractor had much worked. The advantages of the system were greater safety, as well as more efficient use of the power of the tractor, as the implement and tractor worked as one unit. The system could also plough the whole field, without the need to leave a wide headland as with a trailed plough; there was a saving of time and effort in demounting implements for road transport. Finally, the whole system could be mounted on a small tractor, as it used its power more efficiently than the conventional system, and thus find a market on the small family farm which had so far not been tapped.

Ferguson began by selling his system in the USA, for use on Fordsons in the 1920s, using an American manufacturer to make the implements.

greater tractive power than if the implement had been trailed: in work, the fixed implement pushed the front of the tractor down, and reduced wheel slip. The short wheelbase (69 inches) was only slightly larger than the Fordson, and enabled it to work in small areas. The wheels themselves were fitted with strakes, to aid adhesion, and pneumatic tyres were offered as a option in 1937. The engine was a 2.1 litre Coventry Climax 20 hp petrol engine, giving three forward speeds. In 1937 it was replaced by a virtually identical one made by David Brown. From November 1938, as the approach of war pushed up petrol prices, a kit was available to convert the tractor to run on the cheaper Tractor Vapourising Oil (TVO). The tractor was made using the latest materials, and Harry Ferguson's concern with detail was evident; to ease maintenance, only two sizes of bolt head were used, so that one spanner would fit all the nuts and bolts.

A 1938 Ferguson-Brown tractor at Newark Vintage Tractor Show, 2004. *(Peter Dewey)*

The ending of Fordson production there in 1928 left him without a tractor. After many negotiations with potential backers, he decided to make his own tractor to demonstrate the system. This, the 'black tractor' prototype, was built in 1933 (now in the Science Museum, London). Negotiations with David Brown, a Huddersfield gear-making firm, resulted in David Brown equipping a factory to build the tractor, to be built by a new firm created for the purpose, Ferguson-Brown Ltd. This began production in 1936. The tractor was not cheap, and could not compete on price with the Fordson. Its original price was £224, and each implement, which had to be used to get the full benefit of the system, cost £28. The Fordson at that time cost £140. The rigid coupling of tractor and implement gave the Ferguson much greater tractive power, so that, although only half the weight of a Fordson (16½ cwt, as opposed to the Fordson's 1½ tons) it could pull implements more effectively, without wheel slip. To those who tried the Ferguson-Brown, as it was often called, it was a revelation. One such was Henry Williamson, the novelist, who farmed briefly in Norfolk in 1937–38. In getting the capital together for this venture, he could have economised and bought a second-hand Fordson, but, being a perfectionist, he bought a new Ferguson and five Ferguson implements. This all cost him £360, but it was worth it:[51]

> It was half the weight of an ordinary tractor, built of aluminium and immensely strong steel, and it carried its twin-furrow plow under its tail, on three steel arms that looked like a grasshopper's hind legs. On pulling a lever like the short gear-change lever of a racing car, the twin plows lifted up out of the ground. Instead of lugging a heavy sledge of plows around the field, bibbling, as Bob said, at the corners, this new design of tractor lifted its tail and, put in reverse, moved back to exactly where one wanted to drop implements ... Both Bob and Jimmy were sceptical of its performance. 'You won't beat hosses on that ould sud of a Hilly Piece. No tractor can git up that.' ... I put it in plowing gear, let in the clutch, and opened the throttle ... The little machine went up without the least falter. Its thin spiky wheels pressed the ground lighter than horse-hoofs would have done. Its twin shares bit into the sullen soil and turned it over, exposing a tangle of white roots. I heard Bob mutter: 'Blast, I like that patent', as he stared at it. This was the highest praise from one who regarded many of my schemes with hard-eyed caution ...*

The Ferguson system was indeed revolutionary. It was a tribute to it that the new company managed to sell 1,250 complete sets of tractors and implements by 1938. This, although much more than any other British maker apart from Ford, and an impressive performance in 1937, as the

* H. Williamson, *The Story of a Norfolk Farm* (Faber & Faber, 1941; Clive Holloway Books, 1986), pp. 213–14.

The Ferguson-Brown tractor, (1936–39), pulling a binder. Rated at just 18–20 hp, this tractor had much greater tractive power than the conventional tractor with trailed implements. This was the start of the revolutionary Ferguson hydraulic implement lift system. (*Museum of English Rural Life*)

brief recession in the economy hit sales, did not approach the Fordson level of production. Nor was it to last. Faced with the sluggishness of sales, Ferguson and Brown disagreed on policy. Ferguson wanted to cut the price and increase the volume of output. Brown wanted to make a large tractor. The differences being irreconcilable; the partners broke up. Ferguson went off to persuade Henry Ford to make his tractor in the USA. Brown began to produce his own, larger tractor. By the end of 1939, both had achieved their aim, and another new chapter in the world's tractor industry opened.[52]

The major initiatives in the industry in the 1930s were those of Ford and Ferguson. There were also some smaller ones, which would bear much fruit later on. These were mainly in East Anglia, and were dependent on the use of the motor tractor. The most notable was the firm of Howard Rotovator. This was started at West Horndon, Essex in 1938 by A. C. Howard, who had already founded a firm in Australia before coming to England. The principle of the rotovator was to apply engine power directly to the movable blades of a trailed cultivator with, either from a tractor or suitable power unit, through a rotor and hoe blades of a special design. This action created a tilth and prepared the seed bed, and special applications of the principle enabled the machine to be used for a wide variety of farming and horticultural tasks. The principle was not new, since versions of it had been experimented with in the days of steam, but the steam engine was not a suitable conveyor for it, and only when motor tractors came into use was it a feasible proposition. Most of the firm's growth was to take place after 1945, but it was a portent of the wide use of power implements in farming today. Finally, right at the end of the decade, two east Anglian pioneers began developing sugar-beet harvesting machines; John Salmon, a farmer of Dunmow, and William Catchpole, of Stanton, near Bury St Edmunds. Both of the firms founded by these men were to grow rapidly after 1945.[53]

Conclusion

The quarter of a century from 1914 until 1939 was the most cataclysmic period in the history of the agricultural engineering industry. The First World War gave the industry a financial boost, while at the same time severely weakening its competitive ability. The next twenty years were in effect a continuous depression, as the lost foreign markets stubbornly refused to be rebuilt, and as the finances of the industry's leading firms haemorrhaged away. The fiasco of Agricultural & General Engineers brought an element of farce to the story. It was hardly surprising that the industry shrank. By the mid-1930s its output was slightly less than two-thirds of what it had been in 1914. The onset of war began a period of recovery.

CHAPTER ELEVEN

War work again, 1939–1945

Early preparations

Unlike the period before the First World War, the years preceding what became the global conflict of 1939–45 were marked in Britain by intensive preparations of all kinds. The foundations of the military defence plans were being discussed soon after Hitler became the Chancellor of Germany in 1933. Few concrete results emerged from these discussions until British rearmament began in earnest in 1935–36. By then the government had also taken a view on the importance of planning agriculture and food supply in the event of war. The first evidence of this was the formation of the Food (Defence Plans) Department of the Board of Trade, in November 1936. In wartime this was to become the Ministry of Food, to be responsible for controlling food distribution, prices, and the civilian rationing scheme. As far as agriculture was concerned, the policy was to be the same as that of 1914–18. Production of food was to be maximized (in terms of calories) by persuading farmers to shift away from livestock and towards cereal and potato growing. The Ministry of Agriculture (raised from the status of Board to a Ministry in 1919) first promulgated a target for the ploughing-up campaign in January 1937, when it was proposed that in the first twelve months of a war, 1,285,000 acres of grassland should be ploughed up and planted with cereals and potatoes. This would be only a first target. If the war were to go on longer, then further targets would have to be set for subsequent years. In the same year, the County War Agricultural Executive Committees were reconstituted, along similar lines as in 1914–18. These 'War Ags' implemented the system at farm level. Steps were also taken to provide labour and machinery. A new Women's Land Army was formed in 1938, and farm labour was declared a reserved occupation. In the summer of 1939 a new Agriculture Act committed the government to acquire a reserve of fertilizers, tractors and petrol. Finally, as war loomed in the late summer of 1939, farmers were offered a subsidy of £2 for each acre they ploughed up.[1]

For the agricultural engineering industry the environment was radically different in the Second World War from that of the First. Farming would now be a top priority for the government, and farmers would be allowed to make higher profits. The government would encourage the mechanization of farming, as the current stock of horses and machines would be inadequate for the enormous expansion in output being contemplated. Agriculture would be a strongly protected occupation, and the expanded labour force would need more machines – especially tractors and their implements. The first step towards assuring the supply of tractors was taken as the result of an initiative by the Ford Motor Co. An approach was made by the company to the Food (Defence Plans) Department in September 1938, at the height of the Munich crisis, while air raid defence trenches were being dug in London parks, and 38 million gas masks were being distributed.[2] The company was referred to the Ministry of Agriculture, and Mr Harper, who described himself as the Ford's 'liaison officer' with government departments, paid a visit to the Ministry on 21 September. The note of the meeting recorded that, 'he had called to enquire whether his Company could be of any assistance to this Department in the event of a major war'. He observed that production of Fordson tractors was at that time about 250–300 a week, but this could be raised to about 400 a week in less than three months. This visit was succeeded by one from Mr Daniels, the head of Ford's tractor department, on 29 September. This was the day the Prime Minister, Neville Chamberlain, flew to Germany for the third and final time to solve the Czech crisis; the Munich agreement was signed the next day. Mr Daniels '... came in order to make contact with the Ministry and to let us know that the Company are most anxious to do everything in their power to meet the tractor requirements of the agricultural industry should war break out.'[3]

The timing of these approaches was psychologically astute. Fords would be aware that the government had a policy of ploughing up land in case of war, but since neither it nor the farmers had the necessary tractors, it might be panicked into making a deal with Fords.* These talks failed to produce a result at the time. But with the continued deterioration in international relations, and in particular the takeover by Germany of

---

* The Ministry of Agriculture's estimate of the annual cultivating capacity of the average tractor was 99 acres (not, for some arcane reason, rounded up to 100). In early 1939 some effort was put into calculating how much was in practice dealt with by the tractor stock in each county. The answer was only 60 acres per tractor. The Ministry then worked out hypothetically whether there was a surplus or deficiency of tractors in each county, after future action (unspecified) had been taken to raise the acreage worked by each tractor to 99 acres. The intention was then to move tractors from 'surplus' to 'deficit' counties. How farmers were to be induced to work 99 acres per tractor, and how the machines were to be moved between counties, in the absence of compulsory purchase, was not stated. The obstacles were delicately hinted at by a civil servant: 'It may be that many owners would raise objection to their tractor being used on the farm of a neighbour who has not seen fit to purchase one of these machines. Such objections could be dealt with by requisitioning the tractors for outside work, but it is desirable that this step should not be taken if it can be avoided.' The project was abandoned as unworkable. *Source*: PRO, MAF 58/111, 'Tractors and Implements in War Time. Memorandum by Mr Chambers' (20 March 1939).

the remaining Czech territory in March 1939, the approaches eventually bore fruit. On 24 April 1939 the Ministry of Agriculture talked, at its own request, with Lord Perry, the Chairman, and Mr Hennessey, the Managing Director, of Fords. The discussion was about '... the question of producing and holding on behalf of the Government a stock of Fordson tractors to be used for ploughing up grassland in the event of War.' The result was an agreement between Fords and the government on 30 June, that the company would create, at its own expense, a reserve of 3,000 Fordsons (with their ploughs), to be held against the outbreak of war. They would be held with the Fordson dealers, who would keep them in working order. If war broke out, the government would buy them, at a previously agreed, discounted, price (list price, less 27.5 per cent). If peace persisted, then the government would have no obligation to Fords. The quota was 1,800 Standard Agricultural (with metal tyres and strakes) and 1,200 Land Utility Fordsons (with rubber tyres). Deliveries into stock would be carried out over the following six months. The agreement was given parliamentary sanction in the Agriculture Act of 1939.[4]

This agreement was a landmark in the history of the industry. Ford was taking a gamble, and at some expense to itself. The agreement was good news for the government, which so far had done nothing to assure a supply of tractive power for the contemplated ploughing-up campaign. On the conventional assumption that one tractor could plough and otherwise cultivate 100 acres a year, the 3,000 Fordsons could deal with 300,000 acres. Although this was only a part of the nearly 1.3 million extra acres of tillage contemplated for the first year of the war, the rest could be dealt with to some extent from greater use of the 55,000 tractors already on farms, and of course the farm horses. Also, the outbreak of war would provide a larger market for Ford, which would then raise its output. For Ford, the timing of the original approach was significant. Sales in 1938 had fallen dramatically, and were by the end of the year to be only 10,647, as against 18,698 in 1937. Exports were particularly badly hit. While patriotism was undoubtedly a real motive for the original approach, it is also the case that such a deal would benefit the company at a difficult time. Even by April 1939, output per day was only about 55 Fordsons, which was the same as in the previous September, so there was still a need to boost output. For the future, the deal gave Fords a head start against the competition, and helped to embed the Fordson even more securely into the fabric of British farming, much to the chagrin of other tractor makers.

War with Germany was announced on 3 September 1939. The Ministry of Agriculture accordingly took up its option to purchase the Fordsons. By October, 1,500 of them had been taken over by the Ministry. The balance of the order was delivered to the Ministry by the end of the year. However, distribution thereafter to the county committees was slow, only 1,400 having been delivered by the end of February 1940. Fords had by this time raised tractor production to 100 a day. The Ministry had also placed orders for tractor ploughs with British and North American manufacturers, and these began to arrive in 1940. There had been a temporary shortage of tractor ploughs in December 1939, but that had been remedied by February 1940.[5]

It is important to bear in mind that for the first 18 months of the war the trade in agricultural machinery was subject to little direct influence by the government. There was a sharp decline in exports, and some firms were directed to produce munitions, but the bulk of the industry carried on its private business, subject only to the specific demands of the government to supply certain types of machinery. The buoyancy of the home market overshadowed the demand from the government. Thus Fordson production in the six months from the outbreak of war in September 1939 until February 1940 was about 10,000 (and 11,000 tractor ploughs), compared with a total sale of 10,647 in 1938. This overshadowed the original official deal with Fordsons to supply 3,000 tractors. After the tractor deal, the Ministry turned to consider harvesting machinery. By March 1940 it had purchased 3,500 binders and 275 threshers. The binders came from the only firm still producing them, Harrison McGregor. For the rest of the war it was directed by the government to concentrate on producing binders, although it also made potato planters for the Ministry. The peak annual output of Harrison's binders in the war was about 3,800. By June 1940 the Ministry was distributing these machines to the county committees, and beginning the process of rebuilding its reserve of tractors for the next autumn ploughing. The strength of demand, mainly directly from farmers, assisted by government orders, meant that, as the official historian, K.A.H. Murray, wrote, 'by the end of the first year [of the war] the demand for machinery far exceeded the supply'.[6]

New opportunities in agriculture – Plough for Victory

Tractors

The rapidly rising demand for tractors and machinery was fuelled by wartime conditions. The Battle of the Atlantic against the German submarine threat was not decided in favour of the Allies until late in 1943, by which time many millions of tons of shipping had been lost, and the national food supply was under serious threat. In this situation, the expansion of arable farming was all the more urgent. Nothing was to be allowed to stand in the way of the ploughing-up campaign. Farmers had already been given guaranteed prices before the war. These were now stabilized, and the prices of imported feeding stuffs and fertilizers also fixed.[7] Farmers now had every financial incentive, as well as a legal obligation, to cooperate with the ploughing-up policy.

The area of ploughed land expanded enormously. The average tillage area in 1935–39 in Britain had been 8.4 million acres. In 1944 it reached its wartime peak, 13.7 million acres. At the same time the area of permanent grassland declined, from 17.3 million acres in 1935–39 to a wartime low of 10.8 million acres by 1944.[8] This almost 60 per cent expansion of ploughed area required an enormous expansion in tractive power. As the only tractor producer of any size, the government relied mainly on Fordsons to deliver the higher output necessary if the annual ploughing targets were to be met. It was calculated that some 50,000 new tractors would be needed to meet the demands of increased arable production.[9] Since the pre-war tractor population of Britain had been about 55,000, this would entail a rough doubling of the tractor stock. However, the total number of tractors produced was considerably in excess of this, amounting to a total of 115,816 between 1940 and 1944:

Table 11.1 *Number of tractors for agricultural use produced each year in the UK 1937/39 to 1944*

	1937/39	*1940*	*1941*	*1942*	*1943*	*1944*
3-* and 4-wheeled	14,147	17,906	22,168	24,099	21,599	17,770
2-wheeled†	564	1,064	1,894	2,498	2,937	3,611
Tracklaying‡	229	346	339	459	527	690
Total	**14,777**	**18,989**	**24,062**	**26,645**	**24,623**	**21,497**

* 3-wheeled tractors were mainly the row-crop variant of the Fordson, which had a single front wheel.

† i.e., motor ploughs/cultivators steered by a pedestrian

‡ 'Agricultural Tracklayers produced in United Kingdom: 1939 and 1940 production by J. Fowler (Leeds) Ltd, 1942 1943 and 1944 production by Roadless Traction Ltd.' However, most of the tracklayers were small market gardening types, principally the Ransomes MG tractor.

Sources: PRO, CAB 87/18, War Cabinet, Reconstruction Committee. Official Sub-Committee on Industrial Problems. *Agricultural Machinery and Implements. Memorandum by the Ministry of Agriculture and Fisheries.* Table A.

On these figures, total output of conventional wheeled tractors between 1940 and 1944 was 103,542. It should be said that this total is substantially less than the Ford Co.'s own figures, which show a total of 120,281 tractors produced at Dagenham in the same period. For the whole war, from September 1939 until the end of the war with Germany in June 1945, the company recorded a total production of 137,483 Fordsons. To some extent, the discrepancy may be due to the different uses to which Fordsons were put. The War Cabinet figures above are for agricultural tractors only, and Fordsons (and some other makes) were also used for military purposes in the UK and in foreign theatres of war (e.g., North Africa), and for towing aircraft for the RAF. However, the discrepancy is still notable. Fords estimated that of the 55,000 tractors at work on UK farms in 1939, 35,000 were Fordsons, and that by D-Day (6 June 1944)

The plough policy at work: ploughing some of the parkland surrounding the royal palace of Hampton Court. (*Museum of English Rural Life*)

there were some 144,000 farm tractors, of which 85 per cent (122,400) were Fordsons, implying a rise in Fordson usage on farms of some 87,000.[10]

The expansion of Fordson production was a considerable achievement. From the pre-war level of 250–300 a week in early 1939 output had risen to 500 a week by early 1940. There was a lean period in the summer of 1940, when shortages of steel reduced output, culminating in September, when only 555 were delivered in the whole month, but output had increased to 400 a week by December, and, by 1943, to 625 a week. The 5-day week was abandoned, and 7-day shift working instituted. Dagenham also did munitions work, producing Bren gun carriers, engines and military vehicles. The huge factory complex beside the Thames was an obvious target for the German air force. Partly to reduce the visibility of the site, as completed Fordsons were lined up at the dockside awaiting shipment, the colour scheme was changed. The original 3,000 ordered for the government were in the orange paint adopted in 1936. In late 1939 this was altered to green, and all later wartime Fordsons were of this colour. This did not prevent the factory being bombed several times, and six people were killed while at work. However, this did not materially impede production. The company also ran training courses for tractor drivers at its training school at Borehamwood, and took great pains to ensure an efficient supply of spares. On receipt of a telegram, spare parts were dispatched by the next train to anywhere in the country. In addition, in

1941 a scheme was instituted by which any tractor operated by a county committee which required an extensive overhaul was reconditioned and sold cheaply below the maximum controlled price, to enable the poorer farmer to buy a tractor. The contribution of the company to the agricultural programme, and the wider war effort, was considerable. However, it should be borne in mind that there was spare capacity at the works in 1939, and although the peak output figure recorded by the company (27,650 in 1942) was 48 per cent more than the highest pre-war figure (18,698 in 1937), the company had the assistance of the government in ensuring supplies of raw materials and labour, and was able to institute a longer working week without opposition from the trade unions, since wartime legislation prohibited strikes.[11]

The demand for tractors was clearly greater than that created as a response to the ploughing-up campaigns. In 1939 the Ministry had surveyed the tractor population of the UK, and found it to be 52,450. Since this was not a compulsory return, but only covered 'rather less than 90%' of farmers, this figure was raised to a final 56,200. The next survey in 1940 was compulsory, and showed a large rise, to a population of 75,695. After this, purchases slowed down, but the tractor stock was up to 116,800 by 1942. The final figure at the end of the war in 1945 was 179,846:[12]

A wartime (1941) Fordson, with a Ransomes trailed plough. (*Peter Dewey*)

Table 11.2 *Numbers of tractors on UK farms, 1931–45*

1931	22,000
1937	49,800
1939	56,200
1940	75,695
1941	94,500
1942	116,800
1944	173,400
1945	179,846
1946	203,400

Sources: K.A.H. Murray, *Agriculture* (History of the Second World War, United Kingdom Civil Series) (HMSO, 1955), p. 274; PRO, MAF 58/96. Exchange Requirements Committee. Memorandum by the Ministry of Agriculture and Fisheries. *Supplies of Agricultural Machinery and Implements in the United Kingdom* (9.12.40), p. 1. Figures for June in both years. The 1941 figure is an estimate.

The number of tractors had more than trebled in little more than five years. This cannot be explained merely by the ploughing-up campaign: at root, it was due to farmers taking advantage of war-induced prosperity to modernize their techniques. The purpose of policy, which was achieved, was to reward the farmer sufficiently to lead him to produce the goods desired by the government. Thus prices and supplies of the necessary inputs (labour, machinery, feeding stuffs and fertilizers) were controlled to that end. The policy worked, and farmers had a prosperous war. For many of them, this would have been a welcome change from the hard times of the inter-war period. Between 1938/39 and 1944/45, the value of the gross output of the industry doubled, from £295 million to £586 million, and farmers retained a much larger proportion of this as their own income; their net income rising from £53 million to £198 million over the same period. They were thus well placed to invest in modern machinery. For many of them, this was a matter of buying their first tractor. Since there were some 250,000 farmers in Great Britain, this still left some unsatisfied demand at the end of the war. Wartime farmers were also taking the opportunity to upgrade their existing tractors, and delete obsolete types.

Other machinery

Overall, the position and prospects of the entire agricultural engineering industry had been transformed in the first 18 months or so of the war. The enormous drive to mechanisation was not confined to tractors, although demand for them was the most notable. Even at the higher rate of tractor output of 2,000 a month which Fords had achieved by 1941, the company still had three months' orders on the books. While the main policy emphasis was on growing and harvesting crops, attention was also given to animal husbandry. The production of haymaking machinery (including the hay sweeps developed between the wars) was also encouraged, as was the mechanization of milk production, and the number of milking machines at work rose substantially. In the first 15 months of the war, the Ministry of Agriculture had made total purchases of machinery to the value of £2½ million. This may be compared with a total output of the entire industry of £4,395,000 at the 1935 census of production. In the twelve months between June 1941 and May 1942 the value of UK production of farm machinery was estimated at about £20,000,000. In the subsequent twelve months, the figure was estimated to have risen by a further 10 per cent.[13] As the official historian of wartime agriculture, K.A.H. Murray, noted: 'Figures of such magnitude were completely foreign to the farming world of the pre-war days'. In addition, companies were benefiting from income derived from military work. In less than a decade, the industry had been turned around from a position of failure and bankruptcy to one of buoyant prosperity. The output of the more important classes of machinery is indicated below:

Table 11.3 *Output of machinery types, 1937/39–1944*

	1937/39	*1940*	*1941*	*1942*	*1943*	*1944*
Ploughs: tractor	5,539	8,680	10,495	8,929	8,307	9,738
Ploughs: horse	3,867	14,492	14,162	12,485	10,939	13,963
Cultivators	3,954	7,367	11,540	9,851	11,485	10,647
Corn and fertiliser drills	4,121	7,217	9,436	11,543	14,620	13,697
Fertiliser distributors	3,086	4,058	5,165	7,490	8,945	9,670
Binders	581	926	1,010	798	1,567	2,957
Threshers	355	842	998	1,129	1,117	1,095
Hay sweeps	4,362	4,727	5,310	8,097	9,681	8,250
Hay and straw balers	314	411	534	646	799	944
Hay/corn/straw elevators	1,537	1,998	1,923	2,205	2,660	2,882
Potato planters	143	134	2,860	918	1,649	665
Potato lifters	1,327	2,108	5,221	8,719	10,151	7,834
Milking machines	–	–	–	4,582	5,623	4,888

Sources: K.A.H. Murray, *Agriculture*. History of the Second World War. Civil Series (HMSO, 1955), Appendix Table VIII; PRO, CAB 87/18, Table A; Central Statistical Office, *Monthly Digest of Statistics*, No. 15, March 1947, Table 61.

Government regulation and control

The enormous rise in demand for tractors and machinery came largely from individual farmers, although some was due to that of the County War Agricultural Executive Committees. These operated tractors and machinery, either directly or via contractors, to assist individual farmers to meet their ploughing and cultivation targets. By the end of 1940 some 3,000 tractors, 14,000 cultivating implements and 3,000 harvesting machines had been distributed to the 'War Ags' by central government.[14]

Overall, demand was well in excess of supply, even with the contribution of imports. There ensued a 'scramble for machinery', as Murray expressed it, which was only partially controlled until the third year of the war. In October 1941 partial control was instituted over imported tractors and implements from North America. This applied to private imports, and not to the Lend-Lease scheme, which had begun in March 1941, and separately administered. In February 1942 import control was extended to all tractors and in March 1943 to tractor ploughs and threshing machines. In May 1942 a system of county allocations of track-laying tractors, which were in short supply, was instituted, and this was later extended to imported wheeled tractors. In July 1942 an order had been passed establishing maximum prices for second-hand tractors and some other items of machinery and implements.

Initially exports were encouraged in order to earn foreign currency, but in February 1941, no further export of agricultural implements/machines would be allowed, except where implements for export were held in stock of a type unsuitable for use in the UK; where implements destined for export were incomplete, but needed only relatively little extra labour and material to finish them; or where it could be proved that the war effort of the Allies would be handicapped if some implement or part urgently needed in the export market were withheld. For manufacturers, control had so far been indirect. In particular, labour supply was controlled by the system of national service, and the larger makers had their labour protected against military service. The other main factor of production was iron and steel, and a national list of those firms entitled to supplies, either for manufacturing or for repair work, was drawn up by the Ministry of Agriculture. Thus the constraints on manufacturers were largely the external ones of control of imports and exports, the contribution of Lend-Lease, and the need to meet the demands of the customers, both private and government.[15]

Control of the industry by indirect means, and by using the power of persuasion (with the unspoken assumption that resistance would lead to the imposition of formal, legal controls), was the instrument deployed by the government. This can be seen in the matter of price control. Here, the method used was to persuade the Agricultural Engineers' Association to consult with the Ministry of Agriculture before announcing its policy on prices to be recommended to its members. The policy of the Ministry was succinctly summed up by Mr Parker of MAF in 1942:

> We have hitherto exercised no direct control over the prices of agricultural machinery and implements beyond the fact that being

> fairly heavy purchasers we can criticise the prices charged to us … It has been the practice for the Agricultural Engineers' Association to consult us before trade prices were raised, and a broad justification for the proposed advance was submitted for our concurrence …*

The occasion for this observation was the conflict between the government and the AEA which was coming to a head early in 1942. Sir William Tritton had been unwise enough to complain that Fosters of Lincoln (of which he was managing director) were not getting high enough prices for the machinery and implements which it made to government order, and to threaten the Ministry that unless prices were raised, Fosters would look elsewhere for orders. Since the only possible alternatives were to make machinery for the private sector, or to make munitions for the government, and since the Ministry was determined to control what Fosters and the other manufacturers produced, this was a hollow threat, and caused much annoyance at the Ministry. Sir William was reminded that so far control had been lightly applied.[16] Shortly after this, at the AGM of the Agricultural Engineers' Association on 26 February 1942, a resolution was passed recommending an increase of 7½% in the prices of products made largely of iron and steel, of 10% in the prices of products made largely of timber, and a special increase of 15% in the price of binders. It was noted that, since the war had begun, prices had increased by 32½%, 40% and 45% respectively in these three categories.

This demand caused much discussion at the Ministry. It was felt that the existing system of price control (or, rather, influence) was becoming inadequate. A brief enquiry into some of the products made by the principal firms threw up the perhaps unsurprising result that all the manufacturers had experienced cost rises of roughly the same degree since the war began (25% or, more usually, 30%). The rate of profit was calculated by MAF as at least 10%, and usually much more. This spurred the Ministry to institute a much wider enquiry into manufacturers' costs. The results showed that the degree of control exercised over costs and prices was inadequate. The consequence was the imposition of price and production control in May 1943, when an order was passed [The Farm Machinery (Control of Manufacture and Supply) Order 1943] to enable the agricultural departments (the Ministry of Agriculture and Fisheries in England and Wales, the Department of Agriculture for Scotland, and the Northern Ireland Office) to control the industry more closely. It stated that from 31 July all manufacturers of, and dealers in, farm machinery, had to work under licence from the Ministry of Agriculture. The Order gave the Ministry powers to regulate production and prices of farm machinery, and also gave legal authority to the distributive schemes which had hitherto been run on a voluntary basis.[17]

* Parker, covering minute to file 'Investigation into increased costs of machinery' 12.3.42, PRO, MAF 58/140.

For most of the war, the main instrument of government control was the rationing of supplies of steel and timber. These were quite closely defined, being split into (steel): alloy/non-alloy/tinplate/iron castings; (timber): soft/hard/plywood. The quantities were in total quite large (in 1942 – 90,000 tons non-alloy steel; 1,035,000 cubic feet of timber), and a lot of firms were involved. The total number of firms receiving allocations of iron and steel castings was 742 in England and Wales, and a further 187 in Scotland. However, the firms could be repairers as well as manufacturers of agricultural machinery, and there is no easy way of deciding from the list which was which. To take two random examples, there is no doubt about the Ford Motor Co. of Dagenham, but perhaps Antique Reproductions (Worcester) Ltd was not a machinery manufacturer as the term is commonly understood.[18]

There was only one case in which direct and drastic action was taken by the government, leading to the total control of a company. This was Fowlers of Leeds. The reconstructed company was still not into profit, and the board was wracked by serious dissension. On 19 July 1941 the Ministry of Supply implemented two orders under the Defence of the Realm Regulations (DORA) which allowed it to appropriate the company's

Cromwell tanks, made by Fowlers of Leeds, taking part in the London victory parade, 1945. (*Tank Museum, Bovington*)

assets. The aim was to maximize the production of tanks; the new board was told that the firm must consider itself as a Royal Ordnance Factory. In all, a total of 1,633 tanks, mainly of the Matilda and Cromwell types, were produced by the firm. In November 1944 the Ministry announced that it would sell off twelve companies which it had acquired during the war to approved buyers, and in February 1945 the firm was sold to Rotary Hoes, of Horndon, Essex.[19]

The government decided to rely on Fordson tractors because Ford was the only mass producer of tractors in the UK. Even after the Lend-Lease agreement (March 1941), imports were severely constrained by the lack of shipping capacity. As far as the other potential home manufacturers were concerned, a variety of considerations ruled them out. After the disagreement with David Brown, Harry Ferguson had concluded an agreement with Henry Ford – the famous 'handshake agreement' in October 1938, by which Fords agreed to manufacture a new tractor in the USA, using the patented Ferguson three-point hitch for the attachment of implements. This, the Ford-Ferguson, began to be produced in the Ford works at Dearborn from June 1939. Marketing was handled by Ferguson, in association with the Sherman brothers. Meanwhile Ferguson had taken the opportunity to break off the partnership with David Brown. While this solved Ferguson's problem of finding a manufacturer who would make a tractor incorporating his system, it left him without a tractor for the British market.[20]

Having persuaded Henry Ford to make the Ford-Ferguson in the USA, Harry Ferguson turned his attention to the task of persuading Fords of Great Britain to do the same. This he failed to do, even though Henry Ford was the biggest shareholder in British Ford, and wished to persuade Dagenham to make the Ford-Ferguson. The Dagenham management had considerable operational autonomy, and was unconvinced that it would be wise to drop the Fordson, and to retool at great expense to produce a tractor which would probably cost more to produce than the Fordson, even with the advantages of the Fordson mass-production system. Balked by the management, led by Percival (by then Lord) Perry, Ferguson embarked on a large-scale publicity campaign, in which his tractor was demonstrated to the press and leading figures in the industry and the government. The campaign began with a demonstration on his home patch, Ulster, at Muckamore Agricultural College, near Belfast, on 12 October 1939, and culminated with a demonstration in front of the Ford Motor Co. management and government officials in early May 1940.[21] However, the campaign failed, partly, it is thought, because Ferguson was disliked and distrusted by the Ford board. Some of this may have been due to resentment of a pushy and aggressive outsider. In 1939 Ferguson had tried to get a seat on the Dagenham board of directors, and this attempt was repulsed, one director, Lord Illingworth, writing to Perry as follows:

I am decidedly against Ferguson joining the Board. He would be an infernal pest and no one could do any good to get a new machine going during the war, when there is an efficient one on the market already. It has been explained to him why it is impossible during the war and neither he nor Henry Ford, nor Jesus Christ can alter it.*

Even if the Dagenham board had wanted to replace the Fordson with the Ford-Ferguson, the government would not have permitted them to do so. The Ministry of Agriculture was fully aware of the Ferguson campaign, and was clear in its determination to resist it. Following representations by Harry Ferguson to the Ministry, the arguments for and against adopting the Ferguson system were put in a memorandum by a senior official of the Ministry, Donald Vandepeer. The Permanent Secretary, Donald Fergusson, added a note on the memorandum:

Mr. Vandepeer's minute clarifies the issues which arise in connection with the agitation and intrigue conducted over the Ford-Ferguson tractor by Mr. Ferguson and his friends ... The only question that arises under war conditions is whether by using a few hundred or a few thousand Ford-Fergusons instead of Fordson or David Brown tractors, we could substantially increase the number of acres ploughed-up and cultivated for the harvests of 1940 and 1941. I do not believe that such a result could be expected, however good the Ford-Ferguson tractor may be ... It is clearly in the national interest that we should take a firm stand against this agitation, and I think that by doing so now we shall save worse trouble later on.

A note on the minute reading 'I agree' was added by the Minister, Reginald Dorman-Smith.[22]

Harry Ferguson had a fiery zeal to see his ideas put into practice. But he was also annoyed at what he regarded as the preferential treatment accorded to Fordsons in the matter of the pre-war tractor reserve. This also grated on the only other two producers of tractors of any note, David Browns and Marshalls. Of the two, Browns was potentially the larger source of tractors. Even before the break with Ferguson in 1939, David Brown had laid plans for producing a larger tractor of his own. This, the VAK-1, was announced in July 1939, and the company came back from the Royal Agricultural Show of that year with orders for 3,000. The VAK-1 was larger and more powerful than both the Ferguson and the Fordson, and it had streamlined good looks. Its drawback was that it could not use the Ferguson hydraulic systems, which were protected by patents, and Brown evolved his own hydraulic system. This was inferior to Ferguson's, in that it lacked hydraulic depth control, but was an advance on other competing makes such as the Fordson.[23]

---

* Colin Fraser, *Harry Ferguson: Inventor and Pioneer* (John Murray, 1972; Old Pond Publishing, Ipswich, 1998), p. 123.

Anticipating expansion, Brown leased premises of 400,000 square feet at Meltham (near Huddersfield) and had barely started production there when war began. Wishing to expand production of his tractors, Brown was frustrated by the government, which had other priorities for the firm. These were mainly the production of gears for tanks and other military vehicles (the Brown parent company had been a specialist gear producer). Insofar as the company was to be allowed to produce tractors, these were to be of the type suitable for towing aircraft around the hundreds of wartime airfields. For this task the VAK-1 was more suitable than the Fordson, having more power (35 bhp, as against the Fordsons 25 hp; the Ford-Ferguson was 20 hp). Some models were fitted with a fluid flywheel, which allowed the tractor to move aircraft in a more gentle way, without jarring. The pressure exerted by David Brown on the Ministry led to a report on both the tractor and the Meltham factory by the Ministry's most senior machinery expert, S. J. Wright. He was in favour of the David Brown tractor. He considered that it could be used instead of the Fordson, and in place of some of the smaller imported tractors. It was also a consideration that the factory was in a steep secluded valley, less likely to be bombed than the Dagenham plant (it was in fact not bombed during the war). He supported the firm in its plans for expansion. Wright's view was at one point encouraged by the Treasury, in October 1941, with a view to reducing tractor imports, and thus saving foreign exchange; the possibility of raising output from the current 1,500 a year to 4,000 was floated by a sympathetic Ministry civil servant.[24]

However, David Brown's hope of a large expansion in his wartime agricultural tractor business was thwarted. The Tank Supply department of the Ministry of Supply wrote to the Ministry of Agriculture in January 1942 stating that Browns would in future only be allowed to make any tractors at all on condition that they made them for airfield use. If the firm rejected this policy, then it would be confined to making tank gearboxes. If the firm should meet its gearbox and airfield tractor targets, then it would be allowed to produce agricultural tractors. The letter hinted at bureaucratic politicking on the part of the firm:

> Mr. David Brown will undoubtedly do his utmost to produce agricultural tractors. We have to see that he does not let this interfere with the tank and Government tractor programme. In view of our past experience with him, we are relying on you not to let him play your Department off against ours in such a way as to interfere with that programme.*

Some 5,350 VAK-1, tractors were made from its introduction in 1939 to its replacement by the VAK-1A in 1945, was. Most were for airfield use.[25]

---

* PRO, MAF 58/112. Tank Supply Department, Ministry of Supply, to C. I. C. Bosanquet, Ministry of Agriculture, 29 January 1942.

The other firm to be annoyed at the pre-war Fordson deal, and to use this as a stick with which to beat the Ministry of Agriculture, was Marshalls of Gainsborough. In August 1941, Mr Burton* of Marshalls called on the Ministry to make this point, as a preliminary to asking for help in getting some specialist machine tools:

> Mr. Burton of Marshall's called yesterday and discussed his production programme with Mr. Gaunt and me. As on previous occasions, Mr. Burton opened out with the now familiar move that he was not given a look in with the Government contract to supply tractors in June 1939, and it is quite clear that he is suffering from some kind of complex on the matter. It is to no purpose to explain to Mr. Burton how different the situation was in the period before the war from what it is now, and I steered Mr. Burton off these fruitless recollections. If you see Mr. Burton, however, at any time, he is almost sure to open up in the same kind of way, but I can give you the history of the matter at any time should you wish to be fortified on this point.†

At this time Marshalls was mainly engaged on Admiralty work, and their output of tractors was only about 80 a year. Their tractor was the only diesel tractor on the market, and was rugged, sturdy and powerful. It was appreciated that the Marshall tractor had virtues, chiefly being of high enough power to drive a threshing drum, and it was thought that it would be a good policy to have Marshalls increase output to 1,000 a year. But nothing came of this. The Ministry officials visited the Gainsborough works, and Mr Burton talked of having a programme to produce 750–1,000 a year, but it was clear to the officials that, although this would be a useful way of saving on the import of some of the higher-powered US tractors (Massey-Harris 28/40; IHC; Minneapolis-Moline), this could only be done by using machine tools from somewhere else, and this would deprive the services of other necessary products. It was also noted that, when the officials asked how Marshalls would cope with a large contract to supply tractors, Burton was at a loss to suggest how he would execute it. The final word was again had by the Ministry of Supply, which was responsible for determining the supply of machine tools. It pronounced the project dead, and ruled that the machine tools should go to tank production. By March 1943 it was noted that Marshalls were making only six tractors a week, and couldn't expand production until they got 'the new machine tools'. The firm's tractor production remained comparatively low. In the years 1938–45 the total output of its Model M diesel, was probably no more than 890.[26]

---

* Presumably Mark Burton, general manager at the works, and board member, since the reconstitution of Marshalls in 1936: Lane, *Britannia Iron Works*, p. 116.

† Memo. of 27.8.41 by Mr. Parker, MAF: PRO, MAF 58/121.

In 1942 the Ministry of Supply relented, due to the increased losses of food imports from submarine action, and agreed to allocate the necessary machine tools as and when required. Marshalls took advantage of this to develop an improved version of their diesel tractor – the Field Marshall Mark I. This was only a slightly improved version, having the same engine as the Model M, but running at 750 rather than 700 r.p.m. At the end of 1944 the company agreed with the Ministry of Agriculture to produce 2,000 Field Marshalls by July 1946. No government order was offered, and the arrangement was that this was to be on a purely commercial basis. While accurate production records have not survived, it seems that Field Marshall Mark I production was 2,000 in 1945–47, the programme being delayed by wartime and post-war shortages of materials.[27]

Loss of export markets

As part of a general policy aimed at improving the balance of payments exports were encouraged in the first year or so of the war. Until the signing of the Lend-Lease agreement in March 1941, the British government had to pay cash for orders it placed with US firms, and the government was very concerned about the rapidly shrinking reserve of dollars held by the Treasury and the Bank of England. However, there were some restrictions. There were two lists of products. Export of those on List A required the issuing of a licence, whatever the destination. For those on List B, licences were not necessary if the destination was within the British Empire. Tractors, binders and threshers were on List A from the beginning of the war. Ploughs were on the A list until 15 January 1940, when they were transferred to B. They went back on to the A list on 15 August 1940. The following items were free from export control until 15 August 1940, when they were put on the A list: chaff-cutters, cultivators, drills, grinding mills, harrows, manure distributors, potato lifters, rollers, and hay harvesting machinery.[28]

In practice, exports continued for the first year of the war, although at a rapidly declining rate. By this time, the advent of the Dagenham factory had led to Fordsons being the largest single item of agricultural machinery being exported. Most Fordsons, and many of the tractors produced by the other makers, went abroad. The pre-war exports of Fordsons had peaked in 1937, at 11,039. The 1937–38 recession hurt export sales badly, and numbers fell to 5,517 in 1938 and to 4,786 in 1939. In 1940 they were much smaller, but still noteworthy, at 1,711. Export seems to have been fairly free in the first half of 1940, but from July export was permitted 'only in isolated cases or for non-agricultural types'. It was only in 1941 that they practically ceased; the total exported for that year seems to have been 6. The drastic falling-off in exports in the first year or so of the war is indicated below:

Table 11.4 *Exports of agricultural machinery, 1937–40 (£)*

	1937	*1938*	*1939*	*Jan. 1940*	*July–Nov. 1940*
Tractors	1,036,125	621,547	378,929	92,827	57,000
Ploughs	238,149	207,168	158,417	26,032	28,000
Threshers	52,810	48,767	52,747	3,259	39,000
Other	483,697	409,844	323,638	22,818	93,000
Totals	**1,810,781**	**1,287,326**	**913,731**	**144,936**	**217,000**

Source: PRO, MAF 58/79. Exports: Policy. 1939–42; memo. of 6.2.41: 'Exports of agricultural machinery'.

It was only in the last 18 months or so of the war that the position of the farming industry could allow the resumption of exports. Whereas at the pre-war peak of 1937, the weight of machinery exported had been 18,600 tons, this fell to a wartime low of 3,700 tons in 1942, recovering to 7,600 tons in 1944; the 1945 total surpassed the pre-war level at 24,000 tons.[29]

Armament work

Unlike the experience of 1914–18, the Second World War saw a much more measured and organised move to armaments production; the Gadarene rush to get munitions contracts was not repeated. But many firms had to make room for military work. Well before the war, armament contracts had been undertaken on a large scale. Ransomes had orders on hand to the value of £97,463 at the end of October 1938. It was obliged to cease production of lawn mowers almost as soon as the war began, with the exception of gang mowers for airfields. The released capacity of the mower works was given over to high-grade machine tools, and military work such as aircraft assemblies, gun and bomb trailers and bomb parts. In June 1940 the Ministry of Supply declared the firm a controlled establishment. In October 1940 it was making cast-iron rollers for the Admiralty, and was sub-contractor for the manufacture of '25 co-axial gun mountings' per week for Stothert & Pitt of Bath. At the end of July 1942 the firm was machining the tail units for 200,000 mortar bombs, at 33*s*. per 100 for the Ministry of Supply. Among other things, it produced over the whole war period 800 machine tools, 30 bomb trolleys and 1,500 trailers for 25-pounder guns. The thresher works continued to make agricultural machinery, but also produced armaments parts and aircraft hangars. The engine works made precision components for tanks, aero-engines and naval equipment. The electrical vehicle works produced around 1,200 trucks for use at home and abroad.[30]

Another large firm involved in munitions during the drive to rearmament in the late 1930s was Marshalls of Gainsborough. In 1936 the firm had been taken over by Thos W. Ward Ltd. Under the leadership of Ashley

S. Ward, the revived company was back in profit by 1937, although its only agricultural product, the diesel tractor, was a minor part of its total output, and it was engaged in a variety of official contracts, ranging from Merlin aircraft engine components to gun mountings for the Admiralty (as in the previous war). At the outbreak of war, the company had orders worth over £1 million, of which government contracts accounted for £843,000. During the war, most of the company's work was for the government, and at the seventh AGM in 1943 it was announced that the past year's turnover had been £1,335,000. The most notable product was the three-man midget submarine, whose exploits included the successful attack on the German battleship *Tirpitz* in her Norwegian fjord.

It was not just the larger firms which did war work. Bamfords of Uttoxeter were obliged by the government to concentrate on mowers, for which it had a high reputation, but were also called upon to undertake sub-contract work on radar components, gears, etc. A portion of the factory was even taken over by the Daimler Co. for the production of scout cars. A notable contribution to the war effort was made by Taskers, which had been reconstructed in 1932 after its failure in 1926. The reconstructed firm had abandoned agricultural machinery in favour of road trailers, and was not to get back into the agricultural business until 1947. In the war, the company scored a great success with its 'Queen Mary', a long (55 feet, including the cab) articulated trailer, designed for the recovery and transport of crashed aircraft such as Spitfires and Hurricanes. This version, and a later, larger one which could also transport bomber sections, took the total of Queen Marys produced to 3,834. In total, Taskers produced 19,359 trailers of all types during the war. Bentalls made parts for aircraft, especially for Handley-Page and Halifax bombers. But the firm also managed to double its output of agricultural machinery, and considerably expanded its workforce, to the unheard-of level of *c.*1,000.[31]

Working conditions and practices changed during the war. Factories were encouraged to go over to 24-hour, shift working. As in 1914–18, one of the greatest changes was the employment of women on the factory floor. Bamfords recruited women to work the day shift, and thus release men for night work. By August 1943 Ransomes' total of manual workers was 2,948, of which 812 were women. The war saw a large rise in the total of employees. At the trough of the 1930s depression, in September 1931, the number of Ransomes' employees had been reduced to 1,657, and this had only risen to 1,748 in September 1939, but by July 1945 the total was up to 2,898. Ransomes, being only ten miles from a coast exposed to enemy attack, was particularly vulnerable. To meet this danger, manufacture of components was dispersed over a number of factories, and part-time labour (mainly female) in the surrounding villages was recruited to do sub-assembly work without going into the factory. Further energy went into defence, the company forming its own Home Guard unit, made up of 480 employees.[32]

Contracts for the supply of military equipment affected firms of all sizes. Some had specialisms that were very attractive to the government. Roadless Traction of Hounslow, which had developed half- and full-track caterpillar tracks for tractors and other vehicles before the war, found that its products were in great demand by the Air Ministry for airfield tractors. The most widely used was the half-track Fordson N conversion. This was valuable for mowing airfields without damaging the ground surface, but they were also used for moving aircraft and supplies, hauling fuel bowsers and bomb trailers, and even clearing snow from runways. Many were fitted with winches for recovery work. Some Case tractors supplied under Lease-Lend were also converted to half-tracks. The whole half-track output went to the Ministry, and none for civilian use.[33]

Some firms were too small to participate in war contracts. The firm of Bomford Bros, of Pitchill, Gloucestershire, farmed about 1,000 acres, as well as producing ingenious farm machines and doing agricultural contracting (chiefly threshing); it employed only 33 men, 'divided almost equally between farm, works and contracting'. In the war, all manufacturing ceased, and the works became a repair and maintenance shop.[34]

Imports and Lend-Lease

The enormous expansion of agricultural production during the war was clearly going to have to depend to some extent on imported machinery. This was not due to the British manufacturers being otherwise occupied with government orders, as had happened in the First World War. In 1939–45, firms which could make a contribution to the agricultural programme were firmly controlled to that end. It was rather that in some cases home output was inadequate, and in others that there was a need for types of machines which were either not made in the UK, or only made in small numbers. At first, the expansion of the home tractor output was sufficient to meet national needs. In 1939–40, the main products to be imported on government account were tractor ploughs, to be hitched to the Fordsons. In this, as in all cases, the main source of supply was the USA. By the spring of 1941 the raising of the ploughing-up target to 2,160,000 acres for 1940–41 led to further demands for imported machines.

Demand for tractors could not be satisfied entirely from home manufacturers. This was largely a matter of importing types which were not produced in the UK. Such were the larger, higher-powered four-wheel tractors such as those produced in the USA by Case, International Harvester, Allis-Chalmers or Minneapolis-Moline. These could all be supplied in higher power ratings than the Dagenham Fordsons. Their advantages were that they could work faster, and cope more easily with threshing machines and other barn machinery. The Ministry of Agriculture noted that they: 'are wanted for operating the heavy English type of 4′ 6″ threshing drum and ploughing on heavy land wherever a crawler is not absolutely essential'. Another type was the row-crop tractor. This had a higher ground clearance, often a wider distance between the wheels (some models had adjustable axles to vary the track width), and could perhaps accommodate a toolbar, either at the rear or underneath the machine. They could 'take the place of horses and so increase the output of labour in planting and cultivating potatoes, sugar beet and all root crops'. The ministry's memorandum noted that, 'no efficient rowcrop tractor is yet made in the UK'. One exmple of such a machine was the Allis-Chalmers B Model tractor. Fordson had attempted to market a row-crop version of the N model in the 1930s, with a particular eye on the US market, but this had not been successful.[35]

Finally, there was a demand for track-laying tractors. These were required for heavy work, in which it was essential not to compact the ground, or in which conventional wheeled tractors could not get a suitable grip without damaging the soil structure. They were required above all for reclamation work. A memorandum of the Ministry of Agriculture referred to the usefulness of using the heavier crawlers for land drainage work, the smaller models to be used for cultivating the stiff clays and the wetter and hilly grasslands: 'Approximately 1 million acres of old pasture and rough grazing land will have to be ploughed in 1942/43, including reclamation of bush covered and wet land, much of which is stiff clay where tracklayers are necessary.' In the inter-war period, much land had been allowed to 'go back' – to tumble down out of cultivation – and had either become waterlogged or overrun with scrub and small trees. There was thus much work to be done if the ploughing targets were to be met. Drainage was a priority. Over the years 1940–44, some 4 million acres of land was drained. Much of this was done by tracklayers, and the older steam drainage tackle. The tracklayers were almost all imported from the USA. At 31 January 1944, there were working on British farms 6,243 tracklayers of US manufacture – Caterpillar, International Harvester, Allis-Chalmers and Cletrac. These apart, there were only about 60 others, which were either Fowler or Lanz models. Tracklayers were especially useful in the waterlogged Fens, where previous drainage systems had broken down. Alan Bloom's book, *The Farm in the Fen* (Faber & Faber, 1944) describes the difficulties and achievements of one such enterprise.[36]

The enormous rise in the targets for national food output, and the difficulties experienced by home manufacturers, led to a large rise in imports from the start of the war. In 1939–40, imports were only restricted by shortage of shipping space and foreign exchange. Lend-Lease removed the latter constraint from March 1941. At that time, home output was concentrating on wheeled tractors, cultivating implements and hay mowers, but the demand for tractors, ploughs, disc harrows and combined drills (which sowed seed and fertiliser simultaneously) much exceeded the home supply, so that the initial Lend-Lease programme for the second half of 1941 concentrated mainly on these items. However, the entry of the USA into the war on 8 December 1941 confused all plans and priorities, and the original import programme was unfulfilled. The tractor programme had consisted of orders for 1,100 tracklayers (of which 515 were from the Caterpillar Co.), 2,400 wheeled tractors of 30 hp and above (mainly Oliver, Allis-Chalmers, Massey-Harris and Minneapolis-Moline types), and 6,400 row-crop tractors. These latter consisted of 1,400 Allis-Chalmers, 1,300 International Harvester, and of 1,800 Ford-Fergusons, the latter being the new tractor incorporating Harry Ferguson's patented three-point hitch, which was now being manufactured by Ford at Detroit. The Ford-Ferguson, the Ministry of Agriculture had noted, 'may be classed as a rowcrop machine, having adjustable front and rear axles ... these tractors are specially useful for cultivation in small fields and for the potato crop.' However, the reorientation of US manufacturers to war production, and the losses of machinery in transit due to submarine attack, reduced the imports of tractors, so that only 4,200 arrived. The shortfall was particularly evident for tracklayers, many of which did not arrive until the spring of 1943. In November 1942 the shortfall of imported tracklayers was noted as being 2,600. In May 1942 the Ministry of Agriculture had stipulated that the 1942–43 cropping programme depended on the import of 115 heavy, 400 medium and 2,500 small tracklaying tractors. Few of these had arrived in time for the spring cultivations of 1943.[37]

The build-up of the Allied military effort, in preparation for the invasions of Europe and Japan, meant that after 1942 imports of farm machinery declined. But the contribution of imported tractors to the total wartime supply of tractors was considerable, as can be seen from the table below:

Table 11.5 *Total supply of tractors in the UK, 1940–44*

	Tracklaying, agricultural	*Market garden*	*3- and 4-wheeled*	*2-wheeled*	*Totals*
Home	270*	2,074	103,542	12,004	117,886
Imports	4,735	0	29,329	947	35,021
Totals	5,015	2,074	132,871	12,951	152,907
Percentage imported	94	0	22	7	23

* Agricultural tracklayers produced in the United Kingdom: 1939 and 1940 production by J. Fowler & Co. (Leeds) Ltd; 1942, 1943, and 1944 production by Roadless Traction Ltd.

Source: Public Record Office CAB 87/18: *Farm Machinery Supplies 1937–1944*, Table A.

Between 1937–39 and 1940–44 the total UK supply of tractors rose from 18,950 to 152,907; during these periods a very similar proportion was imported stock: 4,021 (21.2%) in 1937–39 and 35,021 (22.9%) in 1940–44. The supply of certain other products was heavily dependent on imports.

While the supply of horse-drawn ploughs rose substantially (from 3,867 annually in 1937–39 to an average of 13,082 in 1940–44), they were almost all produced in the UK. More reliance was placed on imports for tractor ploughs. The import of these had been 3,749 on average in 1937–39, but this rose to a wartime peak of 18,424 in 1941, although declining steadily thereafter, to 6,423 in 1944. Home production rose from 5,539 in 1937–39 to 10,495 in 1941, with some decline thereafter, to 9,738 in 1944. Apart from tractors and tractor ploughs, the items imported in the greatest numbers were corn and fertiliser drills and binders. The dependence on imports for the latter was very striking; of the total supply of 40,593 in 1940–44, 33,335 (82%) were imported. However, this ratio was only slightly higher than that of pre-war. The dependence on imports for drills was less; of the 86,164 supplied in 1940–44, only 34% were imported, which was somewhat less than the pre-war ratio for 1937–39, which had been 45%. Whereas the USA was the main source for tractors and tractor ploughs, substantial imports of drills and binders also came from Canada, and Australia was an important source of binders. The greatest technical innovation of the war was, at least potentially, the combine harvester. A rarity on British farms before the war, home production in wartime was negligible. Even at the end of hostilities, in 1945, only 107 were produced in the UK. But imports from the USA and Canada rose steadily, from a mere 28 in 1937–39, to 930 in 1944. Even in these small numbers, their utility as labour- and horse-saving devices was apparent.[38]

Profits and taxes

The financial results of the Second World War should have been even more favourable to the industry than those of the First World War. In both wars the industry laboured to produce warlike stores. But in the Second War, there was also a large increase in demand for farming machinery of all kinds. Added to this, there was less difficulty in getting labour, and less industrial strife. The constraints were, in certain cases, the novel one of air attack (which could led to the dispersal of production) and restrictions on supplies of raw materials (chiefly steel). The supply of labour was also controlled, and men in military service had to be replaced with women. But, faced with a burgeoning demand, the industry was in a much stronger position financially than in 1914–18.

If high profits were to be made, they had to be made in the early years of the war. Before 1943, the government did not directly control the prices of agricultural machinery. On the other hand, manufacturers were faced with continually rising costs of labour and materials. The general level of inflation of prices, and of wages, was less than in 1914–18, but sufficient to give concern to the manufacturers, and to cause them to revise their prices annually. In an effort to understand the causes of the inflation of machinery prices, the Ministry of Agriculture devised an index of agricultural machinery prices. This showed, on the starting basis of 100 in 1938, that farm machinery prices had risen only to 103 in 1940, but to 118 in 1942 and to 127 in 1943. Their final level, in 1945, was to be 148. In other words, machinery prices had risen by about a half over the whole war. This was higher than the inflation of the retail price index, which had risen by only 31% by 1945. However, the retail price index was by then not a good indicator of inflation, partly because it was based on an out-of-date spending pattern, and partly because during the war the government subsidised the price of bread, which was a large component of the index. A better standard of comparison would be the Board of Trade wholesale price index, and that had risen by 57% by 1942, and 67% by 1945. Thus the inflation in machinery prices was less than the general rise in prices.[39]

Prior to the imposition of price control in May 1943, the industry was free, in theory, to set its own prices. Evidence from the annual statements of firms shows a clear jump in wartime profits. This is hardly surprising, since they were driving a much larger volume of business. Also, many of the military contracts were on a cost-plus basis, with little control of firms' costs. In the case of the larger manufacturers, the level of profit was in any case historically high well before the war, due to the fact that they were already heavily involved in military orders. The pre-war contracts of Ransomes and Marshalls have been noted above. Listers of Dursley had been making Bofors anti-aircraft shell casings from as early as 1936. In the same year they began supplying diesel generators for portable searchlight installations. In the war, the firm's financial record was as follows:

Table 11.6 *R. A. Lister & Co. Ltd: financial record, 1938–45 (£000)*

	1938	*1939*	*1940**	*1941*	*1942*	*1943*	*1944*	*1945*
Gross trading profit*	245	224	139	132	164	399	399	367
Taxation provision	72	94	150	189	242	212	221	178
Ordinary dividend (%)	17½	16	16	16	16	16	16	16
General reserve	607	632	657	759	784	819	769	919
Government securities	–	–	–	–	122	147	149	149
Tax Reserve Certificates†	–	–	–	–	–	400	100	191

* After providing for taxation

† Purchased against future taxation

Source: Gloucestershire Record Office, Gloucester, R. A. Lister archive, D 3310, 1/14]

The profits of the firm were high, but so were wartime taxes. In the Budget of September 1939, Excess Profits Tax had been introduced. Designed to limit the profits of military contractors, it was calculated to recover for the Treasury the wartime profits of firms above their pre-war levels. Set initially at 60%, it was raised to 100% a year later. The impact was softened to some extent by a 20% post-war credit on the tax paid. This was to be had by purchasing Tax Reserve Certificates, as can be seen in the case of Listers. Despite the tax, however, there was enough money left over to make provision for shareholders' dividends, to meet depreciation charges, and to put some into the reserves. As in the case of some firms in the First World War, high profits for Listers were accompanied by a conservative financial policy, which limited the dividends to shareholders (already at a high level in 1938), and made large provision to meet tax obligations. Large sums were paid to the general reserve each year, and the firm invested also in government securities and Tax Reserve Certificates. In addition, the firm was very conscious that there would be a need to restructure its operations at the end of the war, and the rise in wartime reserves was accomplished with this in mind.[40]

A similar story comes from Ransomes, although it was less profitable than Listers. After providing for Excess Profits Tax, its gross trading profit rose from £105,000 in 1941 to £122,000 in 1945, and its reserve fund (for post-war reconstruction and contingencies) rose from £35,000 in 1941 to £170,000 in 1945. As in the case of Listers, Ransomes was mindful of the need to make provision for post-war readjustment, and careful to make provision for wartime taxation as it occurred. Thus it, and other firms in the industry, avoided the arguments with the Inland Revenue which had been a feature of 1917–19.[41]

Not all the firms making high profits on their turnover were large ones. The small firm (capitalised at £25,000) of Nalder & Nalder, of Challow, near Wantage, Berkshire, was struggling in the 1930s. It made a loss of £120 in 1937. In 1938 there was a profit, but only of £555. In 1939 there was another loss, and the chairman played down the possibility of getting government contracts, remarking that large mass-production contracts were not suitable for firms of this small size, and, 'therefore we have to depend on sub-contractors for suitable items to manufacture in most instances, and such sub-contracts are difficult to obtain at present'. By that time (April 1939) the firm had accumulated losses of £4,890. A further loss, of £951, occurred in 1940. The 1941 figures are not available, but in 1942–45 the works were fully occupied, although the chairman complained of a shortage of labour, and in 1944 it was noted that the works employed only 49 men. Full capacity working was accompanied by much higher profits, which in 1942–45 averaged £1,782 annually, and by 1945 the accumulated losses had been reduced to £1,810. At the 1945 AGM the chairman revealed that in fact the works had been largely employed on government work during the war, making, 'jigs and tools for Aircraft makers, Mine Parts, Minesweeper fittings ... and some of our usual food machinery has been employed for making Army Pills'. Wartime work had been split 60% for the government, 20% food machinery, and 20% agricultural machinery.[42]

High profits were also evident in the case of Wm Elder & Sons of Berwick-upon-Tweed, another small firm (capitalised at around £15,000). The firm's low point in the pre-war period had been 1932, when profits were only £159. But there was then a sharp recovery, and by 1938 profits were £6,035, twice as large as the previous period of high profits in the late 1920s. In 1939 they were £8,227. It is not known if this rise in profits was linked to rearmament work, although bearing in mind that the firm had avoided government work in 1914–18, they may have kept aloof. It is also possible that the rise in profits was linked to the Fordson dealership which the firm acquired in December 1938. In the war, profits were very high, rising in each year to reach a peak of £26,815 in 1943.[43]

Another small firm struggling in the 1930s, which was also boosted by the war, was R. & J. Reeves & Son, of Bratton, Wiltshire. In the early 1930s financial weakness seems to have led to a capital write-down and partial reconstruction, but there was still concern in 1938, when sales reached a low of £10,969, and a bank overdraft of £5,000 resulted In the war, the firm confined itself to producing agricultural machinery, and total sales rose substantially, reaching a peak of £31,304 in 1945. Of the products of its own make, the most buoyant were drills; of the products which it sold on behalf of other firms ('factored sales'), haymaking machinery and tractors grew most. But a large part of the increased sales was also due to repair work, which rose from £1,715 in 1938/39 to £5,327 in 1944/45. It can be presumed that profitability rose commensurately. The firm took the opportunity to repay its borrowings, emerging at the war's end in a much stronger financial position.[44]

In sum, the war was, predictably, a period of high profits, although higher taxation drew more of these off than in 1914–18. But the general buoyancy of home demand for agricultural machinery as well as of government contracts ensured that even small previously struggling firms could survive. Larger firms, provided that they were efficient and had the right mixture of products, finished the war in a strong position, having the financial resources to meet the challenges of post-war restructuring. But not all large firms were efficient. There was the world of a difference between firms such as the Ransomes and Listers on the one hand, and the Fowlers and Marshalls on the other.

The industry in 1945

The war had transformed the size of the industry. At the last pre-war Census of Production, in 1935, the total output of the industry had been valued at only £3.14 million, and employment was only 9,351. Both these figures were lower than those of the previous census, in 1930. But as early as 1941/42, the novel conditions of wartime had transformed the industry. This information comes from a report into the post-war prospects of the industry which was compiled by the Ministry of Agriculture in 1945. It is full of suggestive detail and useful information. In the twelve months from June 1941 to May 1942 total sales of new farm machinery in the UK were estimated to have been approximately £25 million, to which should be added £2.5 million for spare parts, making £27.5 million in all. Of this, some £7.5 million was imported, leaving about £20 million as the product of home firms. The figure for 1942/43 was estimated to be about 10 per cent higher. Exports were negligible, and were mostly of implements from stock built up previously.[45]

A part of the increase was due to price rises (around 30 per cent by 1943/44); part is explained by the higher proportion of expensive machines now forming a large part of the industry's output. The outstanding product here was the tractor: at £175 each, the 27,600 Fordsons produced at Dagenham in 1942 would have cost the farmers £4.8 million. More generally, the ploughing campaigns had led to a large expansion of the industry. Total employment is not known, but the *c.*9,000 employees of 1935 may be compared with the *c.*20,000 of post-1945. The industry had doubled in size; its prices had risen by about one-third; and its product structure had shifted towards what the retailers of the 1990s were to call 'big-ticket items'.

But the structure of the industry had not been transformed by the war. The industry was still dominated by a large number of small firms, some of which had been founded many decades earlier, and some of which had outdated methods and premises. Some larger firms also exhibited these features, although the inter-war period had expunged many of the older firms. The number of firms in the industry was, as ever, debatable. The 1935 Census of Production had noted only 94 'establishments', i.e., units of production, but that had been concerned with the specialist producers only. In the war, the definition was widened, to include at one end of the spectrum those firms whose main business was agricultural machinery, but who made other goods also, and those who mainly made other goods, but also produced some agricultural machinery. All these firms were entitled to the allocations of steel necessary for them to continue in the business, and to them was added the firms who did mainly or wholly repair work. Thus the list of firms in the Ministry of Agriculture report was that of those entitled to steel and iron allocations. It represented a definition of the industry and a listing of the firms within it which was the widest possible. Based on the allocations of iron and steel, the total number of firms was 949. Their sizes can be gauged from their allocations, which are shown in Table 11.7.

While there are some difficulties with this survey (chiefly that some firms may have been repairers rather than manufacturers, and some may have not been included since they were entirely taken up with non-agricultural government contracts), this analysis can be taken as representing the structure of the industry at the end of the war. It shows the structural dominance of the small firm, and the dominance of production by the

Table 11.7 *Steel and iron allocations to agricultural machinery manufacturers and repairers, 1945*

	England and Wales	*Scotland*
Number of firms receiving:		
over 5,000 tons per year	4	0
1,000–4,999 tons	16	2
500–999 tons	20	2
100–499 tons	111	13
50–99 tons	86	10
less than 50 tons	505	160
Total	**762**	**187**

Source: Public Record Office, MAF 58/162, *'Farm Machinery and Equipment in the First Ten Years after the War'*, Appendix I, 'Particulars of Agricultural Engineering Industry'.

large firms. A rough rule of thumb employed by the Ministry was that for every 100 tons of steel or iron supplied, six persons were employed, this figure being lower for the larger and higher for the smaller firms. On that basis, only some 151 firms in England and Wales, and 17 in Scotland, employed more than about six persons.

The small size of the overwhelming majority of firms was one of the principal features of the industry. Other distinguishing features related to its technical profile. There was a large amount of foundry and smiths' work and a relatively small amount of precision machining. Wages were rather lower than in some other branches of the engineering industry. The product mixture showed a multiplicity of types, even among the large firms. Thus Ransomes made 112 varieties of horse plough, and 36 types of tractor plough. Bamfords made 10 types of mower. Marshalls made 10 types of threshing machine. There was only one product which was mass-produced, and that was the Fordson tractor. But the main reason for the multiplicity of types was that they were mostly made by small firms, '75% of whose business has been in the repairing of implements'. At the same time, competition in the industry was keen. This combination of factors had worked historically to proliferate the types of implements on offer. Local firms, mainly engaged on repair work, took advantage of the slack winter season to make implements and machines against the expected rise in demand in the spring. Larger firms, finding that they were losing business to the local firms, added to their lines implements of the local pattern. Thus the industry's product proliferation militated against mass production, and contributed to the technical backwardness and relatively poor equipment of many establishments.[46]

Finally, it might be added that the industry was only marginally engaged in research of professional quality. Traditionally, new products had been evolved by a variety of *ad hoc* methods. A favourite one was the taking out of patents by senior members of the founding family. To this could be added the offering of new ideas to firms in the industry by members of the public, and often by farmers themselves. There was also an increasing amount of publicly funded support for the development of new machines and implements. The Agricultural Machinery Development Board had been putting new machines through technical tests since the late 1920s. In 1940 the National Institute of Agricultural Engineering had been founded. But much of this effort remained extrinsic to the firms themselves, and development of new ideas proceeded haphazardly. The full-time researcher or development officer in a firm was a rarity. In the war, even Ransomes employed only four researchers. Few other firms could match even this level of commitment.

Thus the industry, although buoyed up by the wholly exceptional circumstances of the Second World War, had some serious concerns to address. Whether the post-war world would provide the cushion which would enable the industry to come to terms with its problems, both actual and incipient, was another matter.

CHAPTER TWELVE

A very brief supremacy, 1945–1973

The new post-war world and the long economic boom

Few industries escape from the general economic conditions of their age. Whereas the effects of the First World War had been seriously dislocating for the British economy, ushering in almost two decades of industrial problems and high unemployment, the effects of the Second World War had the opposite result. The defeat of the Axis powers, the political realignment of Western Europe and the start of the Cold War inaugurated a quarter of a century of rapid economic growth and outstanding technological change. In retrospect, this period is now being referred to as 'the long European boom', or 'the golden age'. The world economy, the European economies, and the British economy all grew at historically unprecedented rates. In the case of Britain, the growth rate was even higher than that of the third quarter of the nineteenth century, which historians still refer to as 'the mid-Victorian boom'. This may be seen from an estimate of the rate of growth of output (GDP, or gross domestic product) per head over the long period:

Table 12.1 *Britain: growth rate of real output per head, 1856–1998*

	Real GDP per head (% change per year)
1856–1873	1.4
1873–1913	0.9
1913–1924	–0.6
1924–1937	1.8
1937–1951	1.3
1951–1973	2.3
1973–1979	1.5
1979–1990	2.0
1990–1998	1.7

Source: A. Booth, *The British Economy in the Twentieth Century* (Palgrave, 2001), Table 2.2.

Until the slowdown of the 1970s, high economic growth was accompanied by full employment and rapidly rising living standards. It was an age of prosperity, whose tone was caught well by Harold Macmillan, as Chancellor of the Exchequer in 1956, when he observed to the British people that, 'you've never had it so good'.

However, the good years were those of the 1950s and 1960s. Before then, there were serious short-term problems and readjustments to be made in the industrial and trading positions of the leading economies. The war had devastated the industries and cities of Europe, the USSR and the Far East. It would take many years to rebuild the shattered economies of Western Europe and Japan, which had provided Britain with industrial competition before 1939. Even in 1950, when some degree of reconstruction had taken place, world trade in manufactured goods was still dominated by only two nations: the USA and the UK. This supremacy was to be considerably eroded by the end of the 'golden age', but in 1950 it was still a reality. In 1937, the USA had accounted for 20.2% of world manufactured exports, and the UK 22.0%. In 1950 the proportions were 28.7 per cent and 26.5 per cent.[1]

Post-war readjustment

However, the temporary disappearance of much of the previous foreign competition was not a guarantee of success for British exporters. Britain had many problems, some of which seemed insurmountable. The maintenance of the physical equipment of many basic industries such as coal, the railways and the shipyards had been neglected in the war, and they needed fresh capital to renovate. There had been much destruction of housing, thus making what had been a social problem before the war much more serious. There had also been a lot of damage to industrial plant. But the most serious problem in the few years after the war was the that of the balance of payments. During the war, exports had declined to a very low level, partly due to the cutting off of markets because of hostilities; by 1941, Ransomes exported only to the British Empire and South America. It was also due to government policy, as industries were obliged to turn their productive effort over to military goods. But the level of imports remained high, as Britain continued to rely on overseas supplies for about two-thirds of its food and the entire supply of some important raw materials such as cotton and oil. The result was a large deficit on the balance of trade. The gap was now historically very high. In previous years the gap in physical trade had been more than covered by the value of earnings of services – from banking, insurance and shipping services, to which was added the investment income derived from the large volume of capital invested overseas by British firms and citizens. But by 1945 service income was at a low ebb, mainly due to the loss of ships. The loss of investment income was even more serious, as investments in

Europe had been lost; British citizens had sold their dollar foreign investments (at the behest of the government); and governments had defaulted on their bonds. Finally, the British government had run up considerable debts to certain countries in the Empire, chiefly India and Egypt. These 'sterling balances' remained in the post-war period as a constant threat to the stability of sterling.

The seriousness of the balance of payments problem at the end of the war meant that government policy was directed at containing it in the short run, and reducing it in the longer run. Thus the exchange controls begun in the war were to be continued indefinitely. Controls on imports were to be imposed, while every opportunity was to be taken to promote exports. The aim of the government was to raise the volume of exports to 50% above its pre-war level, with particular attention to the dollar area, in view of the large trade deficit with the USA. Only thus would the loss of service and investment income be counteracted, and the balance of payments problem reduced to manageable proportions. In practice, the government's export drive succeeded very well, and by the early 1950s the balance of payments problem had receded somewhat. This success was achieved with the active cooperation of manufacturing industry, in which the agricultural engineering industry played a leading role.

Agricultural policy and prosperity

At the end of the war, many farmers must have had in their minds the 'Great Betrayal' of 1921, when the government had reneged on its promise to guarantee the prices of farm products. However, their fears of a repetition proved unfounded. Agricultural policy was revolutionised after 1945, and a new system of state support for framing, which continues to this day, began. There were various reasons for this change in policy. In the late 1940s there was widespread fear of a world food shortage, as population growth was outrunning the rate of growth of food production. To produce more of Britain's food at home would benefit the balance of payments. Higher incomes for farmers would enable them to invest in and modernise their farmsteads and equipment. Higher incomes for all those involved in agriculture, including (it was hoped) farm labourers were in any case felt to be necessary on social grounds. The result of all these considerations was the Agriculture Act of 1947. This was the beginning of a programme of agricultural expansion, based on guaranteeing prices to farmers. An annual Price Review system was instituted to maintain the real value of the guaranteed prices. The wartime powers to compel farmers to farm by the rules of good husbandry were continued. Farming tenants were given added security of tenure. Agricultural workers were given a new Agricultural Wages Board, to determine a fair level of wages. The higher agricultural prices (and thus farmers' incomes) envisaged by the Act were designed to provide an extra £40 million a year as a capital injection, to improve agricultural infrastructure and working capital. By 1957, when the rising cost of agricultural support led to a new Agriculture Act, which trimmed price support by 2½%, the value of agricultural support policies was running at £288 million a year. This was a far cry from the 1930s.

The assistance given to farmers was of various types: subsidies, improvement grants, tax-free capital depreciation allowances, and assistance with borrowing from banks. Even before the 1947 Act, the Hill Farming Act of 1946 had provided £4 million in grants to farmers in upland areas – the first time that hill farmers had ever received any special consideration from the state. Finally, the farmer was given a body of excellent technical advice, on any farming subject, in the form of the National Agricultural Advisory Service (NAAS) in 1946, later renamed the Agricultural Development and Advisory Service (ADAS). Specific research into plant breeding was carried out by the National Institute of Agricultural Botany. For agricultural engineering, the earlier foundation of the Oxford Institute for Research in Agricultural Engineering (in 1924) was transmuted in 1946 into the National Institute of Agricultural Engineering. Education at degree level in agricultural engineering was now promoted by the establishment of the National College of Agricultural Engineering, which began work in 1962 at the Ford Mechanisation Centre at Boreham, near Chelmsford, Essex. In 1963 the college moved to Silsoe in Bedfordshire, adjacent to the site occupied by the NIAE.[2]

The stage was set for the most expansive period ever recorded in British agricultural history. Agricultural output, productivity and farm incomes all rose rapidly, as farmers invested in new technology. Dr Brassley has recently calculated that the volume of agricultural output in Britain rose in 1870–1935 by an average of only 0.01% a year, but this began to rise in 1936–45, to 0.5%, and in 1946–65 ran at the unprecedented rate of 2.8%, falling off to 1.4% a year between 1966 and 1986.[3] These rates of increase were even higher than those of the 'agricultural revolution' of the late eighteenth and early nineteenth centuries; during which time the years with the highest growth rate, 1800–30, show a growth rate of output of only 1.18% a year.[4]

This rise in the volume of output was accompanied by a rise in farm incomes. The actual value of gross farm output rose from £818 million in 1946–50 to £2,213 million in 1966–70. Taking inflation into account, it rose (at 1986 prices) from £9,526 million to £12,944 million. Also, in the period after 1945, a much higher proportion of that increased income went on machinery. In 1938–39, only 9.5 per cent of farmers' expenditure went on machinery. After 1945, the proportion varied between 16 and 20 per cent.[5]

Higher farm incomes were one essential factor in the post-war boom in the farm machinery industry. Two others were linked: the continuing exodus of farm workers, and the rise in the availability of power supplies.

The number of full-time farm workers in Britain continued to fall, from 739,000 in 1946 to 342,000 in 1966. The numbers of farm horses also continued to fall, from 519,000 in 1946 to 84,000 in 1958 (when their numbers ceased to be collected). In the place of horses and humans there came the tractor, the baler and the combine harvester.

The tractor boom

The stage was set for an unprecedented expansion of the industry, which was even more rapid than that of the early nineteenth century. Government policy, demand from farmers, and technical changes in farming all worked to the advantage of the agricultural machinery industry. Competition from imports was limited by import and exchange controls. Finally, the exodus of workers from agriculture continued even more rapidly after 1945 than in previously. Between 1945 and 1966 the number of regular whole-time workers in British farming fell, from 712,000 to 342,000, so that farmers had another incentive to replace men with machines.[6]

Even in the 1930s, there had been some pioneering new products and enterprises, such as those of Rotary Hoes, Catchpole, Salmon, Teagle, and of course Harry Ferguson, but the financial stringencies facing many farmers meant that these initiatives had been insufficiently developed or widespread. But by the late 1940s the financial constraint was removed. By this time there was a backlog of new ideas waiting to be exploited, and older designs had become obsolescent. The stage was set for a wave of innovation.

In a lecture given in November 1946, W. H. Cashmore, a senior expert on agricultural engineering, drew attention to the possibilities of further mechanisation in farming. The tractor was at the head of his list, since recent improvements, 'have been directed towards making the tractor more versatile so as to replace the horse more completely'. At this time, there were still 519,000 horses kept for farm work in Britain, compared with only 203,422 tractors. The developments which he had in mind were pneumatic tyres, low-powered and row-crop models, and half-tracks. In grain harvesting there was much potential for the use of the combine harvester and the grain drier, particularly on smaller farms. Beet harvesters on the US model should be improved so as to cope with the wetter British conditions. Potato harvesters could be improved, particularly as regarded the disposal of haulm and the separation of stones and clods from the crop. Ploughs could be developed in direct-mounted and reversible versions (reversible ploughs saved labour in marking and setting up ridges). Rotary tillage had a future, as yet largely untried.[7]

The greatest new opportunity was presented by the tractor. Even after its very rapid spread during the war, there was much potential still in the market. There were still some 250,000 farms in Britain, and only some

Combine harvester

The combine harvester dates back to the large fields of California in the 1890s; at that time it had to be pulled by as many as thirty horses. Its widespread diffusion had to await the coming of the petrol or diesel engine. It made its appearance in England in 1929, with Clayton's self-propelled combine. Until after 1945, most were trailed rather than self-propelled. The modern ones are all self-propelled, being essentially a large diesel tractor with a reaper and threshing machine built around it.

The cutting mechanism of the combine is similar to that of the reaper. The crop is cut with a reciprocating cutter bar. At each end of the bar is a large divider, whose function is to separate the crop being cut from that being left for the next round of cutting. Grain lifters can be fitted to the cutter bar to lift laid or tangled crops. The crop is cut while being held against the cutter bar by the large revolving reel at the front of the machine. The cut crop is conveyed up into the threshing mechanism by a series of revolving reels (auger, feeder and pusher), and into the concave, where it is threshed against the stationary beater bars. Thereafter the progress of the grain and straw is similar to that in the threshing machine, the straw being ejected from the back of the combine, either to lie in the row for later baling, or being baled directly by a trailed baler. The grain falls to the bottom of the combine, into an auger, which then raises it to the top of the combine, where it passes through the final cleaning cylinder, and then into the grain storage tank. This is emptied by an exterior auger in a long tube, which can be extended out sideways from the combine, and operated while it is still moving, ejecting the grain into a trailer being pulled alongside the combine by a tractor.

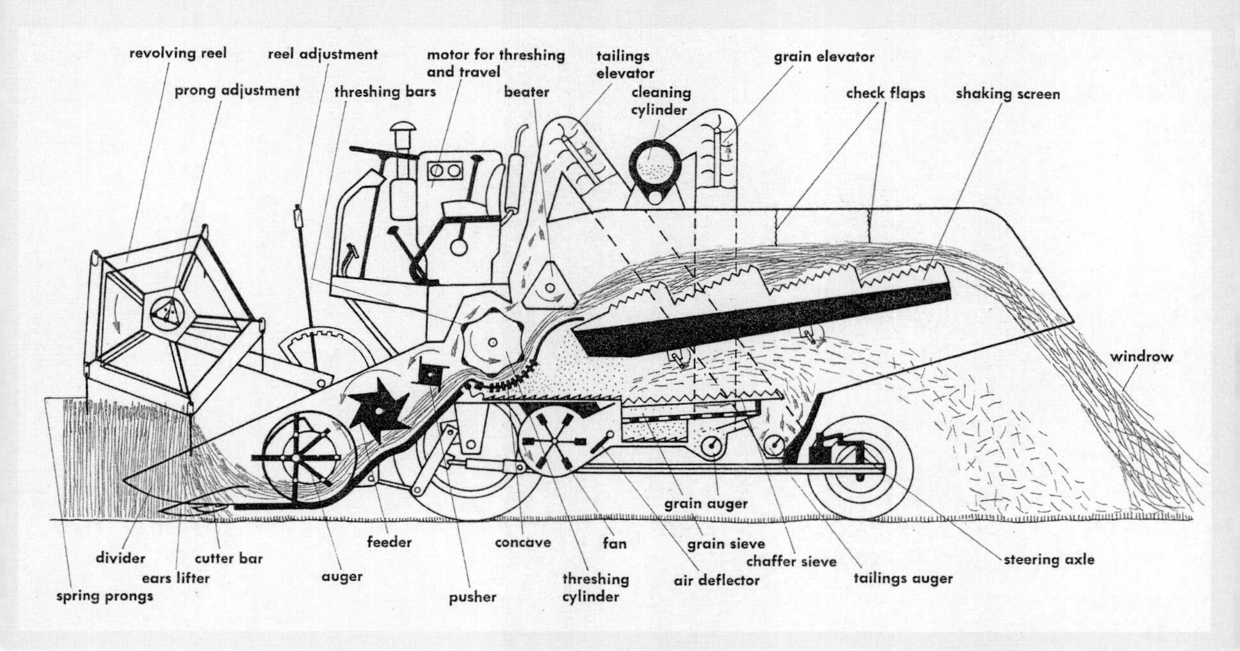

200,000 tractors. There was scope for several tractors on the large farm, and many small farms probably still did not have a tractor. The great majority (around 80 per cent) of the tractor stock was Fordsons, and they were technologically and/or physically aged. There was also scope, as Cashmore pointed out, for a greater variety of power sizes.

Even before the war finished, attention was being given by Ford to a replacement for the Model N. The Ministry of Agriculture was interested, and assisted Ford with the capital cost. The Ministry's specification was that it should be capable of pulling a three-furrow plough on any type of land; be more powerful; use less fuel; and dispense with the rear wheel worm drive. The result was the Fordson E27N Major. Although looking different from its predecessor, it had much in common with it. The engine was much the same, although more powerful (19/28 hp). The tractor came in four versions (with a diesel variant in 1948), and offered a hydraulic implement system as an option. The most important mechanical innovation was the replacement of the worm drive to the rear wheels with a conventional spiral bevel and differential. The Major was a simple, reliable and popular model, which continued in production from its inception in March 1945 until 1951, by which time over 200,000 had been built. The peak year was 1948, when 50,000 were made.[8]

Fordson Major tractor, at the Festival of Britain, 1951. (*Museum of English Rural Life*)

The Ferguson TE-20 tractor, ploughing the steep slopes of a Cornish farm, 1956. (*Museum of English Rural Life*)

Hard on the heels of Fordson came Harry Ferguson. His tractor, the Ford-Ferguson, constructed in the USA by Ford since and marketed there by his own organisation, had been highly successful. Its production, from June 1939 until the end of 1944 had been approximately 200,000. Only 10,000 of these had come to Britain, as part of the Lend-Lease agreement, and Ferguson had failed to persuade the Ford Motor Co. to manufacture the tractor in Britain. But he was determined to press ahead and produce his own tractor in the UK, which he anticipated would outsell the Fordson. Casting around for a manufacturer in Britain, his agent was put in touch with the Standard Motor Co., of Coventry. This had a factory at Coventry, which had been devoted to making aircraft parts and was now standing idle. Standard had been thinking of entering the agricultural market, having in mind some sort of motorised farm cart, and the idea of having a ready-made and successful product to sell appealed to Sir John Black, the managing director. Black and Ferguson first met at Claridges Hotel on 3 August 1945. On the 14th Sir John confirmed to Sir Stafford Cripps, the President of the Board of Trade, that Standard wanted to enter into an agreement to produce the Ferguson tractor, and suggested the use of the Banner Lane factory at Coventry for that purpose. Since the factory, one of the 'shadow' factories built at the start of the war, was owned by the government, official permission was required. On 20 August the agreement between Standards and Ferguson was signed; Standard was to manufacture the tractor for the next ten years, while marketing, design and research was to be handled by a new company named Harry Ferguson Ltd.

There were two more obstacles: the need to have a sufficient allocation of steel, supply of which was still controlled; and permission to purchase dollars. The former was done by Ferguson going personally to see Stafford Cripps, who gave permission for enough steel to produce 200 tractors a day. The latter was agreed with the Treasury in September 1945, permission being given for acquiring $0.5 million for machine tools, $5 million for importing engines made by the Continental Co. in the USA (until 1947, when Standard would produce their own), and $3 million for various implement parts. Standard spent some £3 million tooling up the Banner Lane factory before it was ready for production. The first tractor came off the line on 6 July 1946. Full production was not under way until early October. By the end of the year only 316 had been produced. But the factory was ready for mass-production. It employed 3,500 people, and used the latest American methods of flow-line production and assembly. In 1947, in spite of the difficulties of the cold winter and power cuts, over 20,000 were built, of which over a quarter were exported. The TE-20, or 'little grey Fergie' was on its way. When production of the TE-20* ceased in October 1956, it was with no. 517,651, which far surpassed the production of the Model N Fordson.[9]

Concurrent with the development of the British Ferguson line was the start of Harry Ferguson's own tractor factory in the USA. Towards the end of the war, Ford there had become disenchanted with the Ford-Ferguson arrangement, on which they were losing money. Harry Ferguson's US marketing company, however, was making good profits. Having considered various alternatives, including taking over the Ferguson marketing company, Ford decided to go it alone, and announced at the end of 1946 that they would produce their own tractor. This would be an improved version of the Ford-Ferguson, and marketing would be done by a new Ford-controlled company, Dearborn Motors. This left Ferguson with no tractors to sell. To make things worse, when the new Ford came out it clearly used Ferguson's patents. Ferguson's response was twofold. First, he moved quickly to set up his own Ferguson factory in the USA. Secondly, he sued Ford for the enormous sum of $251 million. Ferguson eventually won the case, although he was awarded just $9.25 million in damages. The impact of all this on the British industry was peripheral, save that Ferguson used the Banner Lane factory to supply his American customers until the time that his new factory in Detroit could produce enough to do so. This created a large, although brief, export trade, amounting to 17,000 Fergusons in 1948–49. In October 1948 the Detroit factory commenced production of the Ferguson TO-20† producing 12,859

* TE stood for 'Tractor England'.

† TO stood for 'Tractor Overseas'.

in the first full year. By the time production ceased in 1954, 125,959 had been built.[10]

By the late 1940s, Ford had been joined by Ferguson as the only other mass-producer of tractors in Britain. But there was one other significant producer. David Brown's first tractor, the VAK-1, had been introduced in 1939, and most of the 5,350 produced had done duty as airfield tractors. Released from this obligation, Brown had moved onto the next stage of his development of an agricultural tractor. This, the VAK-1A, came out in 1946. It had a hydraulic implement hitch, but could not copy the other Ferguson feature, the hydraulic depth control, and performed this function by means of a trailed depth control wheel. Some 3,500 were sold before its more advanced replacement, the Cropmaster, appeared in 1947. This was the firm's most successful model, with 59,800 being sold in its long production run to 1953. It proved very popular from the start. Output was reported to be 3,500 in 1947, and substantially more in 1948. In 1949 it became the first full diesel tractor produced in Britain. It introduced a number of items as standard which had been optional extras before then, such as swinging drawbar and electric lights. In this period the firm introduced a 2-speed PTO, 6-speed gearbox, coil ignition and direct injection diesel engine. The firm's other notable innovations included Livedrive

The David Brown Super Cropmaster tractor. (*Museum of English Rural Life*)

Two successful Marshall products: the Field Marshall wheeled tractor (*Museum of English Rural Life*); and the Track Marshall crawler tractor. (*Peter Dewey*)

(dual clutch, allowing live hydraulics and/or live PTO) in 1957, and the Selectamatic hydraulic system, which could be used at front or back, to power attachments such as loaders or diggers. The company built up its own dealer network, and moved into production of ploughs, ridgers and cultivators. By 1960 tractor output was up to about 10,000 a year, and to 30,000 by 1970.[11]

Marshalls had also decided to stay in the tractor market. Although relatively few Model M diesels had been produced, it had been well received by farmers, and in 1944 the company announced its successor, the 38 hp Field-Marshall, building a new flow-line production track for it at the Gainsborough factory. The publicity emphasised mainly its fuel economy, simplicity (only 66 moving parts) and long life. It used low-grade fuel, and could easily pull a three-bottomed plough on heavy land. One of its more exotic characteristics was that it was started with a special explosive cartridge, which was inserted into the engine casing and then given a sharp tap with a hammer.* The output figures for the Mark 1 Field-Marshall are not known, but the production of the Mark 2, which appeared in 1945, was about 7,000. In spite of much support from satisfied customers, and a fair degree of export success, the Field Marshall was an acquired taste, not helped by the price. In 1945 the Mark 1 cost £550, and the Mark 2 cost £840. These were much higher prices than those of the Ferguson or (even more so) the Fordson. The final straw was the introduction of electric starting, but the expense of this killed the sales. The last Field Marshall was built in 1957. It successor, the MP6, which became available in 1956, was a bold move, having an engine of outstanding power for the time (70 hp). Designed (and priced, at £1,450) to appeal to the export rather than the domestic market, it failed commercially, and only 197 were sold. Marshalls left the wheeled tractor market in 1961, and did not re-enter it until 1984. But the old Fowler factory in Leeds carried on making 'Challenger' crawlers, and Marshalls were making the VFA crawler in the early 1950s; this was developed into the Track Marshall in 1954 (although not released until 1956). The Track Marshall was successful, and variants were made into the 1980s.[12]

The resurgence of the British-based producers, new and old, was accompanied in the late 1940s by the decision of leading North American manufacturers to acquire a manufacturing presence in Britain. There were various reasons for this. Although a considerable number of US tractors had come to Britain in the war, the imposition of controls at the end of the war, and the resurgence of the British tractor industry, had almost halted US exports thereafter. Producers who had formerly exported from North America found they had to abandon the British and other non-dollar markets, or acquire their own production facilities in

* The firm was obliged to discourage the rumour that ordinary 12-bore shotgun cartridges would perform the task, as well as the type supplied by the company!

non-dollar areas. The expected devaluation of the pound sterling (which eventually happened in 1949) and the continuation of the British tariffs under the Imperial Preference system made Britain an attractive location for manufacturing.[13]

The first to do so was International Harvester. It had earlier acquired factories in Germany and France, and had produced tractors in Germany since 1937. In 1938 it purchased a site at Doncaster, and in 1940 was ready to produce tractors there, but the factory was requisitioned by the Ministry of Supply. In 1946 the factory was returned to the company, and implement production commenced. In 1949 tractor production began, with the first Farmall. Massey-Harris had begun building its MH 44 tractor and the self-propelled MH 722 combine harvester at its factory at Trafford Park, Manchester, in 1948, and subsequently at its new factory at Kilmarnock. This was largely an assembly process, the components being imported from Racine. Around 17,000 MH 44s were made, and a further 11,000 of a Perkins-engined variant before production ceased in 1958.[14]

Allis-Chalmers produced the Model B, designed for the small farm, some of which had been imported under Lend-Lease, at its factory at Totton, Southampton (at the rate of about 1,000 a year), from 1948, although it was even then considered old-fashioned. Production soon moved to Essendine, Lincolnshire, and the firm stopped production in Britain in 1970. The firm's effort has been described as having a 'very thin product line and a miniscule market share throughout'.[15] The final addition from overseas was Minneapolis-Moline, which, in conjunction with its British importers, Sale Tilney, formed a company in 1946 to manufacture in the UK. Premises were acquired in several counties (mainly Winnersh, Berkshire, and Alton, Hampshire), and the M-M UDS tractor was produced, and also an unsuccessful combine harvester, made for M-M by Harrison, McGregor Guest at Dowlais, south Wales. The UDS was expensive, costing £1,050–£1,200 at a time when the Ferguson retailed for £325, and production in 1948 was described as 'only a few hundred' in a report by PEP (Political and Economic Planning). In 1949 MM (England) went into receivership.[16]

There were also many attempts by smaller or non-specialist firms to produce tractors throughout the post-war decades. One of the most successful was Ransomes' small market garden crawler, the MG-2, which dated from 1936. Of the specialist producers, Roadless, with its four-wheel and crawler conversions, and County, with its four-wheel drive conversions (both companies concentrated on Ford conversions) were also successful, although the numbers involved were not large in comparison with the total output of Fords. Roadless, which had begun in business in 1919, had converted about 3,000 Fordsons to four-wheel drive by the end of the production run of the Fordson Super Major in 1964.[17] But many small producers were not very successful. The BMB President (1950–56), a small market-garden type; the Garner; the Newman; the Bristol (now

The Nuffield Universal M4 tractor. This was an export model, fitted with a petrol engine. (*Museum of English Rural Life*)

made in Yorkshire); the Bean, and in the 1950s the Turner Yeoman, a larger but expensive tractor, all had short runs. The Turner was in production for six years, but sold only 2,131.

Ford, Ferguson and David Brown apart, the only major domestic incursion into the tractor market was the Nuffield, from Morris Motors, which began in 1948. This, the Universal M4, was of 42 hp, and was regarded as a viable alternative to imported USA models. After the merger between Morris and Austin which created the British Motor Corporation, the Nuffield became a BMC. BMC also tried its hand at the small tractor market, with the BMC Mini Tractor, of 15 hp. Launched at the 1965 Smithfield Show, it did not sell well. The larger Nuffield, uprated to 60 hp was caught up in the merger that created the British Leyland Motor Corporation in 1967, and the Nuffield became a Leyland. In 1981 Leyland was sold to the Nickerson organisation. Over the life of the Nuffield/BMC/Leyland tractor, annual production averaged *c.*7,500 between 1948 and 1964. In 1964 an entirely new factory was completed, and output climbed to about 16,000 by 1969. By the end of the 1960s, the major tractor producers were all offering diesel engines, a range of types in the range 30–*c.*65 hp., and gearboxes with at least six forward and usually two reverse gears.[18]

There was also a market for the smaller, two-wheel tractors, controlled by a man steering the shafts, which were largely used by smallholders, and for market garden work. One such example was the British Anzani Iron Horse, which sold for £130 in 1949. Brockhouse Engineering made three types, the largest being the BMB Plow-Mate. The Howard Gem Rotavator began production in the 1940s, and an improved version was still being sold in the 1990s. There was also a type of tractor intermediate between the 2-wheel and medium 4-wheel. This was the small four-wheeler, with a small engine, usually of 1–3 hp, and often evocative names. In the 1940s and 1950s there were on the market such products as the Howard Bantam, the Wolseley Merrytiller, the Barford Atom, and the Coleby Farmer's Boy.[19]

By 1950 the industry had been transformed. The advent of the mass-produced tractor transformed farming, and with it the machinery industry. The buoyancy of the times created a whole new range of small and medium companies, and even the older ones benefited. Ransomes agreed to manufacture a range of implements in the late 1940s to fit the Fordson. The range of products designed to be used with the tractor expanded. Tractors themselves were being given more powerful engines; diesel engines were becoming more common from the early 1950s; live power take-offs and hydraulics were becoming standard. By the early 1960s, attention was being given especially to better gearbox systems, with higher road speeds attainable. By the 1960s cabs were being improved, and new regulations in 1970 led to the use of reinforced safety cabs, which could withstand a roll-over accident.

The main features of tractor development in these decades may be illustrated from the history of Ford tractors. In the later 1940s the firm's main product was the Fordson Major, the E27N, whose engine was a direct descendant of the Fordson N engine, producing 30 hp. It had three forward gears and one reverse, a conventional clutch and rear axle drive; there was the option of a diesel engine. In 1951 the New Major was launched; this was the first production tractor to have a diesel engine as standard, although petrol versions were available. It proved very popular, and was the tractor which really popularised the diesel engine on the British market. During the 1950s the range was expanded. A successful new model was the Dexta, a small, three-cylinder Perkins-engined diesel, designed for the small tractor market. It was in this decade that the County and Roadless four-wheel drive and tracked option became popular. In 1964 Ford tractor production moved from Dagenham to Basildon, and Ford standardised its worldwide tractor models, with the 'Thousand Series'. The first of these, the 2,000–5,000 models, had live power take-off and hydraulics, differential lock and eight forward speeds. The later 6000 series had 66 hp, while the model 8000 was the first Ford tractor to have an 100 hp engine. In 1971 the largest model was the Roadless 115 conversion, with a Ford 115 hp engine.[20]

Growth and structure of the industry

There was a a sharp rise in the output of the industry, in large measure because of the impact of the tractor, and of new products which included combine harvesters, beet harvesters, potato harvesters, hay and straw balers, fertiliser distributors, manure distributors, machines worked by the power take-off (e.g., saw benches), tractor-mounted winches, and purpose-built trailers. The rise in exports was also remarkable. In 1946 total production (of tractors and machinery) was worth £26.3 million, of which £6.1 million was exported. There followed an unprecedented boom. By 1967 the value of total production was £208.8 million of which £126.1 million was exported. This enormous success can be appreciated better in graph form:

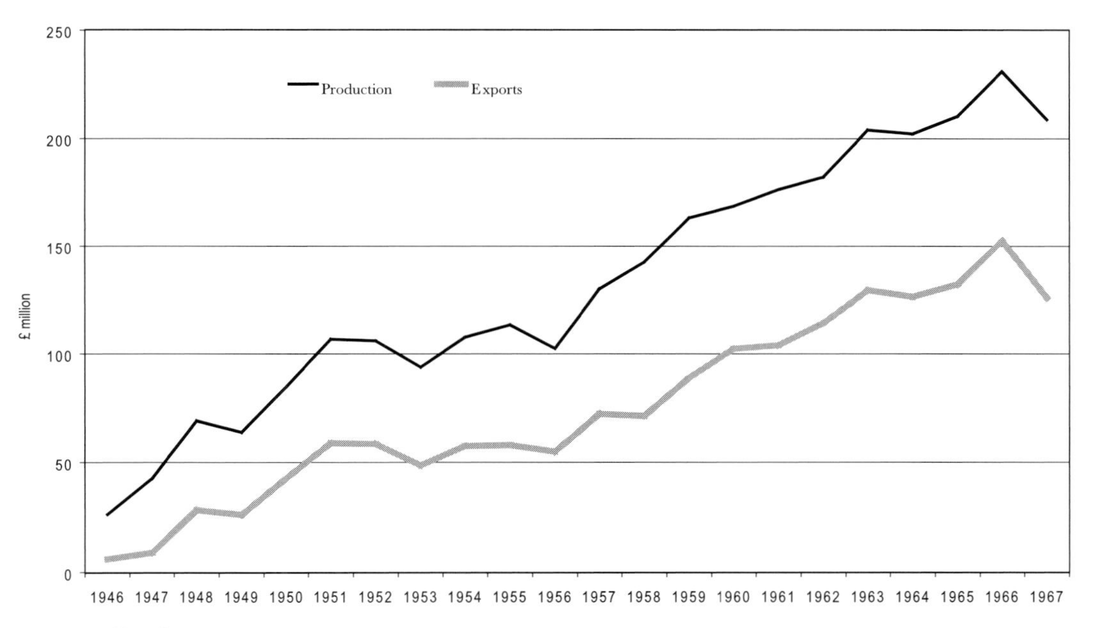

Graph 1. Value of production and exports, 1946–67. (Central Statistical Office, *Monthly Digest of Statistics*)

The tractor played the leading role in the rise of exports. Even in 1935, about half of the value of all machinery exports had been due to tractors. This had become true again by 1947, when tractors and parts accounted for £5.17 million out of the total value of agricultural machinery exports of £9 million. Whereas in 1913 about 70% of all agricultural machinery had been exported, the lean years after 1918 had seen the proportion fall considerably, to about 25% in 1935. By the 1960s the proportion had been almost restored to the pre-1913 level, representing 64% in 1963.[21]

By the early 1950s, the industry had expanded enormously and had become a leading exporter. Thereafter, growth was slower, and the figures owe something to inflation. but the size of the industry had been transformed. At the 1948 Census of Production, the value of agricultural tractor sales was put at £31,302,000, with the following numbers sold (to home and foreign markets):

Table 12.2 *Tractor sales, 1948*

	No.	*Value (£000)*
3- and 4-wheeled	101,860	28,698
Track laying	3,386	1,368
Market garden, wheeled	16,727	1,236
Totals	**121,973**	**31,302**

Source: 1948 *Census of Production*, 3/1/10, Table 8.

By then, tractor sales had overtaken the rest of the industry, accounting for 56% of the value of the total industry sales. The value of the sales of the rest of the industry was put at £23,830,000. This, the machinery section proper, employed 20,247 people, in 142 establishments. Those employed at the tractor manufacturers would have been additional to this number.[22]

By the end of the 1960s, the value of the industry's output had multiplied many times over, although some of this was due to inflation. In 1968 the industry produced 145,381 tractors, and 17,070 'skidded units'.* The total value of the sales of tractors, engines and parts came to £192,785,000. The value of other agricultural machinery produced came to £56,606,000. Thus the total value of the industry's output came to £249.4 million. By now, tractor production accounted for 78% of all sales.[23]

The fortunes of a leading non-tractor machinery manufacturer, Ransomes, Sims & Jefferies throw some light on how firms developed in this expansionary period. A comparison between output in 1948 and 1967 shows the following:

Table 12.3 *Ransomes, Sims & Jefferies: value of output, 1948 and 1967*

	1948		*1967*	
	£000	*%*	*£000*	*%*
Agricultural implements/farm machinery	2,205	65	5,145	55
Lawn mowers/grass machinery	538	16	2,426	26
Vehicles/engineering	627	19	1,832	19
Totals	**3,370**	**100**	**9,403**	**100**

Source: Rural History Centre, University of Reading, Ransomes Sims & Jefferies archive, TR 6 RAN (Add.), Annual net sales analyses 1946–1976.

The firm had increased the production of grass machinery (especially gang mowers for sports grounds and golf courses), but agricultural

* skidded units =engine + bodies, packed for shipment, minus the wheels.

implements still predominanted. The largest single group of products were harrows, worth £1,186,000, but the firm had had some success in the new field of combine harvesters (£573,734). It may be suggested that the buoyancy of the agricultural market, combined with the firm's creativity in the matter of the combine, sufficed to keep it going at a satisfactory, though not particularly dynamic rate; its growth rate was similar to that of the industry as a whole. Along the way it had entered into agreements with or acquired some other firms. The deal whereby Ransomes agreed to make cultivating implements for the Fordson Major tractor ran from 1945 to 1955, and boosted sales substantially. In 1951 Ransomes acquired the firm of D. Lorant of Watford, which had imported Claas low-density balers and resold them under the Lorant name; Ransomes rebranded them as Ransomes products. In 1968 Ransomes was to acquire Catchpole Engineering of Stanton (Suffolk) and Johnsons Enginering of March (Cambridgeshire). But its financial record was not particularly out of the ordinary. On average in the years 1958–67, which were buoyant for industry in general, the firm earned profits before tax of 12.9% of the value of the equity investment of the ordinary shareholders, and paid dividends usually of between 3 and 4 per cent.[24]

Expansion of the industry was not accompanied by great structural change, much of which had taken place before the war. Of the pre-war names, the only one to undergo a radical transmogrification was Fowlers, which the new owners, Rotary Hoes, sold on to Marshall & Co. in 1947. Marshalls trebled their issued capital to £1,500,000 to finance the purchase. The merger brought the manufacture of the larger (40 hp) diesel wheeled and tracked tractors under one organisation. One of its attractions was the acquisition of a modern foundry at Spotborough, Doncaster. In 1968 Marshalls was taken over by the Thos Ward group of companies, and in 1975 was sold on to British Leyland. In 1982 the firm was bought by Charles Nickerson. Harrison, McGregor took over the firm of H.T.A. Ltd at Dowlais, in 1945, being renamed Harrison, McGregor & Guest.

The largest single change was in 1953, when Harry Ferguson sold his interests to Massey-Harris, in exchange for $16 million of Massey-Harris shares. Although Ferguson took on the post of chairman of the board of the new company, he resigned in July 1954 to pursue other interests in the motor industry. The merger meant that Massey gained a ready-made British factory. Standard continued to make Ferguson tractors for M-H, until M-H bought Standard's tractor interests, including the Banner Lane factory, in 1959. M-H also acquired an excellent product, and the North American Ferguson operation. The new company, initially known as Massey-Harris-Ferguson was known as Massey-Ferguson after 1958, and rapidly added a range of cultivation equipment, balers and combine harvesters to the existing Ferguson tractor and implement products. The trend to diesel-engined tractors was consolidated when Massey-Ferguson bought Perkins Engines, of Peterborough, in 1959.[25]

Change was more commonly achieved through capital expansion. In 1945, for example, Ransomes increased its issued capital by £250,000, to £950,000, adding a further £250,000 in 1948, and debenture stock was increased from £250,000 to £800,000. The firm took advantage of the rise in demand to build a new foundry in 1950 on a new site at Nacton, outside Ipswich. The rather smaller firm of Bentalls had expanded considerably during the war, but the now much higher levels of personal taxation meant that it could no longer rely on an injection of family capital. The overheads were also higher. These considerations induced the firm to turn itself into a public company in 1946. With the extra capital, it moved back into feed preparation machinery, and developed grain drying and silo systems. It also expanded its production of combustion engine valves, and recaptured its former trade in coffee-plantation machinery. In 1961 the firm was bought by Acrow (Engineers) Ltd, whose main business was in adjustable steel props for use in construction work.[26]

There were also many new entrants to the industry, often pulled in by the winning of sub-contracts from the major manufacturers. Ferguson had sub-contracting arrangements with more than 20 firms which made the range of implements attached to the Ferguson tractor. David Brown, though, made their own implements, and Ford used Ransomes to make theirs. In 1955 David Brown took over Harrison McGregor & Guest, and used the premises of HMG to manufacture the David Brown range of tillage equipment, such as ploughs, cultivators and ridgers. To these lines were added other products such as front loaders, rotary tillers, trailed combine harvesters, manure spreaders and pick-up balers. At the time of the takeover, HMG employed 514 people, but its export trade was very small.[27]

Although the industry had grown rapidly since the end of the 1930s, its structure had not altered perceptibly. Most counts of firms give a total number of around 1,000 after the war. This would include repairers as well as manufacturers. In late 1947, 986 firms of farm machinery makers or repairers received allocations of steel from the Board of Trade. But most of these received small amounts (900 received less than 10 tons each, and only two received over 5,000 tons). The vast majority of firms remained small, and one consequence of this was that they tended to rely on: '... a small range of implements requiring little manufacturing skill'. In 1946 there were listed 101 makers of harrows, 85 of horse-drawn hoes, 90 of rollers, and 59 of horse-drawn ploughs. There were 22 makers of threshers, and 18 of mowers, although only 3 firms made binders, and 2 made combine harvesters. Even in 1960, the industry directory known as the 'Green Book' had a list of firms totalling 1,210, and in 1970, 1,109. The size of the average firm in the industry may be gauged from the fact that the employment in 94 of the principal establishments was 9,351, and in 1946 68 of the principal firms employed 26,474.[28]

A picture of the structure of the industry in the early 1970s is afforded by the 1973 Census of Production. This survey was limited to manufacturers, and the total number of 'establishments' was 531; there were 523 firms. They had levels of employment as:

Table 12.4 *Farm machinery industry establishments by size, 1973*

Size group Employees	*Establishments*	*Enterprises*	*Employment (no. of employees)*
1–10	313	309	1,463
11–19	104	100	1,513
20–49	56	56	1,729
50–99	20	20	1,521
100–199	20	20	2,854
200–299	3	3	804
300–399	6	6	2,029
400–799	3	3	1,633
750 and over	6	6	9,037
Totals	**531**	**523**	**22,583**

Source: Business Monitor, *Report on the Census of Production 1973: Agricultural Machinery (except tractors)*, PA 331 (1973), Table 4.

Thus the typical firm was still small, and well over half of all firms employed fewer than ten persons. It may be presumed (although there is only anecdotal evidence) that the small firm was synonymous with family ownership; however, some large firms also were still family-dominated, and some of the newcomers were to remain so; JCB up until the present time, and Dowdeswell until the 1980s. Even by the end of the post-war boom at the end of the 1960s, the verdict on the industry of Political and Economic Planning in 1949 still applied:

> In summary, the industry consists of something like 1,000 firms, the great majority of them small, scattered, of limited technical development, and operating with relatively little capital.*

To the total employment shown here should be added those for the tractor manufacturers. There were 29 establishments making tractors, run by the same number of firms. Total employment in them was 28,126, of whom 20,192 were operatives. But most of the tractor employees (26,741) were employed in just five establishments, run by four firms, each of which employed more than 500. Thus the total employment in the whole industry of machinery and tractors was 50,709, or about five times the pre-war level of the industry. However, it is notable that the vast majority of firms were very small, some three-fifths of all firms employing on average about

* Political and Economic Planning, *Agricultural Machinery: a report on the industry* (P.E.P., 1949, p. 11.

five people. It was usual at the production censuses to omit the smaller establishments (fewer than ten employees). However, the 1973 census made an adjustment for these smaller units, and also for establishments whose returns had been unsatisfactory. It is thus probably the best analysis of the structure of the industry in this period. The larger machinery companies (excluding tractor producers) certainly dominated the industry's output, as they had done at the end of the nineteenth century, but there was still scope for the small firm. In 1973 it was estimated that the top 6 machinery companies accounted for 40 per cent of total sales, and the top 16 some 60 per cent, but the proportion accounted for by companies with fewer than 20 employees was about 20 per cent.[29]

New opportunities, new products

In late 1947, the Ministry of Agriculture conducted an enquiry into the industry. The ostensible reason for this was the concern which had been expressed as to how far the industry was well placed to fulfil its plans for expansion, in home and export markets. It was more basically, it seems, concerned as to how to allocate the scarce raw materials equitably between the makers of farm machinery. In particular, there was a need to ration the still-scarce supply of steel. The enquiry can be read as a commentary on the efficiency of the industry, and also as a means of distributing resources in a politically acceptable way. It is also a useful view of the state of the industry and its constituent firms.

The person asked to conduct the enquiry was Sir Alan Gordon-Smith, an industrialist. He was seen by the Ministry as someone who could ask searching questions, while still having regard to diplomatic niceties. He was described as a 'Grand Old Man' who would be plain-spoken, while not ruffling the feathers of the manufacturers. His remit covered the whole industry, both tractor and other machinery sections. It was:

> to report on the adequacy of existing capacity in the Agricultural Machinery Manufacturing Industry to achieve the Industry's programmed expansion of Production for the home and export markets and on the problems involved in the supply of materials, etc., required; and to make recommendations.*

The report was issued in two versions. The first was openly published and circulated. The second version (the 'Appendices') was marked 'strictly confidential', and elaborated some of the veiled criticisms of the first version. In the course of the enquiry, Sir Alan visited 65 firms, which accounted for 82% of the materials allocated to the industry by the Ministry. His first version covered, firstly, high-powered tractors (over 40

---

* Gordon-Smith Enquiry. Report on the Agricultural Machinery Manufacturing Industry. April 1948. Section 1.1 Terms of Reference. PRO MAF 58/237.

hp). It investigated Wolseley (part of Morris Motors, making the Nuffield tractor), Marshall, Marshall-Fowler, Minneapolis-Moline and Massey-Harris. The report gave them all a favourable verdict, as being efficient and capable of meeting their targets. Next, the medium-powered tractors. These were those made by Ford, Standard (making the Ferguson tractor), David Brown, Allis-Chalmers and International Harvester. These were approved, but it was noted that the programmes of Allis-Chalmers and IH would require the import of engines and transmissions, and would also compete with the British firms for steel. Thus the recommendation was that these two firms should be directed into producing only the higher-powered tractors, since there was already ample capacity in the UK for the production of the medium-power types. The horticultural tractor makers (ten firms investigated) could probably not all fulfil their programmes, and the report hinted at an excess supply in the market. The makers of trailed implements (mainly ploughs and other cultivation machinery) had made much progress in modernising their works, but some smaller firms were still 'producing under the more old-fashioned conditions'. The agricultural machinery section (making the larger machines, such as combine harvesters, threshing machines, balers, manure spreaders etc.) was on the whole working 'on sound production methods'. Finally, the dairy appliance sector was considered to be able to meet its targets. Overall, the report recommended that the firms' requests for steel should be granted, with a reduction of only 4 per cent. Sir Alan projected output of £62 million, of which £25 million could be exported.*[30]

The rather generalised and anodyne nature of the first version of the Gordon-Smith report was abandoned in the second, 'strictly confidential' version. This was more critical in tone, although since it was also published and presumably circulated within the industry it still had to be circumspect. The tractor makers were praised, although the note on Marshalls: 'Progress is being made in the reorganisation of the Gainsborough factory on modern lines', was civil service code for 'badly run'. Much attention was paid to the sub-contractors working for Ferguson, of whom the best, such as Pressed Steel, which made the mowers, could fulfil their targets, although there was a shortage of high-carbon steels. David Brown's management was 'capable and progressive'. The management of the horticultural tractor firms was clearly mixed, although J. Allen of Oxford was found to be a 'highly efficient and well-managed firm'.

In the trailed implement section, there was some variation. Bamfords was described as an 'old family business run on sound lines by men of practical experience'. But Bedford Plough and Engineering (which had taken over some of the Howard plough business after 1932) was 'using antiquated methods and there appears to be a marked lack of modern manufacturing experience'. Bentalls had old premises and plant, but

* This was close to the 1949 outturn, as shown in Graph 1 on page 269.

was planning to modernise. Blackstone, Lister and Harrison McGregor were praised. Fisher Humphries had been reorganised and was now 'reasonably well run'. But Dening of Chard was 'a remarkable mixture of good and bad', having a good design department, but poor production planning and layout of work. W. N. Nicholson of Newark was doing its best, but was an 'old-fashioned family business housed in rambling buildings and using antiquated machinery, the bulk of which, however, is in fairly good repair. There is no real production planning.' Ransomes had difficulties over labour supply, but should be supported, since it was the largest producer of implements in the country. Taskers was 'reasonably well run', but was still at an early stage in its re-entry into farm machinery. Catchpole was hampered by working in old farm buildings, and the firm lacked experience of quantity production. Minneapolis-Moline, which was trying to produce combines as well as tractors, was hampered by the head of Sale-Tilney, who although a good salesman, was reluctant to plan production beyond his current order book, and he 'did not inspire confidence'. The firm's Essendine factory was highly unsatisfactory, although the one at Wokingham was better, and its management sound. The dairy section, consisting of Listers, Alfa-Laval, Simplex and Gascoignes (Reading) was efficient.[31]

However, there were also some unpublished notes, compiled as part of the Gordon-Smith enquiry, which throw further light on some of the leading firms. The expansion plans of Ford and David Brown were noted; to that end the latter had completely reorganised the Meltham factory. But Marshalls was given a mixed review. The firm had recently taken over Fowlers (mainly to acquire the Spotborough mechanised foundry), and planned to expand diesel wheeled and tracked tractor output. But the output of wheeled tractors had only just got up to 1,000 a year. The firm had to be supported, since it was the only domestic supplier of crawlers, which were urgently required. But, 'the management of this firm has never been very impressive and it is difficult to be certain of their plans which are liable to disconcerting changes from time to time.' Minneapolis-Moline, having been pressed by Sale Tilney to manufacture in the UK, had acquired factories which were 'scattered and relatively unco-ordinated'. The use of Harrison, McGregor & Guest's factory in south Wales had not helped coordination. The firm had been advised by the Production Efficiency Service of the Board of Trade, and to some extent had taken the advice. However, 'the whole organisation is rather weak'.

Some of the more worrying comments were made in the case of Ransomes. Although essential as the main supplier of tractor ploughs, especially the heavy sort, and of cultivating equipment, and thus deserving every support, 'the management is, however, old-fashioned and the firm is not organised or equipped to overcome the very serious difficulties at the present time in regard to supplies of materials and components'. During the war the Ministry of Production had conducted an enquiry into the firm, in order to see if tractor plough production could be increased. This had resulted in production of the lighter ploughs being put on a flow-line basis, but the heavy ploughs were still made in batches. The greatest individual problems were faced by Denings of Chard. The earlier published report ('Appendices', p. 18) had detailed the main ones – poor production planning and layout of work. The firm had definite potential and will to improve. But the notes add to this picture – over-expansion in the war, weak capital structure, poor reception of their products on the home market. So far they had kept going with the assistance of Edwards, a machine tool manufacturer, which had a large interest in Denings. But the management 'appears to be rather weak and over-burdened'.[32]

Thus the industry entered on the post-war world with a mixed record. That new firms entered after 1945 is not in doubt, although their number is unknown. The number of old firms failing is also not clear, although some examples can be found. To some extent, Harrison McGregor & Guest can be considered to have failed, since its product range and factory methods do not seem to have altered much until the takeover by David Brown in 1955. Dening of Chard failed in 1951, after a large order for the Argentine was cancelled, and the firm was taken over by Beyer-Peacock. The final closure came in 1965, due to problems with the Beyer group itself, allied to a sudden fall in machinery orders. Thomas (originally Mary) Wedlake, which dated from 1785, closed in 1947. But there was opportunity enough for those firms with less of a legacy from the past, or a new product to promote.[33]

The advent of new products also provided an opportunity for some old firms which had so far not developed on a large scale to grow. Such, for example, were Parmiter, Twose, and Bomford. Parmiter, of Tisbury, Wiltshire, dated back to 1885, but in 1947 still employed only nine people, making root lifters, horse hoes and horse rakes, with a turnover of £13,000. Expansion followed, beginning with a purchase of the neighbouring redundant gas works. In 1955 the firm took over the harrow trade of Bedford Ploughs (of Bedford). By 1967 the turnover was up to £200,000, and three designers were employed. In 1976/77 turnover was up to £3,552,000, and there were nearly 150 employees. Twose of Tiverton expanded rapidly after the war, until in 1980/81 sales reached £7.7 million. Bomford had a turnover of only £47,000 in 1947, but by 1973 this had risen to £1,789,921. In 1958 employment had been only 63, but by 1971 was up to 129. Before the war the firm had been involved in contract steam ploughing and cultivating, but after 1945 made its name with a range of ingenious mowing and hedge cutting machines.[34]

Taskers, although carrying on the development of the articulated road trailers, upon which the firm was now mainly dependent, went back into the agricultural machinery market. This began with an association with Rex Paterson, a farmer and inventor, who had evolved the 'buckrake' for picking up hay from the field, for the making of silage. Quick to realise the potential offered by the Ferguson hydraulic lift, Paterson used it to mount a buckrake at the rear of the tractor. Taskers made the product, and the Tasker-Paterson buckrake sold very well. The peak year was 1953, when 7,500 were made. The firm also produced the Fertispread, for accurate distribution of granular fertiliser. This was one of the earliest efforts to solve the problem of efficient and even distribution of artificial fertilisers, whose use was now about to rise enormously. With these bases of expansion, the firm grew to a peak of over 600 employees.[35]

The new products covered most farming operations. In cultivating, the old trailed ploughs which still were dominant in the late 1940s were now succeeded by the mounted ploughs of the Ferguson and other types, and the even newer reversible ploughs. Ransomes offered all three types. Other manufacturers included Fisher Humphries (a firm with a long but unstable history), David Brown, McCormick (from its new factory in Doncaster), and from Scotland, Pierce and Sellar. At the end of the 1960s Roger Dowdeswell, a Warwickshire farmer, invented a three-point hydraulic hitch for crawler tractors, and the Dowdeswell Co. was born. A further Dowdeswell innovation was the 'chisel' plough, designed to rip the ground rather than carve a furrow slice, and by the 1970s these were being made by other firms such as Bomfords, Bamford, International, Massey-Ferguson, Parmiter and Twose.[36]

Cultivation equipment showed the influence of the Ferguson revolution. Ferguson supplied a full range of mounted implements to go on its tractors – cultivators, harrows, tillers. As other firms developed their own three-point hitches, so most cultivation implements became mounted rather than trailed. The disc harrow became more popular, and the pioneering work done before the war by firms such as Rotary Hoes evolved to become the new power harrow, which made a lot of headway. In the early 1960s the reciprocating bar power harrow became popular, and the models were usually imported, from such firms as Amazone and Vicon. Powered cultivators were developed by the Howard Rotovator Co. of West Horndon (as Rotary Hoes became). Seed drills were still of the cup feed type pioneered by Smyth of Peasenhall. Smyth ceased business in 1967, although the Smyth drills continued to be made by Johnson's Engineering at March, Cambridgeshire. But the days of the cup feed drill were numbered; it relied on gravity, and could be inefficient on uneven ground. It was replaced by the force feed drill, capable of sowing seed and fertiliser together. In the 1960s some development of cultivator drills was made. These had a sowing mechanism, set behind a spring tine cultivator – the precursor of the direct drill, of which more was to be heard in the 1970s. Root drills showed a similar line of development, with PTO* and precision versions. Trailed mechanised dung spreaders took the place of men, carts and shovels. Fertiliser spreaders also came on the scene, usually

* PTO = power take-off point (via a shaft from the tractor's engine).

The mechanisation of an ancient and back-breaking task; the Massey-Ferguson trailed dung spreader, *c.*1958. (*Museum of English Rural Life*)

in the form of a large hopper on the back of a tractor, with a spinning distributor revolving at its foot. Weed control was done either by the traditional hoe, or the new type of tractor-mounted sprayer.

At the end of the Second World War harvesting was, in the main, still done by the binder, and threshing by the threshing machine. They were about to be replaced by the combine harvester. This, together with the tractor, was the most revolutionary of all the mechanical revolutions in British farming after the war. In one pass across the land, it cut, threshed, and packed the grain into sacks or inboard collection tanks, exhaling the cut straw onto the ground. The advent of the combine spelled the end of the Edwardian harvest scene of a binder cutting the crop, and men, women and children raking and gathering the cut sheaves into stooks, there to stay for some days until they had dried, when all would lend a hand to get the stooks on to wagons, and thence into the stackyard, to be gathered into ricks, for later thrashing in the winter. With the combine, there remained only the task of collecting and baling the cut rows of straw, and this was performed increasingly by the trailed baler drawn by a tractor; formerly, balers had been large, stationary, steam-driven affairs.

There had been few combines on the land before the war. When the first census of machinery was taken, in 1942, there were 940. By 1946 this had risen to 3,253. The next two decades were years of enormous expansion. The first machines were dependent on a tractor – usually trailed behind it, but sometimes designed to fit round the tractor in a 'wrap-around' fashion. In 1959 there were 19,470 tractor-drawn combines on farms. By then there had been a large growth in self-propelled machines, of which there were 28,700. In later years, these grew to dominate the market, and trailed combines almost disappeared. By 1965, the last year in which trailed ones were separately enumerated, their numbers had reduced to 12,090, and there were 45,860 self-propelled combines.[37]

The majority of the combines in use were made by US firms, albeit in Britain. The first Massey-Harris combines arrived in 1942, and by 1948 there were 1,500 of them at work. Allis-Chalmers also made a wrap-around model, the All-Crop 60, for more than ten years, at Essendine. Earlier, the same factory had been used for the short-lived Minneapolis-Moline G8, from 1947 to 1950. The International Harvester Co. had made its first combine in Britain at Doncaster in 1943, and the firm became one of the two biggest in the trade, along with Massey-Ferguson. Ransomes imported a Swedish Bolinder-Munktell trailed machine in 1953, and in 1954 began to make them under licence at Ipswich. In 1958 they produced their own design, and made 8- and 12-foot cut machines until ceasing combine production in 1976. The largest stillborn project was Harry Ferguson's; following the merger with Massey-Harris, his attempt to build a Ferguson combine took three years and cost over £1 million, yet only eight prototypes were the result, and it never went into production.[38] Massey-Ferguson later made combines, with a range of eight types from Kilmarnock by 1970. The number of smaller manufacturers and imported models made this a very competitive market. By 1965 British farmers

Massey-Ferguson model 400 combine harvester. Produced in the UK between 1963 and 1966. (*Museum of English Rural Life*)

could choose from 32 different home and imported combines made by 10 companies, at prices ranging from £805 for the Aktiv trailed, power-driven bagger to the 14-foot cut self-propelled Bamfords Landlord for £4,000.

The use of the combine to harvest grain turned attention to the question of drying and storing cereals. Unlike the bagged grains of the 1950s, by the 1960s grain was increasingly stored in tanks on the combine, It could then be transferred via an auger system into a large cylindrical silo in the barnyard, or first be fed through a grain drying machine, if it had been cut in wet conditions, or had a too-high moisture content when cut. Firms such as Bentalls developed their own products, such as its 'Inter Continental Vertiflow Dryer', along with a range of round and square silos, with systems of grain ventilation, conditioning and in-bin drying.[39]

Two further products deserve mention. The baler (for hay or straw) was not new, but was usually a large stationary machine, working from stacks in a barnyard or behind a threshing machine, packing the material into high-density bales. Manufacturers included Fisher Humphries, Dening, Opperman, Ransomes, Jones in Wales and Ross in Scotland. The change was now to smaller machines, making lower-density bales which could be trailed behind a tractor, often in conjunction with a metal-framed sledge to collect bales in groups. Thus the manual labour of carting hay from the field was ended. The McCormick B-45 from International Harvester was the first British-made medium-density pick-up baler. Production

A Massey-Harris trailed baler, *c.*1955. (*Museum of English Rural Life*)

began at Doncaster in 1950, and in the next decade more than 37,000 were made. International Harvester became prominent in the business, as was New Holland, a US company which began production in Britain in the mid-1950s, the first machines leaving the factory at Stroud in 1955. At least 13 companies were making pick-up balers by 1958. International and New Holland had been joined by David Brown's Albion baler (named in deference to the name of the mowers formerly produced by Harrison McGregor), and Massey-Harris. Bamfords entered the market in 1954, and was very successful. Other makers included Allis-Chalmers, Claas (imported), Welger (imported) and Jones. The latter, of Mold in north Wales, became a subsidiary of Allis-Chalmers in the 1960s. The Jones baler interests were sold again in 1971, to Bamfords.

Finally, much work was done to ease the burden of harvesting sugar beet, which in the post-war period covered more than 400,000 acres. Firms such as Catchpole, Standen, and John Salmon developed sophisticated machines. In 1946 only about 1% of the crop had been harvested by machine, and back-breaking labour was the lot of the diminishing number of workers who could be recruited for the job. But by 1965 95% of the crop was harvested mechanically. In 1968 Catchpole had been bought by Ransomes, which sold the Catchpole machines under its own brand names.[40]

These apart, there were many products which were not new, but were now produced on a larger scale, with more than a nod to mass production. Such, for example, were the mowers produced by Bamfords, an old firm now in the throes of rapid expansion. In 1958 it became a public company, with a capital of £1 million, and in 1959 it began to make a combine harvester. The company also diversified into the newly developed PTO mowers for hedge maintenance, which are now universal. In the late 1950s it became evident that the old factory would be inadequate, and a major new factory on modern lines was built, with 250,000ft^2 of space. It took over three years to build, and contained a semi-automated foundry and machine, fabrication and press shops. In 1966 the firm took on the agency for the Norwegian Kverneland ploughs, which proved very successful. In 1967 it became closely associated with one of its largest customers, Fredk H. Burgess, although it stopped sort of a merger. In the same year the company rejected an offer for its share capital from J. C. Bamford (Excavators) Ltd, run by a member of the family, John Bamford, which made the by-now famous JCB excavator. In view of the later history of JCB this may have been an error.[41]

Innovation was not confined to field work. In barn work, the most sweeping innovation was the spread of the milking machine. In 1942, when a census of machinery was first taken, there were 29,480 installations on farms in Britain. The movement peaked by 1961, when there were 119,435. Increasing shortage of skilled labour willing to work early in the morning, late at night, and at weekends, was one factor driving

Milking parlour, 1870s

The history of the development of the milking machine goes back to the late nineteenth century, and workable machines were developed before 1914, but take-up by farmers was slow. Early systems needed investment in improved buildings, water supply and dung removal, and had not solved the problems of infection of the cow udder. The use of mechanical milking rose during the Second World War to about half of all British herds, and by 1963–64 it was up to 97 per cent. Early systems used portable buckets for each cow; between the wars some farmers used the portable outdoor bail; by the 1960s, cows came to the machine, in a herring-bone or rotary 'parlour'. The systems improved in cleanliness over time. In the era of hand milking, one man could look after and milk about a dozen cows; the use of bucket milking machines raised this to about twenty. The outdoor bail enabled a man and boy to milk about 50–60 cows. The herringbone milking parlour allows a man to milk a herd of about 80 cows; not much more than with the outdoor bail, but in conditions of greater cleanliness and convenience for transporting the milk off the farm.

In a milking machine a small motor (usually electric) drives a vacuum pump, producing a suction which is transmitted by a pipeline to the milking units. This suction is continuous, but the teat-cups are double-walled, with metal bodies and rubber linings. A device called a pulsator alternately connects the space between the metal walls and the rubber linings, first with the atmosphere and then with the suction. When the space between the liners and the metal walls is connected to atmospheric pressure, the liner collapses, squeezing the teat. Reconnection to the suction releases the pressure on the teat, and it is in this phase that most of the milk is ejected. This intermittent squeeze and release is essential to machine milking. The massaging action of the liner on the teat helps the circulation of blood in the teat, which would be impeded by the application of a continuous vacuum. The massaging action also has some effect on the 'let down' of the milk from the udder.

There are various types of milking parlour in use at present. A common one is the herringbone parlour. This allows the cows to walk to their milking stalls, and after milking to walk out of the front of the parlour. The cows are previously prepared for milking (i.e., washed and made clean) and given their feed of concentrates. The operator stands in a pit some 30 inches below the level of the cows, and attaches the suction cups to each cow's teats. The milk is either collected in a covered bucket for each cow, or taken away via a pipeline to a bulk tank. After each milking all equipment has to be thoroughly sterilised.

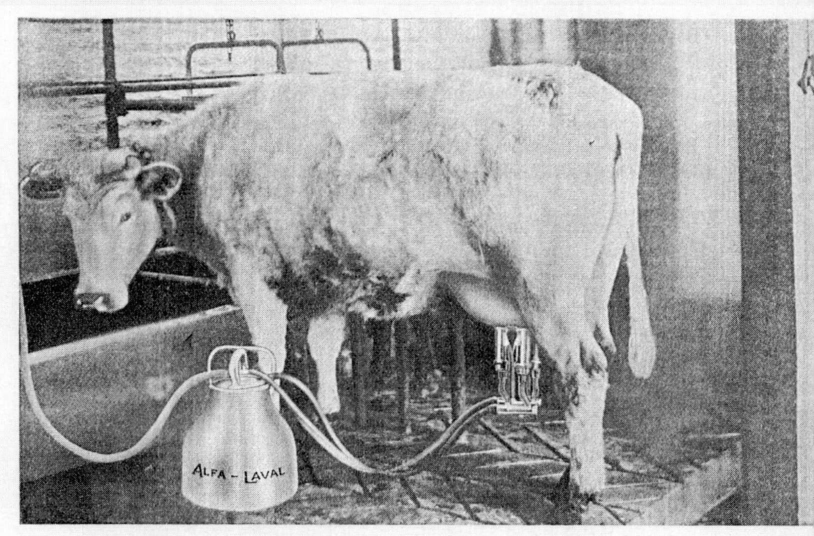

Milking Machine (Cat. No. 203).

By courtesy of Messrs. The Alfa-Laval Co., I

this process; another was the spread of electricity supply to farms. At the 1943 Farm Management Survey, it had been shown that a third of all farms did not have mains electricity. This had been rectified by the end of the 1950s. There were also improvements in the pulsing system, more accurately to replicate the action of a sucking calf, and greater attention was paid to hygiene. A significant development was the beginning of bulk milk collection in 1955. By 1972 there were some 22,700 bulk collection tanks on farms, accounting for 31% of milk producers, and 1,100 road tankers were in use for bulk collection. About half of the latter were owned by the Milk Marketing Board, and the rest by private haulage companies. The owners of larger herds also began to experiment with different installations. For some time the rotary platform was favoured, but in the 1960s the herring-bone milking parlour came into prominence.[42]

The adoption of such a wide range of new products was little short of a revolutionary process, which was in many respects even more dramatic and far-reaching than that of the 'agricultural revolution' and the period of intense mechanisation of the mid-nineteenth century. In 1945 horses did the ploughing; the corn harvest was brought into the farmyard on wagons, having been previously piled in individual stooks in the field; steam did the threshing; haystacks and the rickyard were essential features of the farmstead; the autumn saw large numbers of labourers lifting potatoes and beets. By 1970 the tractor and the combine harvester dominated the scene; few people worked directly on the land; the machines did the the great majority of work.

The export boom

The export performance of the industry was remarkable, and made a notable contribution to the UK balance of payments. In 1946, with output reckoned at £26.3 million, exports were only £6.1 million. But from 1948 onwards, about half of the national production was exported. In that year production was £69.2 million, with exports of £28.2 million. In 1967, production was £208.8 million, and exports £126.1 million. The beginnings of the export boom were based on the entirely altered world trading conditions brought about by the Second World War. The destruction of the war and the slow recovery of the defeated and occupied lands meant that Britain had a presence second only to the USA. In 1950 it was estimated that the world production of tractors was 989,000, the largest producers being the USA (498,000), the UK (150,000), and West Germany (58,000). In 1950, the value of world exports of agricultural machinery totalled £234 million. The USA supplied £125 million, the UK £45 million, Canada £29 million, and West Germany £15 million. World tractor exports were 211,436 units, of which 95,760 were from the USA, 86,881 from the UK, 12,536 from West Germany, and 8,346 from Canada. The UK thus accounted for the production of 15 per cent of the world's tractors, but sold 41 per cent of the world tractor exports. It accounted for 19 per cent of the world exports of agricultural machinery. Imports into the UK were not particularly significant, being some £3.2 million of machinery (just 389 tractors in all).

The UK export trade in 1950 was heavily oriented towards the Commonwealth, with European markets the next most important. Of the total export of machinery (including tractors) worth £45 million, the Commonwealth countries accounted for £23 million. The largest Commonwealth markets were Australia (£9.6 million), New Zealand (£3.5 million), the Union of South Africa (£3.0 million), India (£2.3 million), and Canada (£2.2 million). The largest non-Commonwealth markets were: South America (£3.0 million), Sweden (£2.7 million), Denmark (£2.0 million), France (2.0 million), and the Irish Republic (£1.3 million). French North Africa and the USA each accounted for £1.1 million of machinery.[43]

It is clear that the UK began the post-war decades with a flying start in the world export trade. Its favourable geographical position, largely intact manufacturing sector, and trading links with the Commonwealth led it to have a much larger presence in the world agricultural machinery trade than might be deduced from production figures.

In the ensuing decades, the major developments were that exports continued to grow, although not as fast as those of the other world producers, and there was a growing penetration of the home market by imports. An analysis of the position in 1968 was carried out by the European Commission:

Table 12.5 *UK agricultural machinery sales and trade, 1968 (£ million)*

	Tractors	*Equipment*	*Total*
Total sales	140	76	216
Exports	100	27	128
Imports	4	15	19
Trade balance (= exports *less* imports)	96	14	109
Value of UK market	44	64	108
percentage of sales exported	71%	36%	59%
percentage of home market supplied by the home industry	91%	77%	81%

Source: Heath, J. B. *et al.*, *A Study of the Evolution of Concentration in the Mechanical Engineering Sector for the United Kingdom* (Commission of the European Communities, Brussels, 1975), Table 9.1.

Thus by the late 1960s, while the tractor section was still very successful as an exporter, sending somewhat over two-thirds of its production overseas, the rest of the industry was much less engaged in exporting. There was a corresponding distinction to be drawn as regards import penetration. Whereas the tractor section supplied about nine-tenths of the home market with its machines, the rest of the industry only supplied about two-thirds of the home market demand, the rest being satisfied from imported goods. Overall, while the industry as a whole exported about three-fifths of its products, the home market relied upon imports for a fifth of its supplies.

The growing import penetration reflected the recovery and development of continental European manufacturers since the end of the war in 1945. But growing imports also reflected the structural and technical weaknesses of the industry, which had been apparent in the late 1940s. There were too many firms, and the average size was too small, to compete effectively with large-scale producers located in Europe. The result was that there were too many firms doing more business than could be financed comfortably, so that they were obliged to continue producing models which were obsolete, since the firms could not afford the necessary research and development expenditure. Thus many foreign firms had been able to gain entrance to the British market. Nor were the failures confined to the small producers. A notable element in the import bill was that of combine harvesters; the failure of Massey-Ferguson to produce a high-capacity machine had allowed Claas (West Germany) and Clayson (Belgium) to make a big inroad into the British market. In 1966 imports of machinery (excluding tractors) had been worth £12 million, but £5 million of this was accounted for by combines. Finally, British manufacturers were increasingly acting as import agents ('factors') for foreign firms. This was explained as the consequence of two lines of thought. Firstly, if a firm had ceased to produce a line, it imported a similar product in order to maintain customer goodwill. Secondly, since there was a tendency for British firms to follow the market rather than lead it, firms would try a new imported machine initially, to see how the market responded, before engaging in full-scale manufacture. In all, it added up to a gloomy prognosis for the future competitiveness of the industry.[44]

Conclusion

In the decades of the great European Boom, from 1945 to 1973, a range of new products, most notably the tractor and combine harvester, had revolutionised work on the farm. As the supplier of these products, the industry had risen to new heights of output, and there had been many examples of successful enterprise. Internationally, the British industry had begun with a flying start, and most of this lead was still intact by the early 1970s. But it owed much to the location of the international producers within Britain, and their high propensity to export. In 1973, the three biggest producers in the UK were the tractor makers – Massey-Ferguson, Ford, and International Harvester, with sales of £190 million, £100 million, and £60 million respectively. Of the rest, only six companies had sales worth £10 million and over – Howard (£40 million), New Holland (£25 million), David Brown (£20 million), British Leyland (£20 million), Ransomes (£15 million) and Alfa-Laval (£10 million). Two of these (David Brown and British Leyland) were also tractor makers. It was notable that two out the three leading multinational firms (M-F and IH) were full-line producers, selling the full range of farm equipment, and Ford was a specialist tractor producer. All three multinationals thus had considerable competitive power, which on the whole was not available to the British-based firms except for small specialist producers. On the non-tractor side, there was a clear rise in foreign competition on the British market. This rise was all the more noticeable, since it had been negligible in the late 1940s. But the conditions at that time had been wholly unusual, and could not be expected to last for very long. The resumption of competition had been delayed, but it was about to accelerate. The good times were drawing to a close.[45]

CHAPTER THIRTEEN

Coping with the competition, 1973–2000

The new economic environment: de-industrialisation

Various events at the beginning of the 1970s marked the end of the great post-war economic boom. We have already seen that productivity was growing more slowly as time went on; that there was greater penetration of western European markets by imports. In addition, domestic inflation in the industrial countries was growing, harming competitiveness; there was also greater industrial unrest. These trends were reinforced by the ending in 1971 of the fixed exchange rate system which had been operating since the end of the war, and the linked devaluation of the US dollar. Inflation was driven up further in the early 1970s by world shortages of wheat and oil. Finally, the inflationary tendencies of the 1970s were ratcheted up by the announcement by the Organisation of Petroleum Exporting Countries (OPEC) in September 1973 that the posted price of oil would be quadrupled. The result of these events was that world economic growth slowed down sharply, and inflation accelerated to unprecedented rates. One of the worst performers was the UK, where retail prices grew at a peak rate of 25 per cent in 1974–75.

The rest of the 1970s was thus a decade in which economic growth slowed down and almost stopped, but this was accompanied by high inflation. A new phenomenon – 'stagflation' – the combination of stagnant economic growth and high inflation – was witnessed for the first time. By the end of the decade, the twin pressures of world recession and a tight incomes policy by the UK government had brought inflation down to about 8 per cent, and economic growth had resumed, albeit at a slower rate than that of the post-war boom.

The 1980s began with great political change: the election of a Conservative government in 1979, headed as Prime Minister by Margaret Thatcher. Shortly after this, another sharp slowdown in the world economy began. The recession of 1979–81 hit British manufacturing particularly hard. Unemployment rose to levels unprecedented in the post-1945 world. Government policy towards unemployment had now changed. Before the late 1970s, the response to rising unemployment had been to ease financial and economic policy, to reduce interest rates and to increase public spending to stimulate growth and reduce the social costs of unemployment. In the inflationary post-1973 world, this was no longer contemplated, as to do so would re-ignite inflation. By now the fear of hyperinflation was well entrenched in the national psyche. Thus the manufacturing sector was left to cope alone with the recession. The effects of this were all the more serious for the UK, since its accession to the European Economic Community in 1973 was now exposing the weaknesses of British manufacturing to Continental competition. The balance of trade in manufactured goods deteriorated rapidly; firms closed; unemployment rose substantially. There occurred another new phenomenon – 'deindustrialisation' – a term coined to describe the decline of the manufacturing sector in the older industrialised countries. It should be emphasised that, although this process seems to have begun earlier in the UK than in other industrialised countries, and thus had gone farthest by the end of the 1980s, it was a general phenomenon, affecting the USA and UK in particular.

As the 1980s wore on, the non-manufacturing sector (especially financial services and property) developed something of a boom. But manufacturing did not share in this. For industry, the most that could be expected was to reduce the scale of its operations, cut costs, and hope to weather the new storm. However, this was made more difficult when the late 1990s 'boom' collapsed, with sharp rises in unemployment, interest rates, and bankruptcies. The recession of 1990–92 was almost as sharp as that of 1979–81. In both, national output fell by about 2 per cent. This was an average for the whole economy, but the fall in manufacturing was much greater than this. By the late 1990s, some recovery was discernible, and economic growth did resume, but the previous two decades had been ones of great difficulty for manufacturing.

As far as UK manufacturing was concerned, the relevant problems were: the slowdown in economic growth, which restricted market growth; a greater reliance on exports; the greater penetration of imports, especially from Europe. In short, the economy grew more slowly, was more dependent on foreign trade, and moved strongly away from manufacturing towards services. In 1951–73 the annual average rate of growth of real national income (GDP) had been 2.8%, which was historically the highest recorded. In 1973–79, this fell sharply, to 1.5%. In 1979–98 it recovered, but only to 2.1%. In foreign trade, the opening up of the economy to competition led to a large rise in both exports and import penetration; between 1970 and 1997 manufacturing exports rose as a proportion of domestic production from 16% to 30%, but the share of the domestic market for manufactured goods taken by imports rose from 14% to 33%. Manufacturing output as a proportion of all national output fell from 47% to 28% between 1950 and 1996.[1]

Changes in home demand

Changes in UK farming and its environment also affected demand for tractors and farm machinery. Even before the accession of the UK to the European Economic Community in January 1973, British farming was supported by the state to a large extent. This was chiefly through a system of deficiency payments, although there were also investment and production grants. In 1973, the system was altered to the price support mechanism of the European Economic Community. Administered as the Common Agricultural Policy (CAP) of the EEC, this relied on a policy of setting minimum prices for each of the major farm products. This was backed up by a policy of market intervention; if a surplus appeared likely, the authorities would buy up the surplus, in order to keep the market price up, releasing the surplus on to the market at a future date. Since the efficiency of most EEC farms was lower than that of most British farms, this resulted in support prices being set at levels which in most cases were higher than those which the British farmer had received via the deficiency payments system.

Partly for this reason, and partly as a consequence of the inflation of the time, farmers' incomes rose sharply in the rest of the 1970s. This was especially the case for cereal producers. However, this rise did not keep pace with general inflation, so that the real value of farm incomes was being eroded continuously. The rate of growth of current farm incomes slowed appreciably in the 1980s, as the accumulation of surplus farm products in intervention stores built up (the butter 'mountain' and the wine and milk 'lakes' came in for particular criticism). The result was that policy changed, in order to reduce surpluses and public spending on the CAP. The first major change was the introduction in 1984 of milk 'quotas', which limited the production of milk on individual farms. The next round of reforms (named the 'McSharry Round' after the current European agricultural commissioner) in 1992 cut most support prices, and introduced 'set-aside' of arable land, whereby farmers were paid a modest amount to take land out of cultivation for a certain period. The result of these reforms was that the high income days of the late 1960s were not repeated, and the real income of farming fell from the early 1970s more or less continuously, although it stabilised in the later 1980s, before rising rapidly in the early 1990s, to fall precipitously after 1995. Briefly, in 1995, farmers' total real income (at 1970 prices) had slightly exceeded the level of 1970. But thereafter, the effects of the BSE crisis, and then of the foot-and-mouth epidemic, reduced farm incomes very sharply. In 1990 the total income from UK farming had been £2,779 million; in 1995 it was up to £6,180 million, and in 2000 it had fallen to £1,540 million.[2]

The other main tendency which affected the machinery market was the structural change in farming. Over time, farm holdings were being amalgamated, and thus reduced in number. This was due to the desire to obtain economies of scale, which in turn could only be achieved fully by the use of larger and more powerful machines. Within individual holdings, such economies were being obtained by grubbing up hedgerows in order to create larger fields. This process went furthest in the cereal-growing districts of the eastern and east Midland counties of England. In 1951 there had been 451,804 agricultural holdings in Britain; in 2000 these had halved in number, to just 233,200.[3]

The industry's output

These factors operated on the overall tractor and machinery market. In 1970, the total value of the industry's sales had been £402 million. In 2000 it was £1,754 million.

However, these figures are at current prices, and so distorted by the high inflation of the period (especially the 1970s). When they are deflated by a suitable price index (based on 1907 prices), it is shown that the real value of sales in 2000 was lower than that in 1970:

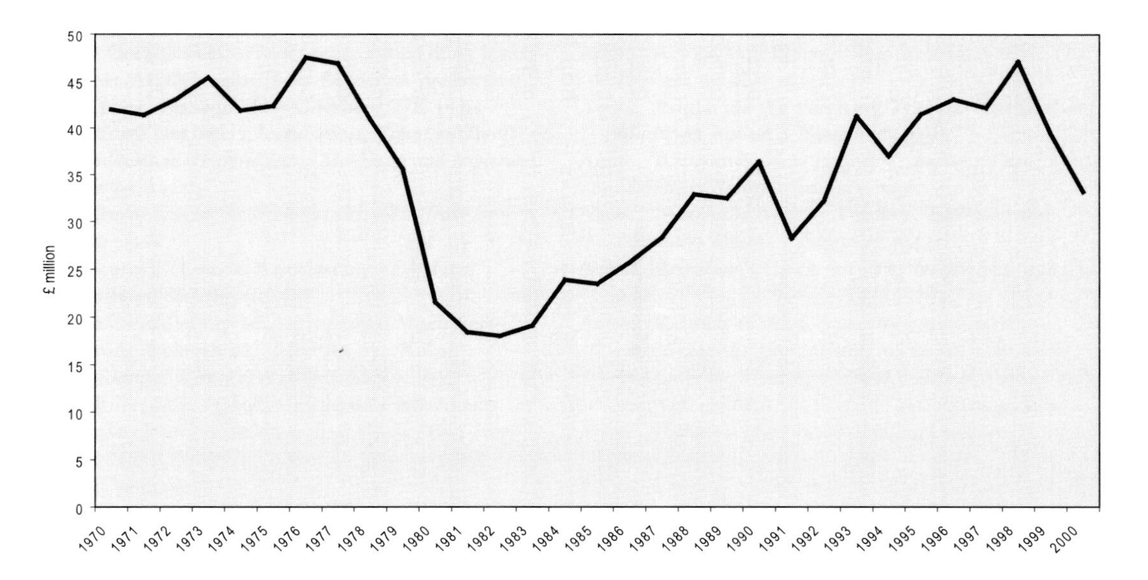

Graph 2. Real value of production (£ million), 1970–2000. (*Source*: see Appendix)

Thus the last thirty years of the twentieth century are revealed as peculiarly difficult for the industry. The recessions of the late 1970s and early 1980s cut a devastating swathe through the industry. From the peak in 1973 to the trough in 1982, total real output fell by about 60%. What remained was only a large rump. However, recovery did take place, and by the early 1990s real output was back fairly close to its 1970s level. The late 1990s saw a renewed decline. The relative importance of tractors, machinery and factored goods to the total industry sales (in current prices) is indicated below:

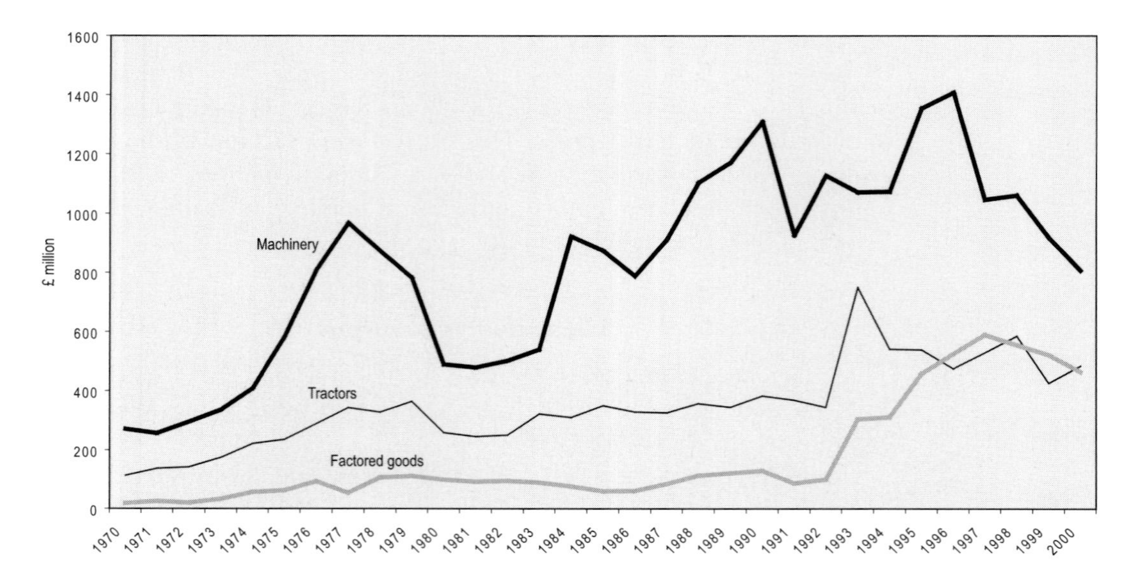

Graph 3. Value of production (£ million, at current prices), 1970–2000. (*Source*: see Appendix)

The peak of sales was in the early to mid-1970s. By the mid-1980s, the real value of sales had fallen by slightly over a half. The crisis in the industry led to many firms closing. Some famous names disappeared, including Taskers, Wallis & Steevens, Garrett and Fowler, although many of them had long since ceased to have much connection with farming. There was some recovery thereafter until the mid-1990s, but it was erratic. In the last few years of the century, sales fell anew, so that at the time of writing (November 2003), their real value was little more than that of the later 1960s.

The economic recessions of post-1973, especially that of 1979–81, had serious consequences for the economy as a whole, and agricultural engineering in particular. Growing import penetration and slowly growing exports had been noted for many years. In addition, the structural and technical weaknesses of the industry were apparent as early as 1969.[4] A decade or so later, these were still in existence. A report by the Royal Society in 1982 into the research and development work of the industry repeated the conclusions of previous observers – declining competitiveness, low investment in R&D (amounting to about 1.3% of turnover), many small companies with limited horizons (unlike the European industry, which had experienced a period of rationalisation), and a low proportion of qualified manpower in the labour force.[5]

A notable feature of this period was the rise in the sales of factored goods – goods not made by the manufacturer, but bought in from other manufacturers, either home or overseas. In 1970, such goods had accounted for only 4% of all sales. In 2000, they accounted for 26%. The type of arrangement commonly found is exemplified in the case of Massey-Ferguson, bought by the Canadian Varity Corporation in 1986 (and sold on to AGCO in 1994). In 1988 M-F formed a joint venture with the Dutch firm of Greenland NV, then the largest manufacturer of farm implements in the world. The new company, Massey-Ferguson Implements (UK) now dealt under the M-F name with products mostly supplied from the Greenland factories in Europe.[6] For the industry as a whole, machinery sales were almost unchanged as a proportion of all sales (28% at both dates), but tractors fell from 67% to 45%. In one sense, it could be said that the industry was filling the gap left by declining tractor sales by becoming agents for other manufacturers. Employment had fallen proportionately, from a total in 1979 of 40,800 in the machinery and tractor sections together, to 23,700 in 1983.[7]

The British industry had become more exposed to foreign competition. This led to a shrinkage in its place in the national economy. By the same process of internationalisation, other markets were being opened up to British producers. While these international developments had led to a shrinkage of the industry, compared with its size during the 'golden years' of 1945–73, they also presented opportunities. The spread of multinational corporations offered the benefits of rationalisation and an efficient product at a competitive price. The disadvantage was that decisions could be taken quickly which would have adverse consequences on the host economy in the short run. Such was the decision by AGCO, announced in June 2002, to close the Massey-Ferguson Coventry factory in 2003. The high sterling exchange rate and high production costs were cited as main reasons behind this decision, together with the trend to higher-powered tractors, which has reduced the number of tractors being sold. Production will be relocated to Brazil (for the smaller tractors) and Beauvais, France. It remains to be seen whether the acquisition of the Case-NH Doncaster factory by Landini in 2000 will lead to its closure. The logic of these moves is that the British agricultural machinery industry, which has come to be largely dependent for its output and exports on the tractor, will shrink further.[8]

The maturity of the market

By the 1970s, the great movement of mechanisation and modernisation which had dominated the history of British farming since the 1940s was slowing down. The wartime boom had raised the number of agricultural tractors to 203,000 in 1946, although there were still more horses than tractors on farms. The subsequent boom more than doubled the tractor stock, and by 1959, there were 505,000 tractors in use. For the remainder of the century, the tractor stock was to remain at about the half million mark; in 1996 it was 496,975. By the mid-1970s, horses had ceased to be used on farms, and the tractor was universal. On the larger farms, it was not uncommon to have two or three tractors. There was clearly a limit to how many tractors could be sold in the future. The market was becoming largely a replacement market, although the continual technical innovation, and continuous rise in power per machine, to cope with the larger farms and fields, would serve to underpin the market.

British farmers were buying fewer tractors, and the export market was shrinking. In 1948 total sales of tractors by British manufacturers had been 121,973. The peak of the post-war boom came in 1963, with sales of 241,933. Thereafter there was an almost unbroken decline, and by 1988 sales were down to 79,274. Most of these went for export, since in that year there were only 22,521 tractor registrations on the home market. But by 2000, total tractor production was down to 48,086, and the home market had shrunk to just 10,422 tractors, For the industry, the decline in the numbers of tractors sold was to some extent offset by the rise in the size of tractors. In 1986 the average horsepower of new tractors was 82 hp, and by 2000 this had risen to 120 hp.[9] The long-term history of tractor production may be shown graphically:

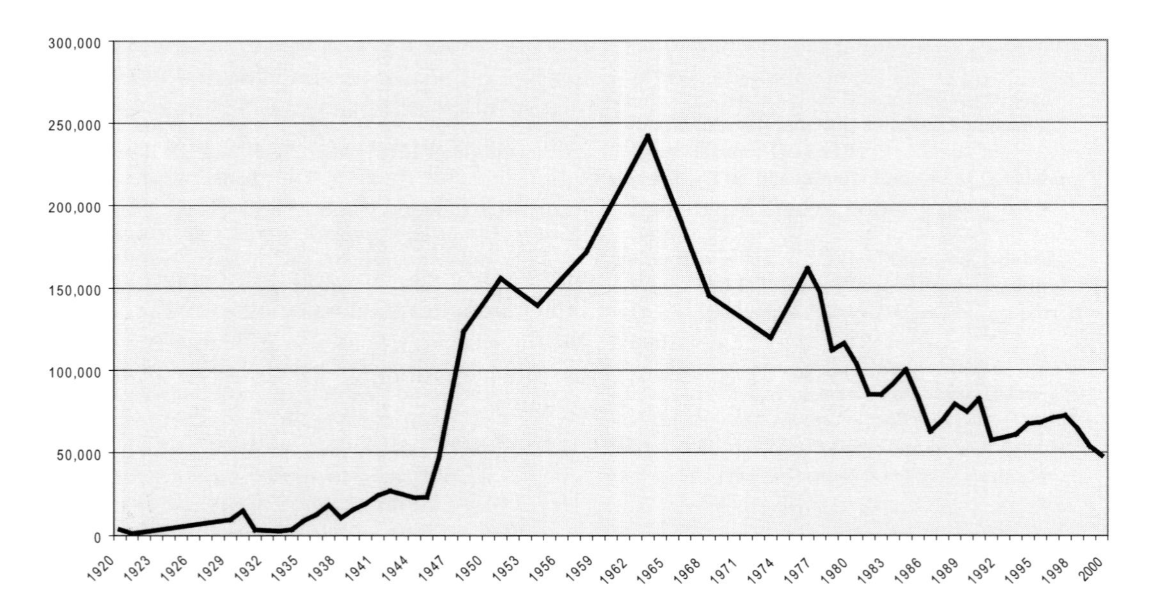

Graph 4. Annual UK tractor production, 1920–2000.

The other large item of machinery for many farmers was the combine harvester, and this was also becoming much more a replacement market. At first, combine sales held up better than tractors. In 1976 the annual registrations of combine harvesters was 2,622. Registrations rose to a peak of 3,273 in 1984, but fell back to 2,623 in 1986. Thereafter the home market shrunk rapidly, sales in the early 1990s being little more than 1,000 machines, although this did rise somewhat, to 1,300, in 1996. By then, the typical combine produced between 150 and 300 hp, and cost over £100,000, compared with *c.*£20,000 for a medium tractor.[10] The mechanisation of the cereal harvest, with its associated investment in grain dryers, storage silos and barns, could not be economically justified by the smaller farmer, and this encouraged the employment of contractors for combining and other specialist arable operations. This, and the greater size and cost of the machines, limited the sales of combine harvesters.[11]

In the case of both tractors and combine harvesters, the market was clearly maturing, with most sales by the end of the century being replacements. The industry was also constrained by the longevity of the product, rather as it had been in the case of steam engines and threshing machines before 1914. This was rather more pronounced for tractors, which were notoriously long-lived, even when apparently neglected. Attention to changing the engine oil and antifreeze sufficed to keep the aged machine running, even if the 'tinwork' (as the trade called the engine covers and mudguards) looked in bad condition. Combines deteriorated more quickly, since they had a more complex mechanism, and servicing could be easily omitted or skimped between harvests, but since they were essentially a more complicated version of a larger and more specialised form of tractor, the same remarks applied. In 1996, 25% of combine harvesters and 35% of tractors in use on British farms dated from 1982 or earlier[12]

New products and new versions of old products

Tractors

The main changes in tractor technology followed on from those of the 1960s. The most obvious development was the dominance of the diesel engine, which could supply the necessary power, and was now reliable, easy to maintain, cheap to run, and long-lasting. Along with this, engines got bigger. Sometimes turbocharging was used to get extra power; the first tractor on the British market with a turbocharger was the Ford 7000 of 1971. There was much attention paid to improved transmission systems to utilise this extra power. Synchromesh gears and clutchless powershifts became standard on many of the larger tractors; the day of fast gear changes on the move dawned. The market for small (less than 20 hp tractors) shrank. Power take-off speeds were raised to *c.*1,000 rpm, and the pulley belt disappeared. Larger machines had problems of adhesion, and soil compacting; to obviate these, wider and multiple tyres were used. Adhesion while using a rear implement was also enhanced by adding temporary weights to the front of the tractor. Four-wheel drive served similar ends, and now was produced in the manufacturers' factory rather than being added on by a specialist company (such as County or Roadless). Steering was assisted by hydraulic and hydrostatic linkages, and brakes were also now power-assisted. Much more attention was paid to driver comfort, with padded cabs to absorb noise, insulated roofs (some water-cooled), fully enclosed cabs, heating and air conditioning. A typical top-of-the-range tractor of the late 1980s included four-wheel drive, 24 forward and 24 reverse speeds, synchronised on-the-

By the 1990s tractors had become much bigger and more powerful. *Above*: Massey-Ferguson 3125 at the 1992 Royal Show. (*Museum of English Rural Life*)

Below: Ford 7840. A popular early 1990s model. By this time a glassed-in cab with roll-over protection was standard. (*Peter Dewey*) move gearbox changes, automatic powershaft de-clutching in overload situations, automatic differential lock and front wheel drive engagement, and automatic wheelslip control with a radar sensing system. The final change of the 1990s was to the appearance of the tractor. Manufacturers downswept the front end of the engine hood, so that vision of the driver was improved. This was the only major change to the look of a modern tractor since the Fordson of 1917.[13]

The enclosed cab improved working conditions for the farmer immeasurably. It also made it possible to use radios, for relaxation, and in the 1990s the spread of the mobile phone ended the isolation of the ploughman. Finally, the use of electronics developed rapidly. Already used in engine and gear control systems, and in controlling draft, in the 1990s they were used in conjunction with GPS (global positioning systems) to pinpoint the position of the tractor in the field. This technology could be used in combine harvesters to decide on how much seed and fertiliser to sow in particular parts of the field, the onboard computer drawing its own map of the yield density over the whole cropped surface. In the next year's sowing, this data would form the basis of the pattern of seed, fertiliser and weedkiller delivery.

Ford had introduced their 8000 and 9000 models in 1968, and these formed the basis of their tractors for the next two decades. This was part of a process whereby the major North American firms unified their tractor range on a worldwide basis. Massey-Ferguson and International Harvester followed. By the 1980s Ford was offering a full range of models, from a 13 hp compact model, through the mainstream models of 80–120 hp to the giant 325 hp tractor based on the North American Steiger, a firm which specialised in these large machines. In 1987 Ford also acquired Versatile of Winnipeg, Manitoba, which made similar machines. In 1986 Ford had bought New Holland (formerly the tractor section of Sperry), and from then on its tractors were badged as Ford New Holland. In 1991 Fiat acquired Ford New Holland, and merged it with Fiat Geotech to produce N. H. Geotech. In 1993 this company was renamed New Holland (again), and a new range of 24 models in three different ranges was announced in 1994. The largest ones were the articulated Versatile 82 series, with 12-speed transmissions. The Versatile 70 range has power output of between 170 and 240 hp.

The story was similar at the other major producers. In 1984 Case-IH was formed. Its product range for the late 1980s included the Model 685L, at 69 hp, with an 8 forward and 2 reverse gearbox, and hydrostatic steering, and the 96 hp Model 1594, with a four-range hydrostatic transmission. The Massey-Ferguson range of the 1980s was enhanced with models of 100–190 hp. By the late 1990s, there were only three major tractor producers in the UK – Case New Holland (in Basildon), Massey-Ferguson (Coventry), and Case-IH (Doncaster).[14]

The history of the smaller manufacturers was more varied. David

A typical 1990s tractor; the Case 7130. (*Museum of English Rural Life*)

Brown had been absorbed by Case (itself owned by Tenneco) in 1972. Marshalls carried on. The old Steam Plough Works in Leeds inherited from Fowlers was finally closed in 1974, and demolished the following year. In 1982 Charles Nickerson had purchased the wheeled tractor division of British Leyland, and with it the Marshall interests. Track Marshall became a holding company. Production of wheeled tractors was resumed at the Gainsborough works, and in 1984 the Field Marshall name was resurrected, being the company's first venture into wheeled tractors since the demise of the MP6 in 1961. The turbocharged Marshall 100 offered 116 hp and four-wheel drive. At the same time the Track Marshall crawlers sold well. But strong competition in wheeled tractors and what was in effect a dated Marshall product developed from the Nuffield range limited sales, and the firm went into receivership again in 1985. The company owed its bankers and unsecured creditors £12.5 million, and 170 of its 280-strong workforce were dismissed immediately. The wheeled tractor division was bought by Bentall Simplex, then renamed Marshall Daimler. Track Marshall was bought by Hubert Flatters, and sold after his early death in 1987 to the TWR group of companies (in 1990). The new owners kept Track Marshall production and development going, producing a revolutionary crawler in 1989, with rubber tracks which made it suitable for road work, but at £95,000 the TM 200 fell foul of the new farming recession of the early 1990s. The deterioration in orders resulted in the decision to make machines on a built-to-order basis, in a small part of the Gainsborough factory. The long Marshall saga was almost over.[15]

Roadless continued with its four-wheel adaptions of (mainly Ford) tractors, and diversified into forestry and construction variants, but competition from Ford and other major manufacturers now producing their own four-wheel drive models spelt the end. Peak production had been 300 sets in 1975, but only 10 in 1982. The company went into liquidation in 1983.[16]

A notable entry into the tractor market came from the JCB company, makers of tractor-based excavators. It set up a separate company, JCB Landpower, to develop its concept of a High Mobility Vehicle, better known as the JCB Fastrac, which was launched at the Smithfield Show in 1990. The turbocharged version developed 115 hp. The 1998 model had 54 forward and 18 reverse gears. Its novelty was that it was fast on the road, with the power to haul a 14-ton load at 45 mph. Unloaded, the Fastrac could travel at 75 mph. It also had the full range of large-tractor field functions, and proved popular, especially for farmers with scattered fields or holdings.[17]

The JCB Fastrac, shown here with spraying equipment. (*Museum of English Rural Life*)

Field machinery

Ploughs had to be stronger to cope with more powerful tractors and faster ploughing speeds. Opportunity was also taken to increase the number of ploughshares on the plough. These influences married with another innovation, the reversible plough. The concept of this was not new. The Kentish turnwrest plough of the early nineteenth century was reversible, and Ransomes and other makers had supplied them from time to time. But this time the concept produced a quite novel instrument. The pioneer of this was Roger Dowdeswell, a Warwickshire farmer, who began by developing a three-point linkage for his crawler tractor, and then produced a fully mounted (i.e. secured to the tractor only at its forward end, the rear end being suspended above the ground, without benefit of a rear land wheel) reversible plough for his new linkage. Production soon outgrew the farm workshop, and moved in 1975 to the Blue Lias works, at Stockton near Rugby. The resulting ploughs, of Dowdeswell and other makes, which now dominate the ploughing scene, might have anything up to 5–7 shares for a mounted plough, and up to 10 for a semi-mounted plough with landwheel.

There were also short-lived fashions, such as for chisel ploughs (a heavy-duty cultivator), stubble ploughs, shallow ploughs, and push-pull ploughs. The use of mounted ploughs led to the use of extra demountable weights at the front of the tractor, to balance it. In the 1990s, a promising innovation in mouldboards was the square mouldboard, which sometimes had plastic facings to extend its working life.

Other cultivation machinery was revolutionised by the universal adoption of the PTO (power take-off) and three-point linkage. By the 1970s powered cultivation machinery was in common use, the reciprocating bar and rotary power harrows being widely used on arable farms. In the 1980s the number of companies making power harrow/drill combinations increased rapidly. A novel innovation was the use of direct drilling, which became popular from the 1970s. In this, the land is not ploughed, but the seed is injected via a drill directly into the land, from which only the stubbles have been cleared. The necessary planting furrow is opened up by either coulters or harrow tines. The greater power of tractors made it feasible for the business of sowing and cultivating to be conducted at one pass: the direct drill opens up the land; the seed is inserted in the soil; the fertiliser is applied, and the power harrow or cultivator covers the sown seed. Thus several machines can be carried at once on the tractor. In root planting, the higher-power tractors led to the use of planting arms of great width, and the planting was now electronically controlled. Weedkiller sprayers also increased the length of their arms, which could be folded up for road travel.

The advances in technical development were notable in the case of combine harvesters. The developments noted above in tractors were all applied to the combine, which also grew in size. By the 1980s the 200 hp

A large New Holland combine harvester at work. (*Museum of English Rural Life*)

combine, delivering 18 tons of grain an hour, was common. Delivery was speeded up by the use of rotary flow systems. By the 1990s, combines were equipped with satellite mapping systems, by which the monitoring system created a yield map of the field, which could be used to help crop management the following year. The storage of cereals was developed further by the use of augers and storage silos, and the large-volume grain drier became an essential part of the arable farm.

The other major harvest field development was in baling. By the end of the 1960s, the trailed straw baler was commonly seen at harvest time. The new developments were in the speed of work, and the size and shape of bales. 'Big balers' were first introduced by the Howard Rotovator Co. in 1972. The resulting cylindrical bales contained 9–14 cwt of hay or 7–9 cwt of straw, and were at least four times heavier than bales had been previously. The new bales proved very popular, and in the 1980s they were joined by another type of baler, which produced bales of similar size, but cubic in shape. Along with the increase in size and weight of bales, handling systems also developed, usually to fit at the front end of a tractor, using its hydraulic power. The larger capacity of the balers, and the greater power of tractors, enabled baling to be carried out at much higher speeds than formerly. A trailed wrapping system was developed, which wound a wide band of black PVC around the hay. When cut green, this was a shortcut to silage-making in the bale.

Higher tractor powers also affected animal feed systems. Silage making was revolutionised by the use of high-speed forage harvesters, which projected the cut grass over the back of the tractor into a high-sided trailer. Once in the yard, handling was eased by the increasing use of loaders. These might be hydraulic tractor attachments, but they could be also purpose-designed loader/excavators, such as those supplied by JCB or Matbro. In addition, trailed feed mixers arrived; these mixed the roots, silage etc. as they were fed in, and the result turned out in the field for the animals to consume.

Root harvesting was part-mechanised by the 1970s; by the end of the century, it was being entirely performed by complex machines. Thus the back-breaking work of lifting and cleaning sugar beet or potatoes came to an end. Toward the end of the century, feed preparation vehicles, into which rations were loaded, came into use. They were used to distribute feed to the fields, for consumption by sheep or cattle on site.

The growing of vegetables and soft fruit was speeded by the increasing use of large irrigation devices. These might comprise a sort of walking overhead gantry, which could be towed across the field, or static hose assemblies, which were used to spray the surrounding area before being moved to a new location.

Finally, care and maintenance on the farm was much eased by the use of the tractor and its attachments. Ditching and drainage could be done by a tractor using either a forward scoop or a back-mounted digger shovel. Hedge cutting was now universally done with a reciprocating cutter bar mounted on telescopic or flexible arms at the rear of the tractor. Keeping down weeds and thistles in grassland was now done at high speed, usually with a chain or flail mower.

In the milk industry, the technical changes pioneered in the previous period came to fruition. Milking was entirely mechanised; the milking parlour was now usually herring-bone or similar; the milk was collected in a bulk tank, from which it was fed to the bulk road tanker when it called.[18]

Changes in the global machinery business

Although world economic growth slowed down sharply in the 1970s, and there were two sharp recessions, in 1974–76 and 1979–81, farmer incomes in northern Europe and America were insulated in the first half of the 1970s, as inflation pushed up the price of farm produce. In Britain, this was augmented by the effects of joining the European Community (1973), and there was something of a boom in farm incomes. The price of farm land rose substantially, provoking a scramble for it among institutional investors. But by the end of the decade, farm incomes in North America and Europe had been badly hit by oversupply and the sharp industrial recession of 1979–81. The result was that sales of farm machinery and tractors fell heavily in 1979–82, and soon even the biggest corporations were in serious financial difficulties.[19]

The first to suffer was International Harvester. Enormous financial losses between 1980 and 1983 made it impossible for the company to continue independently. In 1984 Tenneco, which already controlled Case, bought the IH agricultural products division, renaming the result Case-IH. The Doncaster plant, along with the former IH plant at St Dizier, France, continued to be the main production source for British and European tractor manufacture. In 1986 Case-IH acquired Steiger, the maker of giant four-wheel drive tractors in the USA. In 1999 Case-IH was acquired from Tenneco by Fiat, and the new owners immediately sought a merger with New Holland, which took effect in 2000. One of the conditions attached to the merger by the European Commission was that the new company, Case New Holland, should divest itself of the Doncaster plant. This was done in January 2000, with the announcement that the plant was being sold to Landini S.p.a. of Italy. Production of Case-IH tractors would continue, being effected by the new Landini management, which may at some later date make its own tractors at the plant. For the time, the Case-IH line continued, albeit IH tractors had re-assumed the McCormick name as a marketing tool, so that McCormick tractors were once again on the market.[20]

Allis-Chalmers also got into trouble in the early 1980s. In 1985 it was taken over by a West German company, Klockner-Humboldt-Deutz, which renamed the tractor division Deutz-Allis. Production of tractors at the old Allis-Chalmers base in West Allis, Wisconsin, stopped in 1985. In 1990 the company was sold to Allis Gleaner (AGCO) which renamed the tractors AGCO Allis. Since then, AGCO have made no fewer than 17 acquisitions, most notably of Massey-Ferguson in 1994, and of Fendt, the largest German tractor maker, in 1997. In just seven years, AGCO had gone from revenues of $200 million to $3.2 billion. In 2002 it was announced that the Coventry factory at Banner Lane, which had begun by making the Ferguson tractor in 1946, would close in 2003, production for Europe being centralised at AGCO's French factory at Beauvais.[21]

Ford held aloof from the merger process until 1986, when it acquired New Holland, becoming Ford New Holland. In 1987 it bought Versatile, the other North American maker of giant tractors. In 1991 Ford New Holland was acquired by Fiat Agri, and renamed N. H. Geotech. This changed its name back to New Holland in 1993. New Holland had the right to use the Ford name until 2000, by agreement with Fiat.[22] At the end of 2003, Fiat owned 91% of CNH shares. After the end of the millennium, the process of concentration has carried on, with the purchase by Claas of Renault Agriculture in the spring of 2003.

Thus by the end of the century, many of the historic names and factories of the industry had disappeared, as companies failed, were taken over or sold out of the business. By 2000 the only major tractor manufacturer to have survived, and to have retained its independence and original name, was John Deere.

The rise of foreign competition

In 1968 UK producers had still ruled their home market. Home tractor manufacturers supplied 91 per cent of the home market. Import penetration was greater for other farm machinery and equipment, but the home producers still accounted for 77 per cent of the combined home market for tractors and other machinery. Some 71% of tractor production and 36% of equipment production was exported.

This position was to change dramatically in the next three decades. As noted above, both the national export ratio and the import ratio rose rapidly from 1970 onwards. However, the rise of the import ratio was more rapid, implying a deterioration in the trade balance. As early as 1978, it was noted in a report on the non-tractor machinery industry that import penetration was well advanced. This was particularly evident in harvesting machinery. The decision of Ransomes to cease manufacturing a combine harvester in 1974 had left the UK as the only major country in the EEC not to make them, and the resulting penetration of the British market by Claas (Germany) involved large sums. Import penetration by 1976 was as follows:

Table 13.1 *Proportion of the British agricultural machinery market accounted for by imports, 1976*

Product sector	*percentage imported*
Grain harvesters	79
Balers	43
Forage harvesters	77
Root harvesters	65
Haymaking equipment	79
Cultivators and hoes	52
Mowers	49
Mouldboard ploughs	52
Fertiliser distributors	53
Drills and planters	26
Manure spreaders	15
Average all sectors	54

Source: Department of Industry, *The Agricultural Engineering Industry* (1978), Annex D.

This 1978 DoI report was at a loss to explain the high degree of import penetration. To some extent it had been assisted by British manufacturers selling imported machines as factored goods. Thus one of the two largest companies, Howard Machinery, derived 23 per cent of its turnover from factored imports. It was possible that the fragmented nature of the industry had impeded product rationalisation. There had been little evidence of mergers between companies making similar products, which would have given scope for such rationalisation. The exceptions were that recently the Wolseley Hughes group had acquired five companies (Vapormatic, Sparex, McConnel, Kidd, Parmiter), and Jones Balers had been taken over by Bamfords, which had allowed some rationalisation. But even then, these two latter companies still marketed independently. Perhaps the most useful contribution to this debate was in the section of the report which gave the views of a sample survey of machinery dealers. There were about 1,000 dealers, and their views and policies were very influential in forming farmers' decisions. The dealers considered that the availability of foreign machines was better, especially those factored by a UK manufacturer or reputable importer, and the foreign manufacturers were more innovatory, their machines being more in line with contemporary requirements. In general, the view was that imported machines were marketed more aggressively. There was some concern that the initial cost of imported machines, and of imported spares, was higher than for the native product, but this was offset to a large extent by better availability and performance.[23]

In the next two decades, import penetration proceeded apace. By 1996 the value of total tractor sales had risen to £2,880 million. Of this, exports were worth £1,136 million. Exports accounted for 39 per cent of output. The home market was worth £2,081. Imports supplied some 16 per cent of the home market. There was thus a large positive trade balance in favour of the home tractor producers. The position was somewhat different for the machinery manufacturers. It was a much smaller sector than tractors, recording sales by home manufacturers in 1996 of only £660 million. Exports accounted for 43 per cent of output. But import penetration had developed to the point at which imports now exceeded exports. Over half (57%) of the home market was supplied by imports. There was thus a negative trade balance in machinery. But taking tractors and machinery together, there was still a positive trade balance, of some £591 million.[24]

The trade position of the industry had altered enormously. The main reason for this was the rise of competition from European producers. This includes competition from the multinationals established in Europe. In 1995, 83% of the total £293 million of tractor imports, and 75% of the total £428 million of machinery imported came from the European Union. Most of the machinery imported was field machinery, the largest items being combine harvesters (£37 million, mainly from Germany and Denmark), ploughshares (£26 million) and balers (£20 million), and £27 million of parts. The major sources of tractors were Germany (£130 million), France (£76 million), the USA (£32 million) and Holland (£20 million).[25]

Readjustment and restructuring

It is difficult to discover the extent to which the size structure of the industry altered in this period, because the methodology employed at the (by then, annual) censuses of production changed, making direct comparison between the 1970s and the 1990s difficult. At the 1973 census 'establishments' (units of production) and 'enterprises' (controlling businesses) were noted. In 1993 the categories had changed to 'businesses' and 'enterprise groups'; the distinction between these latter categories was not clear, especially since identical numbers were returned in both categories. The results themselves give rise to concern: it is unlikely that the number of businesses or enterprises in the industry doubled in twenty years of difficult trading; and it is especially unlikely that the number of businesses of fewer than twenty persons had more than doubled in number. The uncertainties surrounding the data concerning the small businesses make further analysis of them unprofitable. What can be safely concluded is that the great majority of businesses/enterprises remained small, about four-fifths of them employing fewer than 20 people, and that there were only about 100 firms employing more than 20. The bulk of employment was in the few large establishments or businesses. In 1973 there were 18 enterprises employing 200 or more persons, comprising 60% of the total industry workforce, and in 1993 there were 14 businesses/enterprise groups employing 200 or more persons, amounting to 58% of the total industry workforce.

Table 13.2 *Industry establishments by size, 1973 and 1993*

1973				*1993*			
No. of employees	*Establishments*	*Enterprises*	*Total employment*	*No. of employees*	*Businesses*	*Enterprises*	*Total employment*
1–19	417	409	2,976	1–19	1,104	1,104	2,900
20–99	76	76	3,250	20–99	70	70	3,000
100–199	20	20	2,854	100–199	11	11	1,400
200–799	12	12	4,466	200–1,499	11	11	4,100
800+	6	6	9,037	1,500+	3	3	5,800
Totals	**531**	**523**	**22,583**	**Totals**	**1,199**	**1,196**	**17,200**

Source: Central Statistical Office/Office for National Statistics, *Censuses of Production*, 1973 PA 331 Table 4; 1993 P.A. 29.3 Table 4.

The censuses also show that the employment in the industry shrank, by about 24 per cent, a large fall over quite a short period. Some of this shrinkage was undoubtedly due to the increased mechanisation of factories, with computer-controlled machine tools and robotics playing a larger role. Some of it was due to the increased pressure of competition, both internal and external, and the sharp recessions of the early 1980s and 1990s.

Increasing competition was being felt from the later 1970s onwards. To some extent, the industry had had a good decade. The accession to the EEC, with its higher support prices, the high inflation of farm product prices, and the land price boom were all good for the industry. Cereal farmers especially were buying new machinery, attracted by high profits and generous depreciation allowances to set against income tax. Thus farmers were probably investing more heavily in new machinery than was warranted by the long-term outlook. But in 1979–80, farm product prices slumped, and the land boom came to an end. The latter had allowed some profit to be made from land sales, and some of this had been reinvested in new machinery. The resultant recession was weathered by farmers simply postponing further replacements, and making their machinery stock last longer. The result was that by 1980–81, the industry was in a severe sales and profit squeeze. In 1977 the industry had sold in the UK tractors worth £954 million and other machinery worth £236 million By 1979 this was down to £710 million and £189 million respectively. Exports of both categories together in the former year had totalled £621 million, and in 1979 they were down to £572 million. A business report covering 60 of the leading firms in the industry for the period 1979–81 and including tractor and machinery manufacturers, concluded that by 1979 nine of the 60 were making losses. The Agricultural Engineers Association reported a fall in employment amongst its 227 member firms of 5,842 in 1980, and a further fall of 1,856 in 1981, making a drop of roughly 14% of the total employment in the industry in 1980–81. Such a sharp fall had not been experienced since the recession of post-1919.[26]

This report also analysed the financial performance of the 60 companies concerned. In order to get a picture of the industry at the beginning of the 1980s, it may be useful to list them. The value of the leading firms' sales (the four major tractor manufacturers, and the ten largest machinery firms) in 1978/79 is given in Table 13.3.

There had been some recent changes in the industry at the time of the report. Massey-Ferguson had decided to close the Kilmarnock factory, and concentrate production at Marquette, in France. British Leyland had ceased tractor production at Bathgate, having sold the tractor operation to Marshalls of Gainsborough. Massey-Ferguson and International Harvester were 'working through refinancing programmes'. But the owner of David Brown, Tenneco, was disinclined to make a large capital injection.

Most firms were independent, but there was one large holding company, Wolseley Hughes, whose subsidiaries were Archie Kidd, Parmiter, McConnnel, Bruff and Wolseley Webb. The largest and best-known non-tractor manufacturers were Howard Machinery (formerly Howard Rotovator) and Ransomes. To these one might add Alfa-Laval, since its

Table 13.3 *Leading agricultural machinery companies, 1978/79*

	Sales (£ million)	*Main products*
Tractors		
Massey-Ferguson (UK)	313	Tractors
Ford Tractor	175	Tractors
International Harvester	149	Tractors and combines
David Brown Tractors	95	Tractors
Machinery		
Howard Machinery	78	Tillage machinery
Alfa Laval	35	Dairy equipment
Ransomes Sims & Jefferies	34	Grass cutting, tillage machinery
Twose of Tiverton	18	Tillage
County Commercial Cars	14	4-wheel-drive tractors
Fullwood and Bland	12	Milking equipment
Simplex of Cambridge	11	Storage
Craven Tasker	10	Trailers
Bentall	10	Driers
Weeks Associates	9	Grading equipment

Source: ICC Business Ratios, *Agricultural Equipment Manufacturers* (1980), Table 4.

sales were of a similar size to Ransomes. The rise of Twose of Tiverton to a leading place is notable.

The typical firm was small, and in 1978/79: 'As a general rule the smaller companies are motivated by one person endowed with either a good eye for implement design or a capacity for profitable marketing ...'[27] In the mid-1990s the typical firm was still small. There was even a number of small tractor manufacturers. In 1995, 30 were returned at the census of production, 5 of these having an annual turnover of less than £49,000. Fifteen more had turnovers of between £100,000 and £499,000. The remaining ten had turnovers in excess of £1 million. The turnover of most machinery companies remained very small. Of the 1,440 returned at the 1995 census of production, 445 had a turnover of less than £50,000, and a further 240 had a turnover of between £50,000 and £99,000. Thus 48 per cent of firms had turnover of less than £100,000. At the other end of the scale, 225 companies had a turnover of over £1 million, but they only amounted to 16 per cent of all firms.[28]

The fall in farm incomes and the general economic recession of the late 1970s and early 1980s had a serious effect on the profitability of firms. Even in 1978/79, the rate of profit of 51 of the largest companies had been only 4.0% of their sales (£50 million on sales of £1,246 million). In the following year this fell to 2.1%, and in 1980/81 a loss of 3.2% was recorded among these 51 firms (a loss of £41 million on sales of £1,252 million).

The greatest losses in 1980/81 were suffered by the tractor manufacturers. International Harvester of GB lost most, at £20 million, followed by David Brown, at £13 million, and Ford Tractor, at £9 million Massey-Ferguson (UK) was slightly in profit, making just £600,000. County lost £1.9 million. Some of the larger machinery manufacturers were in profit. Ransomes posted a profit of £2.3 million, 4.7% of its sales. Alfa-Laval made £1.8 million. Elswick-Hopper made £0.7 million. Vapormatic made £0.8 million. Hayters made ££0.7 million. Twose made £0.2 million. Howards may have been in profit, but the last accounts were in 1979, showing profit of £1.2 million. But there was a long list of loss-makers, the largest losers being Hestair (–£1.3 million), Lister (–£0.54 million), Wright Rain (–£0.4 million) and British Lely (–£0.6 million). The consequence of this squeeze on firms' financial positions was that some long-established names disappeared from the industry, as firms either failed, were taken over by other firms, or moved out of the farm machinery industry. By the later 1980s Ransomes had moved out of the sector altogether, and into grass care machinery. Wilders also left the sector. County Cars went into receivership in 1983, and although the name was revived, the successor firm ceased tractor manufacture in the early 1990s.[29]

But by the middle of the last decade of the twentieth century, the structure of the industry was remarkably unchanged over the previous half century. Although the tally of the total number of firms in the industry fluctuates alarmingly according to the particular census of production, this may be due to the definitions adopted by the census takers, rather than reflecting a real change. Shrinkage in employment and the failure of some famous company names had not been accompanied by structural change. In 1994–96, the list of leading companies showed some changes, but there were plenty of familiar names (see Table 13.4).

Newcomers in 1994–96 were JCB Landpower, which opened a subsidiary factory in Georgia (USA) in 2000, Fermec, Ifor Williams, Lowman, and Permastore. The only large firms present in 1978/79 not to be on the 1994–96 list were Howard and (though a smaller firm) Twose. Ransomes was not to be long on the list as an independent firm, being taken over by a US company, Textron, in 1998. Ruston's produces a range of balers, livestock handlers, drills and ploughs, under the Reco name, and the Mascho range of tilling and seeding equipment. Dowdeswell was not big enough to be listed in 1978/79, but had a medium firm-sized turnover in 1994–96. By this time it had also moved into the earthmoving machinery business, although still making the ploughs (up to 11 furrows) with which it began. It employed some 250 people in 1997. Bomford Turner was the product of the amalgamation of Bomford & Evershed and Turner International, of Alcester. The united company is medium-sized, at 177 employees in 1997, and makes grassland machinery and hedge cutting equipment. It was taken over by Alamo Group (TX) Inc. of the USA in 1993. Alfa-Laval Agri, located initially at Brentford, Middlesex, and

Table 13.4 *Leading UK agricultural machinery companies, 1994–96*

Company	*Year ending*	*Turnover (£ million)*
Tractors		
New Holland UK	December 1995	754
Massey-Ferguson	December 1995	460
Case United Kingdom	December 1995	318
Machinery		
Ransomes	September 1995	186
Fermec International	September 1995	95
Alfa Laval Agri	December 1995	22
Claas Holdings	September 1994	48
Fullwood and Bland	December 1994	48
Ifor Williams Trailers	March 1996	33
Lowman Mfg	December 1995	21
Permastore Holdings	April 1995	16
Dowdeswell Engineering	June 1995	15
Ruston's Engineering	October 1996	14
Bomford Turner	December 1995	12
Parmiter	July 1996	6
Total		**2,574**

Source: Keynote Market Report, *Agricultural Machinery* (1997), Table 25.

latterly at Cwmbran, Gwent, is owned by a Netherlands holding company. Ifor Williams Trailers produce a range of agricultural and leisure vehicles, including horseboxes and boat trailers.[30]

While there were some newcomers at the top end of the scale, there were also some further down, whose origins were fairly recent, or at least post-1945. Among these may be mentioned Allman (sprayers); Alvan Blanch (crop processing; from 1952); Downs of Glemsford (Suffolk), which had been in business since 1850, but did not begin its current specialism of potato grading machinery until 1965; R. J. Herbert Engineering (making potato graders, since 1972); Haith Tickhill of Doncaster (*c.*1950 – grading and packing systems); McHale (Northern Ireland; bale wrappers since 1989); Marston (Marston, Lincolnshire, started as blacksmiths in 1966; now manufactures a range of trailers, manure spreaders, etc.); McConnel (mowers; since 1935, but reformed in 1945); Moore Uni-Drill (Northern Ireland, since 1973); Parmiter (since 1885, but small until the 1940s); Pirie (sprayers); PortAgric (Mark Cross, East Sussex, from 1989); Ritchie (Implements) of Forfar (livestock equipment and latterly bale handling; founded 1870, but small until post-1945; Shelbourne Reynolds (1972 – green crop harvesting); Tong Engineering (potato handling; dates from 'the past fifty years'); and Harry West (manure spreaders, Whitchurch, Shropshire).

There are also some more historic names whose bearers are still successfully in the business of making farm machinery, although the original family may no longer have any connection with the firm. Amongst these is J. B. Edlington, of Gainsborough (1865); Bamfords (sold in 1981, and sold again, to Bensons of Knighton in 1988); George Brown of Leighton Buzzard; Grays of Fetterangus (dating from 1929, but reformed in 1972); Knight Farm Machinery of South Luffenham (Rutland) which makes fertiliser sprays and disc harrows, having begun in a blacksmiths shop in 1921, and in 2002 taking over the Parmiter range of disc harrows; Standen (of Ely), which has made root harvesters since 1936, and took over some Dowdeswell lines in 2000; Teagle (from 1944 – straw shredders and feeders, grass machinery); and Twose of Tiverton (hedge cutting, mowing). Parmiter (of Tisbury, Wilthire), owned by the Wolseley Group since 1978, was bought out by its own management in 1999, the company thus reverting to private ownership. In 2003, Parmiter sold off all its product lines, with the exception of chain harrows, to Shelbourne Reynolds.[31]

Finally, there is the roll call of once-prominent firms which have disappeared. With the notable exceptions of the large failures in the inter-war period noted in chapter 10, there have been few failures in the last two centuries. Nor has there been any particularly well-marked takeover movement, although there have been some large individual sales, most notably that of Harry Ferguson, David Brown, Bentall, and Catchpole (to Ransomes). Ransomes also absorbed Lorant in the mid-1950s. By the late 1970s, some more firms had became absorbed by larger industrial groups; Standen by Tremlett; Simplex by GEC; Stanhay by Hestair; Salopian Kenneth Hudson by Rubery Owen, and Gascoigne Gush and Dent by Thomas Tilling.

The largest sale was that of the largest manufacturer, Ransomes, in 1998. By the early 1970s, its sales of grass machinery had exceeded those of its farm machinery. In 1987 it sold its farm machinery interests to the Electrolux Group, and fell back on grass machinery alone, having re-entered the domestic mower market in 1985, with the acquisition of Mountfield of Maidenhead, and of Westwood of Plymouth. In 1988 and 1989 it acquired several overseas companies, mainly in North America. The North American acquisitions in 1989 were funded by heavy borrowing, and this proved fatal in the recession of the early 1990s. By August 1993 total borrowings were up to £85 million, the trading profit only just covered the interest payments on the debt, and the Ransomes shares were worth only 19p each. In the circumstances, it was hardly surprising that the company succumbed to the takeover by Textron in 1998.[32]

Although there have been some commercial failures in the post-1945 period (the cases of Marshalls and County have been noted above), it is more usually the case that firms ceased trading in the normal course of business, perhaps due to lack of a suitable member of the controlling family to carry on the business, or perhaps because the firm decided to quit the business of farm machinery in favour of something perceived to have more long-term potential. Thus Wedlake ceased trading in 1947, Knapp in 1966. Cooch, British Anzani and Fenton & Townsend of Sleaford are no more. Smyth of Peasenhall ceased trading in 1967. Wallis & Stevens left the agricultural market in 1970–73, selling their agricultural interests to John Wallis Titt, and ceasing to trade in 1981. Taskers, taken over by Craven Industries (a subsidiary of John Brown) in 1968, was sold again in 1983, the new owners closing the Tasker factory in 1984. Hestair was a conglomerate briefly involved in making agricultural machinery, from *c.*1974 to 1983. Wilders of Crowmarsh closed their foundry in the early 1980s, and closed their manufacturing business in the early 1990s, concentrating on being a dealer in machinery, being prominent in the trade as the dealers Lister Wilder, although by 1998 the Wilder family had sold their interests. Elders of Berwick ceased trading some time in the 1980s. Gibbs of Bedfont (Middlesex) had by the 1980s moved over entirely into the motor trade, in which they had been increasingly involved since the 1950s. Howard Machinery ceased trading in 1985, and its lines were taken over by Dowdeswell and Howard Farmhand.

Thus there is ample evidence of a rather natural progression of firms into and out of the industry, as family interests and ability wax and wane. The only clear examples of firms leaving the industry because they could not keep up the momentum of technical development are Smyth (a one product firm whose product had not altered much since 1914) and Dening of Chard, whose product range and production methods were by all accounts distinctly old-fashioned by the 1950s.[33]

The industry at the end of the twentieth century

On the eve of the First World War the industry was at a new peak. In the interwar period, it shrank substantially under the pressure of external events. After 1940 it grew very rapidly, first under the stimulus of war-induced demand, and then under the influence of the revolution in agricultural policy, and the technical backlog which had developed since the 1920s. In this fresh expansion, the tractor industry, mainly of North American ownership or influence, played a crucial role. At the peak of this fresh expansionary phase, around 1970, the industry was in size and export capacity many times larger than it had been in 1914. However, this was not to last. Under the renewed impact of severe recession and foreign competition, the industry's export performance began to wilt. More serious was the larger loss of market share to imports. As a consequence, some leading firms faltered, or were taken over. Also, a lot of the smaller historic names disappeared, partly through the sheer efflux of time and loss of family commitment; the greater part of the industry still consisted of family firms.

The fluctuations in the industry's fortunes, and the impact of technological change, have had surprisingly little effect on the size of the industry in the long term, relative to the engineering industry as a whole. The first occasion on which the industry's output was measured was at the Census of Production in 1907. In that year, the total output of the industry was £4.6 million, which may be compared with the total output of the mechanical engineering industries, valued at £93.2 million. Thus the industry's output was worth about 4.9% of that of the mechanical engineering sector as a whole.

By the later 1940s, the industry was changing its product structure, as the output of tractors rose rapidly. By the late 1950s, tractor output was worth more than the output of farm machinery. The industry had changed its nature, boosted by a new and very valuable trade, encroaching on the motor industry, and exhibiting something of the mass-production nature of that industry. At the peak of the post-war boom, at the end of the 1960s, the industry had become larger in size relative to mechanical engineering in general. By then it was dominated by tractor production, the value of which in 1970 was £209.3 million, that of agricultural machinery being £116.3 million. Taking both together, the value of the industry's output was 6.8% of the total value of the output of the mechanical engineering industries, then worth a total of £4,786 million.

The commercial crises of post-1970, and the rise of foreign competition, have pushed the industry back to a point where its relative size is similar to that of the early twentieth century. In 2000, the value of agricultural machinery output was £544.6 million, and that of tractors £918.8 million, making a total of £1,463.4 million. This represented about 4.6% of the total value of the output of the machinery and equipment industries, which was worth £31,800 million.[34]

The current danger is the over-reliance on the multinational tractor producers, whose decisions can be sudden, and have severe consequences. The closure in 2003 of the Massey-Ferguson Banner Lane factory at Coventry, and the uncertainty surrounding the future of Case/Landini at Doncaster are the obvious examples.[35]

However, the machinery sector is very different from the tractor sector. It is remarkable that the number of firms in the industry since estimates began has always been somewhere between 500 and 1,500, and probably close to the middle of this range. In spite of wars, revolutions, periods of trade restriction and free trade, enormous technical changes, and periods of boom and bust, the structure of the industry has remained almost unchanged. The great majority of firms are small, privately owned and family-run. Some grow rapidly, but most do not. Some explicitly state that they do not wish to grow, wishing to stay in touch with farmers and their own employees. There is a constant inflow and outflow of new and old firms, few of which fail. They serve the regional and national market, and some serve international markets, but always start with a local base of support. There is also a constant flow of new ideas, from the new firms themselves, and from farmers and entrepreneurs. It seems a fair prediction that there will always be a place for a new firm with a new idea, and that there will always be a home market to give it a good start. Whether the industry as a whole will continue to shrink as it has done in the last quarter of the twentieth century remains to be seen.

CHAPTER FOURTEEN

Retrospect

Long-term factors in the development of the industry: technical change, demand and entrepreneurship

The long-term economic and social factors underlying the growth of an industry or of a national economy are many and various, but they may be summed up under the three general headings of technical change, demand and entrepreneurship. As far as technical change is concerned, the roots of the British agricultural machinery industry may be traced back to a long time before the industrial revolution. There had always been *savants* who were interested in better ways to produce food, and 'ingenious mechanicks' willing to bend their minds to assist this. The line of agronomists runs from Pliny via Thomas Tusser to Jethro Tull, Alderman Mechi and Harry Ferguson. The outpourings of such men would, however, have been useless unless they fell on favourable soil. As far as the British mechanical engineering industry was concerned, that soil was the industrial revolution. The ingenuity of the early inventors occurred at a time and in a context that was favourable to the embodiment of good ideas in mechanical form. The working or iron and wood had progressed by the late eighteenth century to the point at which it was feasible to make a workable threshing machine, even if small, static, and powered by hand, water or horse. Likewise the invention of the chilled ploughshare by Robert Ransome was dependent upon a sufficient supply of good-quality and reasonably cheap cast iron. Both these conditions had been supplied by the revolutions in iron-working recently achieved by, among others, the Coalbrookdale company.

The stage was set for the heady days of the first half of the nineteenth century, which was the coming-of-age of the British agricultural engineering industry. By the time of the Great Exhibition of 1851, the international superiority of British technology in general, and agricultural engineering in particular, was not in doubt. To be sure, there were signs of the coming incursion from the USA in the shape of the mechanical reaper, but that was a small cloud on the horizon. The lack of competition from continental Europe may be generally explained by the slower process of industrialisation there, except patchily, and by the preponderance of small-scale, owner-occupied farms, with low or non-existent cash incomes.

In the later nineteenth century, technical advances spread widely. The British government abandoned its efforts to keep it a British secret. The prohibition on the export of machinery was repealed in 1849. But even had this not been the case, the new technical knowledge would have spread anyway, pirating and illegal copying being rife (as it still is). It was therefore not surprising that British farm machinery exports to Europe were subject to much competition by 1914, and manufacturers turned to the exploitation of less developed markets, in Russia and South America. To some extent, there was also a loss of the home market, to North American harvesting machinery. But in the field of steam and allied technology, British was best. Unfortunately for the more conservative British manufacturers, technical change did not halt with the compound steam engine, and the years between the late 1890s and 1914 are a time of lost opportunity. By 1914 Henry Ford had revolutionised the transport industry, and was going to do the same three years later for the tractor industry. The *coup de grace* for the steam ploughing engine took a long time to come, but in 1914 the glory days of steam were numbered. Since the industry had relied so much on steam technology, especially in the biggest export market, Russia, this was serious. It should perhaps be added that the continental European producers were in a similar predicament to Britain, and were equally badly placed to cope with the era of Henry Ford.

Technical change in the inter-war period presents a mixed picture. On the one hand, the older large firms were struggling to survive without changing their technology. The new firms were either finding niche markets (e.g. sugar-beet and potato cultivation) or were themselves already established, and with the capacity to manufacture the tractor. Thus Ford at Dagenham (1932) and the Ferguson-Brown tractor (1936) began a new agricultural revolution, whose full effects were not felt in Britain and Europe until the 1960s. The most recent changes have been built around the technology of the computer and satellite navigation, and are still being worked through. Taking the long view backward, the technological developments since 1800 have been staggering. A leading role in this had been played by British farmers, engineers and entrepreneurs. But technology does not exist on an island, and by 2000 it was clear that it had been thoroughly internationalised.

The second explanatory variable in the story of agricultural engineering has been demand. Without the demand from a market, the best ideas will wither. While the interaction between technical change and demand growth is a two-way process, there is no doubt that the market for the most advanced agricultural machines was growing rapidly from the end of the eighteenth century. The high farming profits of the Napoleonic war period provided a short-term boost. More important in the longer term was the enormous rise in the rate of population growth, the British population quadrupling between the first census of population in 1801 and the last pre-war census in 1911. British farming thus provided the necessary growing market for the machinery manufacturers, as farmers and landowners modernised their methods and equipment. When this process was over, by about 1880, there was the rising demand from the less developed economies in eastern Europe, Australasia and South America to occupy the manufacturers.

The hiatus in the growth of demand came in the period between the world wars. For the industry, the war and its consequences were disastrous, effectively undermining the main basis of pre-war profitability. The Bolshevik Revolution and its consequences killed off the Russian export trade, and rising protectionism did the rest. The industry's leaders must have mused bitterly on the dangers, unperceived before 1914, of being so dependent on exports to a limited number of markets. This skewedness of pre-war markets explains why the industry's output fell more than the generality of British manufacturing firms from the 'older industries' after 1918. By the mid-1930s, it was a much smaller operation than it had been, and many famous names from the prosperous pre-war days were no more.

The revival and subsequent growth of the industry after 1939 is a story of rapid demand growth. Initially, this was spurred by wartime exigencies. After the war it was favoured by the national policy towards farming and towards exports, and benefited from international policy towards the reconstruction of Europe. In the longer run, demand was enhanced by the next phase of the agricultural revolution; the widespread adoption of the tractor in British agriculture up to the mid-1960s, and of western European agriculture thereafter. Since then, demand has grown little in real terms; the home demand is largely a replacement demand; exports have declined and imports have grown. Thus the industry has entered on a new phase of shrinkage, and the present state of demand is not encouraging.

Entrepreneurship is the third leg of the explanatory tripod. It is a nebulous concept, hardly susceptible of measurement. Thus it can be interpreted and described in many different ways. In the nineteenth century Samuel Smiles could write his *Lives of the Engineers* as a sort of contemporary hero-worship. In the later twentieth century economists wrote books with titles such as *Britain's Economic Problem: Too Few Producers* (Bacon & Eltis, 1976), and *Slow Growth in Britain* (Beckerman, 1975). It was widely assumed by the 1980s that something called the 'British disease' existed, and that it was somehow linked to a generalised 'entrepreneurial failure'. What can be said about the quality of entrepreneurship in agricultural engineering?

It is common to approach this subject by contrasting the heady days of Victorian entrepreneurs with the difficult and problem-beset twentieth century. It is also easy to conclude that in the first period the entrepreneurs were more capable than they have been subsequently. However, it may be suggested that both the British and the world economy at any time between 1800 and 1914 were so buoyant that it was difficult not to succeed. There were, it is true, regular periods of trade depression. The 'trade cycle' had become a reality by the 1840s. But this merely meant that manufacturers would lay off a proportion of their workforce and wait for better times. The consequences of the periodic recessions were thus borne by the labour force, and the firm continued in business. In this particular industry, the only documented failures in the nineteenth century are Barrett, Exall & Andrewes of Reading, Brown & May of Devizes, and Crosskill of Beverley, although there were some firms, such as Mary Wedlake, which did not fulfil their early promise.

In the post-1918 period, when the industry was shrinking, and some of the leading large firms succumbed, it is also tempting to adduce entrepreneurial failure, and on a much larger scale. To some extent this must be so. Yet the main evidence comes from the single clearest example of failure, that of Fowlers of Leeds. This occupies such a large place in the story, partly because the firm's history has been told in such detail and so expertly by Michael Lane, and partly because it is an extreme case. After reading Lane's history, it could hardly be suggested that the history of Fowler's entrepreneurship after 1918 was anything other than a story of failure. Yet, given the firm's over-commitment to steam engines and the German and Russian market before 1914, it would be a brave management consultant who could suggest any better ways of saving the firm after 1918. Marshalls managed change rather better, but had the advantage of being less committed to agricultural engineering than Fowlers.

But there is a wider case to be made here. The outstanding failure between the wars was not that of a single firm like Fowlers, but the more general case of Agricultural & General Engineers. This extraordinary experiment went far beyond any other attempt at industrial rationalisation in the 1920s. None of the other 'older industries' such as coal, cotton and shipbuilding, which all faced similar problems to agricultural engineering, ever attempted anything so grandiose. Why was this? Several reasons may be given. The first is that the agricultural engineering industry's exports were harder hit than those of cotton and coal (although shipbuilding was in a worse case). These older industries had not relied on Russia, but on India or Europe and the wider world for their markets, and their markets, although damaged, had not disappeared. Secondly, there was a great fear of competition from the USA. The lesson of the Fordson had struck home, and the British firms, mesmerised by the mass-production methods of 'Fordism', clambered aboard the bandwagon, which they hoped would lead them in turn to the blessed state of mass-production, without stopping to think how it could be applied or whether it was appropriate to their own conditions. Finally they were cajoled by some powerful personalities: the Garretts and Maconochie in the initial stages, and Rowland thereafter. In short, the industry leaders panicked, and were subsequently bullied.

Had Agricultural & General Engineers not occurred, it is likely that the weaker firms would have succumbed more gently over time, as the banks lost patience and called in their overdrafts. Some may have revived, but the wholesale wipeout of 1932 would have been avoided.

After 1939 demand revived. At home, agricultural modernisation and mechanisation hinged around the tractor. Abroad, demand for tractors was enormous, and the British industry was second only to that of the USA in satisfying it. Again, it was a good time to be an entrepreneur. There are few post-war examples of failure, although some products did not find a lasting market. The Opperman Motocart, a 'self-propelled three-wheeled vehicle suitable for light farm transport' of the later 1940s was one such. Of original design, its air-cooled engine was built into the centre of the large front wheel by which the vehicle was also steered. It was really aimed at replacing a horse and cart, but was upstaged by the smaller tractors such as the Ferguson.

More generally, the industry never really succeeded in the crawler market, and the very small tractors did badly. But within the ranks of the agricultural engineering industry, enterprise flourished, new firms came on, and exports rose to new heights. After 1970, demand has fallen, and the internationalisation of the trade has made incursions into the British home market. And after 2000 there are large question marks over the future of the tractor industry in Britain, without which the rest of the industry will be something of a left-over rump. But even after the loss of markets since 1970, the industry's output in real terms is many times larger than it was at the time of the first census of production in 1907.

A part of the British economic decline?

Consideration of the periods during which the industry has shrunk – the inter-war period, and the period since the 1970s – raises the question as to how far the industry's failure on those occasions has contributed to, and should be seen as a part of, the general failure of the British economy.

The debate on British 'economic decline' sometimes presupposes a degree of decline which has not in fact taken place. The British economy has grown in the long run, and still does. Thus there is no question of absolute decline, merely a relative decline in comparison with other industrial countries of similar size. Put simply, the British economy has grown more slowly over the past two centuries than the others, and thus is no longer the largest industrial economy, as it was briefly in the third quarter of the nineteenth century.

A cultural explanation for this relative decline, popular in the 1980s, is that advanced by Martin Wiener in 1981, whose thesis is that after the middle of the nineteenth century the British entrepreneurs of the first generation, who had made their fortunes, succumbed to the lure of high society and bought country estates. They were succeeded, it was argued, by their public school educated sons, who preferred other ways of making money (the City, finance) or preferred to have a career in the public service, as imperial or home civil servants. This newly educated elite turned away from making money, finding it a vulgar pastime.[1] This argument has been investigated exhaustively by F. M. L. Thompson, who found little evidence to support it. In the case of the farm machinery industry, there are some cases of the successful men acquiring estates, but this was as much to try out their products in the field as to join the county set. A more persuasive variant of the Wiener thesis is that propounded by Cain and Hopkins, which proposes that the major economic actors were seduced into a type of 'gentlemanly capitalism', in which active participation in the market and the production of goods and services were beyond the pale for a gentleman. There is little evidence of this in the agricultural engineering industry. The sons of the family on the whole stayed in the business, and so did their sons, if the firm lasted that long. They may have liked to think of themselves as gentlemen, but they were still in trade, or 'business', and likely to remain so.[2]

Another allegation that British manufacturing management was inadequate. There are really two strands to this argument. One is that the typical British manufacturing firm stayed under family control, and remained of small or medium size. Thus few grew to the size at which economies of scale could be achieved, and so batch production, or tailor-made orders for individual customers, remained the norm. At the same time, limited size precluded the possibility of adopting the large-corporation structure, which was pioneered in the USA by such as US Steel, International Harvester, or General Motors. This form of organisation had various advantages which were denied the smaller family firm – larger-scale operations, devolution of responsibility to operating units, a clearer career path and structure, and greater emphasis on long-term planning and research. This is the essence of Alfred Chandler's criticism of the British industrial firm. However, for the British agricultural engineering industry, it is difficult to provide an instance of a firm with a large enough market share to assure it of economies of scale, except in limited product instances, or to set up a corporate structure. For most of this period, the largest firm on the machinery side was Ransomes, and this firm happened to be heavily involved in plough production, which benefited from attention to individual markets, and batch production. The reliance on bespoke orders and batch production was not confined to Ransomes, and it is difficult to avoid the conclusion that, for the agricultural engineering industry, the Chandlerian prescription was inappropriate. It was, of course, another matter when it came to the tractor producers, but they were essentially a part of the motor industry, and their history was very different. They were, or became, offshoots of large multinational corporations. But they were not typical of most of the firms in the farm machinery sector.[3]

However, the other strand to the 'management failure' argument focuses on specific shortcomings. Here the argument is on firmer ground. It has long since been accepted that one of the weaknesses of British manufacturing is the lack of a clear link between school, post-school training, and work. In Germany, this gap was less evident, and the consequence was and is that the average employee has a higher level of training, and qualifications to match. In Britain, the lack of such a link has resulted in a workforce which lacks technical training, and learns a variety of semi-skilled procedures on the job, without formal instruction. This is true for the majority of engineering firms, and agricultural engineering is no exception. A further weakness is the lack of commitment to expenditure on research and development. This means that there have never been many people in the industry specially hired to do this work, and that the majority of new products are drawn from the inspiration of the leading members of the management.[4]

These weaknesses – too many small firms, too little spent on research and development, a shortage of qualified and trained manpower (the industry has not been good at attracting graduates), and in some cases a tendency to follow rather than lead the market – were apparent by the 1980s, and some may still be there. But in common with other sectors of British manufacturing, productivity has risen since the crises of the 1980s and 1990s, and many ancient firms are no longer in the business. The future does not perhaps look as bleak as some pessimists might fear.

Conclusion: a story of continuing change

In the long run, the agricultural engineering industry has made possible the enormous rise in agricultural productivity which the world has witnessed in the past two centuries. In the later eighteenth century, British farm output rose at less than 1 per cent per year. Then it steadily increased, to around 1.8 per cent annually in the mid-nineteenth century. Thereafter there was a period of slow growth, from the 1870s until the 1920s. After that, the growth was enormous; usually at least 2.5 per cent annually. The long-term result of this compound growth was that by the early 1980s, British agricultural output was nearly nine times what it had been in the first decade of the nineteenth century. This enormous rise in production was accomplished by the use of the products of the agricultural engineering industry, as the new machines, implements and tractors replaced horses and men. Of the 800,000 farm horses at work on mid-Victorian farms, nothing now remains except a few kept for ploughing matches. In 1851 there were *c.*1.3 million hired farm labourers in England and Wales alone; in 2000 there were only 120,000 in the whole of UK agriculture. The contribution of the agricultural engineering industry to farming and to human progress is not in dispute. The uncertainties are those concerned with its future as an industry.[5]

It is perhaps most appropriate to see the history of this industry as one of continuing, and on the whole successful adaptation to a world which has ever been, and is still, continually changing. While certain people in the older industries in the later nineteenth century – coal, cotton, shipbuilding – might have thought that they were destined to rule world trade for ever, there is little evidence of this in agricultural engineering. The business archives of Ransomes, Sims & Jefferies, which was for most of these two centuries the largest machinery manufacturer, show that its management was continually on the move, looking personally at potential new markets around the globe, and reassessing the existing trades. It was only in the last two decades or so of the twentieth century that the firm seems to have lost its touch – or, since it moved substantially out of agricultural engineering altogether, perhaps it was just realising the inevitable, and acting on its hunch. On the whole the story of the industry is one of successful adaptation. Individual products might fail once in a while, but this was uncommon. The greatest failures were confined to the 1918–39 period, as former luminaries of the industry desperately tried to find new products. After 1945 there was a quarter-century or so of success, then a diminution of growth, and then a shrinkage. There is plenty of evidence that the industry had structural and technical weaknesses in 1945, which had not been addressed before the post-1970 crises. But there has been nothing in the recent history of the industry remotely comparable to the staggering, catastrophic, industry-wide failures seen after 1970 in the British car, motorcycle, shipbuilding, and electronics industries, and the industry has emerged in a shrunken but efficient state, displaying a lot of entrepreneurial energy.

Notes and references

Notes to Chapter 1. The origins of an industry, 1750–1820

1. Mitchell, B. R., *British Historical Statistics* (Cambridge, 1988), p. 756. The best brief guide to the agricultural revolution is by J. V. Beckett, *The Agricultural Revolution* (Blackwell, Oxford, 1990).
2. Chapman, J., 'The extent and nature of parliamentary enclosure', *Agricultural History Review*, vol. xxxv (1987), p. 28.
3. Grigg, D., *English Agriculture: an historical perspective* (Oxford, 1989), pp. 149–50.
4. Somerville (Lord), *Facts and Observations relating to Sheep, Wool, &c.* (1809), pp. 67, 75.
5. Quoted in Fussell, G. E., *The Farmer's Tools*, p. 44.
6. Fussell, *Farmer's Tools*, p. 58. On the Rotherham plough, see Overton, M., *Agricultural Revolution in England*, (Cambridge, Cambridge University Press, 1996), p. 122; Bailey, M., *One Hundred and Six Copper Plates of Mechanical Machines and Implements of Husbandry, approved and adopted by The Society for the Encouragement of Arts, Manufactures and Commerce and contained in their Repository in the Adelphi Buildings in the Strand ... carefully corrected and revised by Alexander Mabyn Bailey, Registrar to the Society*, vol. I (London, Benjamin White, 1782).
7. Fussell, *Farmer's Tools*, pp. 101–5.
8. Fussell, *Farmer's Tools*, p. 116.
9. Fussell, *Farmer's Tools*, pp. 152–61.
10. Harris, J. R., *The British Iron Industry 1700–1850* (Macmillan, 1988), pp. 31–40.
11. Riden, P., 'The Output of the British Iron Industry before 1870', *Economic History Review*, vol. xxx (1977), pp. 448, 455.
12. Harris, *Iron Industry*, p. 34.
13. Harris, *Iron Industry*, p. 61; Roepke, H. G., *Movements of the British Iron and Steel Industry – 1720 to 1951* (Urbana, Illinois, 1956), p. 45.
14. Blake, S., 'An Historical Geography of the British Agricultural Engineering Industry, 1780 to 1914', unpublished Ph.D. thesis, University of Cambridge (1974), p. 48.
15. Blake, *thesis*, pp. 80, 249; Palmer, F., 'The Blacksmiths Ledgers of Bucklebury', University of Reading Research paper No. 2 (1970).
16. Anon: 'Oldest-established Implement Manufacturer in Britain? Knapp's are now in their Third Century.', *Implement and Machinery Review*, 1 April 1962, pp. 489–90; Fussell, *Farmer's Tools*, p. 169.
17. I have been unable to trace any further reference to Corcoron. It may be that this firm eventually mutated into Bryan Corcoran, 'the well-known London contractors for malting equipment', who manufactured Coleman & Morton's (founded 1848) adjustable screens in 1914 – Booker, J., *Essex and the Industrial Revolution* (Chelmsford, Essex Record Office, 1974), p. 43.
18. Warwick Record Office, Troth & Hillson archive.
19. Lane, Michael R., *The Story of the St. Nicholas Works: A History of Charles Burrell & Sons Ltd, 1803–1928* (Unicorn Press, Grimston, Yorks., 1994), pp. 2–3.
20. Whitehead, R. A., *Garretts of Leiston* (Percival Marshall, 1964, pp. 8–9.
21. Fussell, *Farmer's Tools*, pp. 48–50.
22. For previous four paragraphs, see Weaver, C. and M., *Ransomes 1789–1989. A Bicentennial Celebration* (Ipswich, Ransomes, Sims & Jefferies PLC, 1989), pp. 11–18; Grace, D. R., and Phillips, D. C., *Ransomes of Ipswich. A History of the Firm and Guide to its Records* (University of Reading, Institute of Agricultural History, 1975), p. 1; Brigden, R., 'Agricultural Iron Foundries of the Nineteenth Century', *Foundry Trade Journal*, December 1983, p. 556.
23. Black, E., 'Plenty's taking museum pieces to steam fair', *Newbury Weekly News*, 3 May 1973, p. 22; North, L., *Wishing you a year of Plenty and Progress, 1790–1958* (Newbury, the firm, 1958),

pp. 1–2.

24. Anon., 'Through Five Generations! Cooch & Sons celebrate Their 150th Anniversary', *Farm Implement and Machinery Review*, 1 October 1950, pp. 939–40; Kemp, P. K., *The Bentall Story: commemorating 150 years service to Agriculture* (the firm, n.d.), pp. 4–5.
25. Pointer, M., *Hornsby's of Grantham 1815–1918* (Grantham, Bygone Grantham, 2nd edn, 1978), pp. 2–5.
26. Hueckel, G., 'English Farming Profits during the Napoleonic Wars, 1793–1815', *Explorations in Entrepreneurial History*, vol. 13 (1976), pp. 331–45; Thompson, F. M. L., 'The Land Market in the Nineteenth Century', *Oxford Economic Papers* IX (1957), reprinted in Minchinton, W. E., (ed.), *Essays in Agrarian History* (David & Charles, Newton Abbot, 1968), p. 35; Overton, *Agricultural Revolution in England*, p, 162; Beckett, J., Turner, M., and Afton, B., *Agricultural Rent in England 1700–1914*.
27. Generally on the agricultural revolution in East Anglia, see Wade Martins, S., and Williamson, Tom, *Roots of Change: Farming and the Landscape in East Anglia, c.1700–1870* (Exeter, British Agricultural History Society, 1999), especially pp. 115–19 and the yields noted on pp. 158–69; also Williamson's more recent *The Transformation of Rural England: Farming and the Landscape 1700–1870* (Exeter, University of Exeter Press, 2002).
28. Mingay, G. E., *Land and Society in England, 1750–1980* (Longman, 1994), p. 40.
29. Mitchell, *Statistics*, p. 25.
30. Grace and Phillips, *Ransomes*, p. 1; Kemp, *Bentall Story*, p.6; Booker, J., *Essex and the Industrial Revolution* (Colchester, Essex Record Office, 1974), p. 5.
31. Phillips, A., 'Early Colchester Foundries', *Essex Archaeology and History*, 14 (1982), pp. 102–3; Booker, *Essex*, pp. 5, 7–8.

Notes to Chapter 2. Towards a national market, 1820–1850

1. Mitchell, *Statistics*, p.756.
2. Thompson, F. M. L., 'Land Market', p. 35; Thompson, R. J., 'An Enquiry into the Rent of Agricultural Land', *Journal of the Royal Statistical Society*, LXX (1907), Appendix, reprinted in Minchinton, W. E. (ed.), *Essays in Agrarian History*, vol. II (1968), pp. 57–86.
3. Walton, J. R., 'Aspects of agrarian change in Oxfordshire, 1750–1880', unpublished DPhil. thesis (Oxford, 1976).
4. [Hobsbawm, E. G., and Rude, G., *Captain Swing* (1969); Overton, *Agricultural Revolution*, p. 125].
5. Mitchell, *Statistics*, p. 762.
6. Blake, *Thesis*, pp. 81–2.
7. Norfolk Record Office; archives of these firms.
8. Radcliffe, V., 'Entrepreneurial Success: A Local Case Study: W. N. Nicholson & Sons, Agricultural Engineers of Newark on Trent', *The Nottinghamshire Historian*, No. 54, Spring/Summer 1995, p. 8.
9. Reeves, M., *Sheep Bell and Ploughshare: the story of two village families* (Bradford-on-Avon, Moonraker Press, 1978), pp. 81–3.
10. Suffolk Record Office, Ipswich; James Smyth archive, Sales Number Book 1, HC23/F1/1.
11. Grace and Phillips, *Ransomes*, pp. 1–2.
12. Fort, T., *The Grass is Greener: Our Love Affair with the Lawn* (HarperCollins 2001), p. 113.
13. Rolt, *Waterloo Ironworks*, pp. 35–8; Course, Edwin, 'The Rolt Memorial Lecture, 1994: Engineering Works in Rural Areas', *Industrial Archaeology Review* XVIII, No. 2, Spring 1996, plate 13; Lane, M. R., *The Story of the St Nicholas Works: a history of Charles Burrell & Sons Ltd* (Stowmarket: Unicorn Press, 1994), pp. 3, 5.
14. Brown, Jonathan, *Farm Machinery, 1750–1945* (Batsford, 1989), pp. 20–1. The figure of 50 is my own suggestion.
15. Blake, thesis, pp. 88–9.
16. Whitehead, R. A., *Garretts of Leiston* (Percival Marshall, 1964), p. 11.
17. Goddard, N., *Harvests of Change: the Royal Agricultural Society of England, 1838–1988* (Quiller Press, 1988), ch. 1 and fig. 1, p. 29, table 1, p. 35; p. 47; Pigeon, D., 'The development of Agricultural Machinery', *Journal of the Royal Society of England* (hereafter *JRASE*) series iii, I, (1890), pp. 257–75.
18. Kemp, *Bentall Story*, pp. 10–12.
19. Goddard, *Harvests of Change*, p. 53; Rural History Centre, University of Reading, Ransomes Sims & Jefferies archive, company history, C5–7; illustrated in the accompanying blueprint marked 'fig. 2', which seems to be the 1842 engine.
20. Whitehead, R. A., *Steam in the Village* (David & Charles, Newton Abbot, 1977), p.20; Clark, Ronald H., *The Steam Engine Builders of Norfolk* (Yeovil, Haynes Publishing Group, 1988), p. 15; Lane, Michael R., *The Story of the Wellington Foundry, Lincoln: A History of William Foster & Co. Ltd* (Unicorn Press, 1997, pp. 3–4); idem., *The Story of the Britannia Iron Works: William Marshall Sons & Co., Gainsborough* (Quiller Press, 1993), p. 9.
21. Whitehead, *Garretts*, p. 48.

22. Whitehead, *Garretts*, 23, 27, 57.
23. Collins, E. J. T., 'Power availability and agricultural productivity in England and Wales, 1840–1939', in van Bavel, Bass J. P., and Thoen, E., *Land Productivity and Agro-systems in the North Sea Area, Middle Ages–twentieth Century* (Turnhout, 2000), p. 211.
24. Mitchell, *Statistics*, 541; Aldcroft, D. H., and Dyos, H. J., *British Transport: an economic survey from the seventeenth century to the twentieth* (Harmondsworth, Pelican, 1974), p. 129; Freeman, M., and Aldcroft, D., *The Atlas of British Railway History* (Beckenham, Croom Helm, 1985), p. 15.
25. Dyos and Aldcroft, *Atlas*, p. 134; Mitchell, *Statistics*, p. 541.
26. Generally, see Perkin, Harold, *The Age of the Railway* (1970); also Simmons, J., *The Railways of Britain: an historical introduction* (1961).
27. Grace and Philips, *Ransomes*, p. 3; Rural History Centre, University of Reading, Ransomes, Sims & Jefferies archive, internal history of the firm, TR RAN SP1/1A, period 1840–1850, pp. C5, C14.
28. Blake, *thesis*, p. 18 and figures 21 and 23.
29. Blake, *thesis*, figure 9.
30. Blake, *thesis*, figure 11 and pp. 98–104.
31. Lines, C. J., 'The Development and Location of the Specialist Agricultural Engineering Industry, with Special Reference to East Anglia' (unpublished M.Sc. thesis, University of London, 1961), p. 40.
32. Skehel, M., *Tales from the Showyard: Two Hundred Years of Agricultural Shows in Lincolnshire* (Lincoln, Lincolnshire Agricultural Society, 1999), p. 15.
33. Collins, E. J. T., 'The Age of Machinery', in Mingay, G. E., (ed.), *The Victorian Countryside* (Routledge & Kegan Paul, 1981), vol. I, p. 203.
34. Lord Willoughby de Eresby described his system, in his *Ploughing by Steam* (J. Ridgway, 1850), which contained detailed illustrations. The system employed a railway locomotive, running along rails in the centre of a field, and operating a plough on either side, attached to an endless chain. There is no evidence of it having got beyond the prototype stage.

Notes to Chapter 3. At the works around 1850

1. Berwick-upon-Tweed Record Office, archive of Wm Elder & Sons Ltd, item 4990, *Tweedale Press Supplement*, 1965.
2. Brigden, R., 'Agricultural Iron Foundries of the Nineteenth Century', *Foundry Trade Journal*, December 1983, p. 560; Philips, D. C., 'Howard, James', in Jeremy, D. J. (ed.), *Dictionary of Business Biography* (Butterworth, 1986), p. 373.
3. References for the above paragraphs: *The Engineer*, 20 July 1860, p. 38; Brigden, *Foundry Trade Journal*, 8/22 (1983), p. 560; End page of Howard's catalogue, 1 January 1864; Bedford and Luton Archives Collection, Photograph BR37, p. 2, negative 1782.
4. Hopewell, Charles, *Recollections of Charles Hopewell* (1863), pp. 1–10, unpublished typescript in Ruston & Hornsby archive, Lincoln. Photocopy kindly supplied by Mrs Pat Gregory, Lincoln.
5. Anon., 'Visit to the Stamp End Works, Lincoln', *Bell's Weekly Messenger*, 4 July 1859, p. 3.
6. 'The Druid', *Saddle and Sirloin, or English Farm and Sporting Worthies* (1870), pp. 465–9.

Notes to Chapter 4. A brief supremacy, 1850–1875

1. Ernle, Lord, *English Farming Past and Present* (6th edn, Frank Cass, 1961), chapter XVII, 'High Farming: 1837–1874'.
2. Collins, E. J. T. (ed.), *Agrarian History of England and Wales*, vol. VII (1850–1914), chapter 2A, 'The High Farming Period, 1850–75', pp. 78, 86–7, 98–100, 108–9, 116.
3. Collins, *Agrarian History*, pp. 117–23.
4. Dewey, P. E., 'Agricultural Labour', in *Agrarian History of England and Wales*, VII (1850–1914), chapter 12, pp. 851–2, Appendix I.
5. Pidgeon, D., 'The Evolution of Agricultural Implements, II', *JRASE*, 3rd ser., 3 (1892), p. 252; Whitehead, *Garretts*, p. 89; Shearman, L., *Portable Steam Engines* (Shire Publications, Princes Risborough, 1986), pp. 4–6.
6. Pidgeon, *JRASE* 3rd series, 3 (1892), p. 252; Brown, *Farm Machinery 1750–1945*, p. 42; Brigden, R., 'Equipment and Motive Power', chapter 7C of E. J. T. Collins (ed.), *The Agrarian History of England and Wales*, vol. VII, pp. 509–10.
7. Clark, *Steam Engine Builders of Norfolk*, throughout.
8. Clark, *Steam Engine Builders of Norfolk*, items XVIII, XIX.
9. Lane, *Wellington Foundry*, p. 5; Hill, (Sir) F., *Victorian Lincoln* (Cambridge, Cambridge University Press, 1974), p. 123.
10. Newman, B., *One Hundred Years of Good Company* (Lincoln, Ruston & Hornsby, 1957), pp. 1–8.
11. Rolt, L. T. C., *Waterloo Ironworks: A history of Taskers of Andover, 1809–1968* (Newton Abbot, David & Charles, 1969), pp. 86–90; 2

photographs on p. 52.

12. Bell, B., *Ransomes, Sims & Jefferies, agricultural engineers: a history of their products* (Ipswich, Old Pond Publishers, 2001), pp. 63–4.
13. Whitehead, *Garretts*, pp. 69–70, 129–31; fig. 51 has illustration of the 1879 'peasant' thresher from *The Engineer.*
14. Lane, *Britannia Iron Works*, pp. 16–17.
15. Lane, *Wellington Foundry*, p. 16.
16. Whitehead, *Garretts*, p. 127.
17. Collins, E. J. T., 'The High-Farming period: the Golden Age, 1850–75', in Collins, E. J. T. (ed.), *The Agrarian History of England and Wales*, vol. VII (1850–1914), chapter 2A, p. 76.
18. Haining, J. and Tyler, C., *Ploughing by Steam* (Model and Allied Publications, Hemel Hempstead, 1970), pp. 87–9, 108–11; Lane, *St. Nicolas Works*, ch. 2.
19. Tyler, C., *Digging by Steam: a History of Steam Cultivation by means of the application of Steam Power to the Fork, Mattock and similar implements* (Watford, Model and Allied Publications, 1977), pp. 25–7, 156.
20. Tyler, *Digging by Steam*, pp. 56–7, 84–9.
21. Haining and Tyler, *Ploughing by Steam*, pp. 70–75.
22. The three preceding paragraphs are based on Lane, *Steam Plough Works*, chapters 2–4.
23. Anon., 'Reports of the Committees appointed to investigate the present state of steam cultivation', *JRASE*, 2nd series, vol. 3 (1867), p. 367.
24. Collins, E. J. T., 'Harvest technology and labour supply in Britain, 1790–1870', *Economic History Review* 2nd ser., 22 (1969).
25. Fussell, *Farmer's Tools*, pp. 117–23.
26. *Hampshire Chronicle*, 13 September 1851, p. 2.
27. Hutchinson, Wm T., *Cyrus Hall McCormick* (New York, 1968), vol. II, pp. 411–14, 425; Pidgeon, D., 'The Evolution of Agricultural Implements – I', *JRASE*, 3rd series, vol. 3 (1892), p. 65.
28. Fussell, *Farmer's Tools*, pp. 130–37; Blake, *thesis*, p. 219; Collins, E. J. T., 'The High-Farming period: the Golden Age, 1850–75', chapter 2A of Collins. E. J. T. (ed.), *The Agrarian History of England and Wales*, vol. VII (1850–1914), p. 76.
29. Wigan Archives Service; Leigh Record Office, Lancs., Harrison McGregor archive, D/DY HMG, Profit and Loss account for the Year ended 31 December 1875.
30. 'Crosskills of Beverley', *East Yorkshire Local History Society Bulletin*, Autumn 1982, p. 7; Thomas Nelson and Sons, *An Encyclopaedia of Agriculture* (1914), vol. II, p. 252; East Riding of Yorkshire Record Office, Beverley, DDBD 91/30 *Crosskill's Implement Newspaper*, part I, July 1848.
31. Ipswich RO, Smyth archive, HC23/F2/17, Order Book 1st Jan. 1874–24th April 1875.
32. Brigden, R., 'Equipment and motive power' *AHEW*, VII, p. 509.
33. Ibid., p. 508.
34. Blake, thesis, p. 215.
35. Grace, D., 'The Agricultural Engineering Industry', in *AHEW*, vol. VII, chapter 16A, pp. 1016–17.
36. Lane, *Steam Plough Works*, pp. 50, 114, 136, 155–6.
37. Ipswich Record Office, Garrett archive, HC 30/C8/2, Steam Engines from 1 January 1877; HC/30/A3/3, Sales Account 1879–81; Lincs. Record Office, Lincoln, Amalgamated Power Engineering archive, APE1, Foster of Lincoln, Balance Sheets 1877–1960; Wigan Archives Service, Leigh RO, Harrison McGregor archive, D/DY HMG, Profit and Loss Account for year ended 31 December 1875.
38. Lane, M.R., *The Story of the Britannia Iron Works: William Marshall Sons & Co., Gainsborough* (1993), pp. 10–13; Lincs. Record Office, Lincoln, Marshall archive, Directors' Minute Book, vol. I, 1862–1881.
39. University of Reading, Rural History Centre, Ransomes archive, AD 7/17, Sales 1870–1939; TR RAN SP1/1E, 'Period 1860–1870'; TR RAN 3/5; Grace and Phillips, *Ransomes*, pp. 4–5.
40. Lincs. Record Office, Lincoln, Ruston & Hornsby archive, Hornsby Threshing Machine Register 1854–1904; Hornsby Engine Registers 1873–1913, vol. I; University of Reading, Rural History Centre, Nalder & Nalder archive 334, Synopsis Book 1870–1901; 335, Analysis Book 1870–92.
41. University of Reading, Rural History Centre, Bomford archive, TR BLL, Humphries 1875 Catalogue, P2/A1; Hants. Museum Service, Winchester, Taskers 1873 Catalogue; 1873 steam engine catalogue.
42. Goddard, N., *Harvests of Change: The Royal Agricultural Society of England 1838–1988* (Quiller Press, 1988), pp. 72–3.
43. Goddard, *Harvests of Change*, pp. 71–4; Brigden, *AHEW*, VII, p. 509.
44. Wiltshire RO, Trowbridge, WRO 2574; 2440.
45. East Riding Record Office, Beverley, Crosskill archive, DDX/347/I, abridged catalogue of East Yorkshire and Crosskills Cart and Wagon Co.
46. Cambridgeshire Record Office, Lack archive R77/56; Hunt archive, 773.
47. Clapham, (Sir) J., *An Economic History of Modern Britain*, vol. II (Cambridge, CUP, 1952), p. 217.
48. *Annual Statement of Trade* 1853; Mitchell, *British Historical Statistics*, p. 452.
49. *East Yorkshire Local History Society Bulletin* no. 26, Autumn 1982, p.2.

50. Mitchell, *Statistics*, pp. 453, 482.
51. Grace, D., 'The Agricultural Engineering Industry', *AHEW*, VII, chapter 16A, p. 1010.
52. Rural History Centre, University of Reading, Ransomes archive: typescript notes of the firm's history by C. J. Palmer, TR RAN SP 1/D.
53. Ruston & Hornsby archive, European Gas Turbines (now Alsthom), Firth Road, Lincoln. Lincoln. Ledger: 'Hornsby: Threshing Machine Register 1854–1904'.
54. Munting, R., 'Agricultural Engineering and European Exports before 1914', *Business History*, xxvii (1985), pp. 126–9; Grace, *AHEW*, chapter 16A, p. 1011.

Notes to Chapter 5. Exports to the rescue, 1875–1913

1. Phillips, A. D. M., *The Underdraining of Farm Land in England during the Nineteenth Century* (Cambridge, 1989).
2. Ministry of Agriculture, *A Century of Agricultural Statistics: Great Britain 1866–1966* (HMSO, 1968), tables 42, 46.
3. Afton, B., and Turner, M., 'Agricultural Output', chapter 38 of *AHEW*, table 38.11; Thompson, F. M. L., 'An Anatomy of English Agriculture, 1870–1914', in Holderness, B. A., and Turner, M., *Land, Labour and Agriculture, 1700–1920: Essays for Gordon Mingay* (Hambledon Press, 1991), pp. 232–3, Table 11.6.
4. Trow-Smith, R., *Power on the Land: a centenary history of the Agricultural Engineers Association 1875/1975* (AEA, 1975), pp. 23–47.
5. Grace, D., *AHEW*, pp. 1005–7.
6. Rural History Centre, University of Reading, Ransome Sims & Jefferies archive, TR RAN AD 7/17.
7. Mitchell, *British Historical Statistics*, p. 836.
8. Mitchell, *Statistics*, p. 728.
9. Grace, *AHEW*, p. 1006.
10. Mitchell, *Statistics*, p. 728.
11. Hornsby's had already moved into the export market for their threshers – see table 4.3, chapter 4 above.
12. Lane, *Steam Plough Works*, p. 181; Grace, *AHEW*, p. 106.
13. Grace, *AHEW*, p. 1009.
14. Lane, *Steam Plough Works*, pp. 86–8, 187; Whitehead, *Garretts*, pp. 102–7.
15. Maclean, I. W., 'Anglo-American Engineering Competition, 1870–1914: Some Third-Market Evidence', *Economic History Review* (1976).
16. Wattenberg, B., *The Statistical History of the United States* (New York, Basic Books, 1977), series P. 123–176, p. 684. To get the rough equivalent in British pounds sterling, these dollar figures should be divided by 5.
17. Phillips, W. G., *The Agricultural Implement Industry in Canada* (Toronto, University of Toronto Press, 1956), pp. 10–24.
18. Wattenberg, *Historical Statistics*, series K 1–16, p. 457.
19. Grace, *AHEW*, pp. 1007–8.
20. Munting, R., 'Agricultural Engineering and European Exports before 1914', *Business History*, xxvii (1985), pp. 133–4.

Notes to Chapter 6. A mature industry, 1875–1913

1. Collins, 'Land productivity and agro-systems', in van Bavel (2000), tables 8.3, 8.4.
2. The foregoing is based mainly on Corley, T. A. B., 'Barrett, Exall & Andrewes, 1818–1887', typescript MS, University of Reading, Rural History Centre, p. 1678, 18. pp. 000; idem., 'Barrett, Exall & Andrewes' Iron Works at Reading: The Partnership Era, 1818–64', *Berkshire Archaeological Journal*, v. 67 (1973–74), pp. 79–87; Measom, G., 'The Katesgrove Iron Works of Messrs. Barrett, Exall, and Andrewes …' *The official illustrated guide to the south-Eastern Railway* (1858; repr. 1970), pp. 174–7.
3. Brown, G. P., 'Crosskills of Beverley', *The East Yorkshire Local History Society Bulletin*, Autumn 1982, pp. 7–8; *Victoria County History of the Counties of England*: Yorkshire East Riding, vol. VI, p. 139; Rural History Centre, University of Reading, Samuelson archive, item 79.
4. Taylor, Audrey M., *Gilletts: bankers at Banbury and Oxford: a study in local economic history* (Oxford, Clarendon Press, 1964), pp. 128–9, 165; Langley, J.L., 'Memories of Late Victorian Banbury', *Cake and Cockhorse* v. 2, 4 (1964), P. 55; Hutchinson, W. T., *Cyrus Hall McCormick*, v. II, pp.414, 441; Rural History Centre, University of Reading, Samuelson archive, item 68.
5. Weaver, Paul, 'Brown & May, Devizes, Wiltshire', in *Chippenham & District Preservation Society, Newsletter Supplement* (1984).
6. Rolt, L.T.C., *Waterloo Ironworks: A History of Taskers of Andover, 1809–1968* (Newton Abbot, David & Charles, 1969), pp. 129–47, 150–3.
7. Shearman, Lyndon R., *Portable Steam Engines* (Shire Publications, Princes Risborough, 1986,

1995), pp. 18–23.

8. Clark, *Steam Engine Builders of Norfolk*, Fig. 168 and 170, p. 107.
9. Lane, *Steam Plough Works*, pp. 78–81, 183.
10. Lane, *Steam Plough Works*, p. 261.
11. Rayner, D. A., *Thomas Aveling 1824–1882: His Life and Work* (n.d.), pp. 1–4, 16.
12. See the photo of the Hero engine by Taskers; Rolt, *Waterloo Iron Works*, p. 52.
13. Lane, *Steam Plough Works*, p. 131.
14. Lane, *Steam Plough Works*, p. 137 et. seq.; chapter 9; generally Wilkes, P., *An Illustrated History of Traction Engines* (Spur Books, Bourne End, Bucks., 1974, chapter III.
15. Fussell, *Farmer's Tools*, pp. 196–8; Orwin, C. S., and Whetham, E., *History of British Agriculture 1846–1914* (Longmans, 1964), p. 347; Jannson, T., A *Historical review of the Development of the Milking Machine with special reference to technical advances and improvements* (Alfa-Laval, Tumba, Sweden, 1973).
16. Fussell, *Farmer's Tools*, p. 191 and plate 107.
17. Voelcker, A., 'On the Composition of Cream and Skim-Milk obtained by De Laval's Centrifugal Cream-Separator', *JRASE* 2nd ser., v. 16, (1880), p. 160.
18. Fussell, *Farmer's Tools*, p. 192.
19. Evans, D. E., *Lister's: The First Hundred Years* (Gloucester, Alan Sutton, 1979), pp. 29–72; Gloucs. Record Office, Gloucester, R.A. Lister & Co. archive, D3310/3/5.
20. *IMR*, vol. 34, no. 397 (1908), pp. 69–70.
21. Wailes, R., 'The early history of Akroyd Stuart's Oil Engine', *Transactions of the Newcomen Society*, v. 48 (1976–7), pp. 103–110; Edginton, D., *Old Stationary Engines* (Shire Publications, Princes Risborough, 1996), pp.3–7.
22. Key, M. (ed.), *The Blackstone Collection*: 'Carter Brothers Special', no. 8, March 2003, pp. 1–12; Anon, 'The Works of Messrs. Clayton & Shuttleworth', *Engineering*, (1920), reprinted in *Road Locomotive Society Journal*, vol. 44/4, Nov. 1991, pp. 112–17.
23. Evans, *Listers*, p. 68; Kemp, *Bentall Story*, 25–7, 31; Jackson, R., 'A brief history of E. H. Bentall & Company', *Farm & Horticultural Equipment Collector*, July/August 2000, p. 9.
24. The history of the farm tractor continues to fascinate, and there are many books on the subject. Some are more pictorial than textual, and the quality of the photographs is very high. Even the mainly pictorial works have some useful information on company history, and most blend history and pictures to good effect. Some of the best, on which these foregoing paragraphs are based, are Moorhouse, Robert, *The Illustrated History of Tractors* (Quintet Publishing, 1996) and Williams, Michael, *Great Tractors* (Blandford Press, Poole, 1982). Also to be recommended are Carroll, John, *Tractors of the World* (Anness Publishing, 2000) and Wilkie, Jim, *An Illustrated History of Tractors* (Ian Allan, 1999). Also consulted is Pointer, Michael, *Hornsbys of Grantham 1815–1918* (Grantham; Bygone Grantham, 1977), p. 24.
25. Moorhouse, *Tractors*, p. 24.
26. Dugdale, J. B., 'Miscellaneous Implements Exhibited at Park Royal', *JRASE* 64 (1903), pp. 229–31; Beaumont, W. W., and Hippisley, R. J. B., 'Trials of Agricultural Motors at Manor Farm, Bygrave, Baldock, Hertfordshire', *JRASE* v. 71 (1910), p. 184; Moffitt, J., *The Ivel Story* (Japonica Press, 2003), pp. 111, 162.
27. Rural History Centre, University of Reading, Ivel Motor Co. archive; Senate House Library, University of London, Pollitt Papers, 8/89, 8/108, 8/114, 8/124.
28. C.J. Palmer, *History of the Orwell Works 1771–1928*, unpublished typescript, Reading University, Rural History Centre, TR RAN, SP 1/1, K117–19.
29. Lane, *Britannia Works*, pp. 75–6.
30. Lane, *Steam Plough Works*, pp. 268–73.
31. Lane, *Wellington Foundry*, pp. 106–11.
32. Newman, B., *One Hundred Years of Good Company* (Ruston & Hornsby, the firm, Lincoln, 1957), pp. 86–90.
33. Cawood, Charles L., *Vintage Tractors* (Princes Risborough, Shire Publications, 1980), p. 19; Baldwin, N., 'A-Z of Tractors; Crossley', Tractor & Machinery May 2001, p. 63.
34. Baynton & Hippisley, q.v.; Whitehead, *Garretts of Leiston*, pp. 126, 180–1.
35. Haynes, Walter, *History of Ruston & Hornsby Limited* (Lincoln and Grantham; the firm, 1954; 3 vols, typescript), vol. 2, pp. 21–5). Walter Haynes was a director of the firm. I am indebted to Mrs. Pat Gregory of Lincoln, for the loan of these volumes.
36. Lane, *Britannia Works*, chs. 1 and 2, pp. 73–4; Lincolnshire Archives, Lincoln Record Office, Marshall & Sons archive, Directors' Minute Book 1862–1881; Marshall Sales Sheet 1910.
37. Rural History Centre, University of Reading, Samuelson archive, balance sheet and profit and loss account, 30 September 1888; Lane, *Britannia Works*, p. 19.
38. Wigan Archive Service, Leigh, Lancs., Harrison, McGregor archive, Reports of Directors 1911–13; Shareholders' Meeting Minutes,1892–1910.
39. Rural History Centre, University of Reading, Nalder & Nalder archive, sales figures and

annual balance sheets, 1875–1913.

40. Northumberland Record Office, Berwick-upon-Tweed, Elders archive, BRO 126, Balance Sheet for y/e 31 December 1904; *Tweedale Press Supplement* on the firm's centenary 1865–1965; Box 4, Sales Ledgers 1906–12.

Notes to Chapter 7. At the works in 1913

1. Hill, Sir F., *Victorian Lincoln* (Cambridge, CUP, 1974), Map 3.
2. Dewey, P., 'Farm Labour', in Collins, *AHEW*, VII, pp. 831–2; Whiting, R.C., *The View from Cowley; the impact of industrialization upon Oxford, 1918–1939* (Oxford, Clarendon Press, 1983).
3. Victoria History of the Counties of England, *Lincolnshire* (1906), p. 395.
4. *RLSJ*, q.v.
5. Hill, *Victorian Lincoln*, p. 217; Lane, *Wellington Foundry*, pp. 27, 38–9; *IMR* 2.11.1900, p. 25921.
6. Anon, 'Extension of Blackstone & Co.'s Rutland Iron Works, Stamford', *IMR* 1 Dec., 1904, pp. 918–19; Rural History Centre, University of Reading, Ransomes archive, *History of the Orwell Works*, TR RAN SP1/1K, p. 39; Whitehead, *Garretts*, pp. 28–40.
7. Alford, L.C.G., *The Development of Industrial Lincoln*: a lecture given to the Lincoln Historical Association and the Lincoln Training College (1950), typescript; Anon., 'A Visit to a Lincoln Engineering Works', *The Engineer*, 23 May 1913.
8. *RLSJ*, q.v.
9. Reeves, M., *Sheep Bell & Ploughshare: the story of two village families* (Moonraker Press, Bradford-on-Avon, 1978), p. 86–7.
10. Mitchell, *Statistics*, p. 137.
11. Wigan Archives Service, Leigh, Lancs., Harrison McGregor archive, Directors' Minute Book no. 5, March 7, 1913.
12. Whitehead, *Garretts of Leiston*, p. 193.
13. Hill, *Victorian Lincoln*, p. 219; Lane, *Britannia Works*, p. 74.
14. Evans, *Listers*, pp. 74–80.
15. Lane, *Britannia Iron Works*, pp. 39, 61, 73; Wigan Archives Service, Leigh, Lancs., Harrison McGregor archive, Directors Minute Book no. 5, Nov. 8 1915.
16. Dewey, Peter, *War and Progress* (1997), pp. 7–8.

Notes to Chapter 8. Dynasties around 1914

1. Pointer, M., *Hornsbys of Grantham*, pp. 2–3, 16–18, 22.
2. Phillips, in Jeremy (ed.), *Dictionary of Business Biography*, p. 995.
3. Newman, *One Hundred Years of Good Company*, ch.1 and pp. 22–3.
4. Newman, *One Hundred Years*, pp. 30–1; Phillips, D. C., 'Joseph Ruston', in Jeremy, D. (ed.), *Dictionary of Business Biography* (Butterworth, 1986, pp. 994–5.
5. Glasgow City Archives, John Wallace & Sons. Ltd, archive, history of the firm by Robert G. Bryson, 1957, TD 726/1; Siderunt book of Trustees of John Wallace's estate, containing deed of settlement, TD 733/75.
6. Northumberland Record Office, Berwick-upon-Tweed, archive of Wm Elder & Sons, *Tweedale Press Supplement* (1965), 'Family Links Unbroken in Surge Forward'; Balance Sheet, 31 December 1904; Box 1, Correspondence 1875–1930.
7. Gregory, P., 'Joseph Ruston of Lincoln (1835–1897) – was he an enlightened entrepreneur?', Advanced Certificate in Local History, University of Nottingham, 1997, unpub. Ms., p. 24. I am indebted to Mrs. Gregory for sending me a copy of her thesis; Phillips, D. C., 'Joseph Ruston', in Jeremy, q.v., p. 995.
8. Jeremy, *Dictionary of Business Biography*, entries by Phillips, D. C. (Burrell, Fowler, Howard, Ransome, Ruston, Shuttleworth), and Holderness, B. A. and Phillips, D. C. (Clayton).
9. Thompson, F. M. L., *Gentrification and the Enterprise Culture: Britain 1870–1980* (Oxford, OUP, 2000), pp. 165, 185, 187. On possible 'entrepreneurial failure' before 1914 and cultural attitudes to money, business and industry, see W. D. Rubinstein, *Capitalism, Culture and Decline in Britain, 1750–1990* (Routledge, 1993); B. Supple, 'Fear of failing: economic history and the decline of Britain, *Economic History Review* (1994); M. J. Wiener, *English Culture and the Decline of the Industrial Spirit, 1850–1980* (Cambridge, 1981); see also J. F. Wilson, *British Business History, 1720–1994* (Manchester, 1995), pp. 113–19.
10. Lane, *Britannia Works*, various.
11. Lane, *Britannia Works*, p. 84.
12. Bamfords Limited 1871–1971 (the firm, Uttoxeter, n.d.); Bell, B., *Fifty Years of Farm Machinery; from starting handle to microchip* (Ipswich, Farming Press, 1993), p. 126.
13. Milton Keynes Museum of Industry and Rural Life, E. & H. Roberts archive; Wooton, K., and Loudon, N., 'Roberts of Deanshanger', *Old Glory*, no. 31, Sept. 1992, p. 66–69; Riden, P., *Victoria*

County History: Northampton, vol. V, Cleeley Hundred (Institute of Historical Research, 2002), pp. 229–30.

14. Wiltshire Record Office, Trowbridge, Reeves of Bratton archive, various.

Notes to Chapter 9. War work, 1914–1918

1. Dewey, P., 'Economic Mobilisation', in Turner, J., (ed.), *Britain and the First World War* (Unwin Hyman, 1988) pp. 71–82.
2. GEC – Alsthom, Firth Road, Lincoln, Ruston and Hornsby archive, Hornsby & Sons, Reports of Directors; Ministry of Munitions, *History of the Ministry of Munitions*, vol. XII, pt. VI, p. 1.
3. Rolt, *Waterloo Iron Works*, p. 157; Rural History Centre, University of Reading, Nalder & Nalder archive, items 343, 476; Cambridge University Library, Vickers archive, Doc. 605, 'History of Robert Boby Limited', p. 3; Minute Books No. 1, meeting of 18.7.17.
4. Wigan Archives Service, Harrison McGregor archive, Directors' Minute Book no. 5; Northumberland Record Office, Berwick-upon-Tweed, Wm Elder & Sons Ltd archive, Box 1, Correspondence 1875–1930.
5. Whitehead, *Garretts*, pp. 187–8; The Long Shop Museum, Leiston, *Guidebook to the Museum*, p. 9.
6. GEC-Alsthom, Ruston & Hornsby Ltd, the firm's archive, Firth Road Lincoln, folio, 'Record of Aeroplanes'. I am indebted to the then archivist, Mr. Ray Hooley, for his very kind assistance.
7. Lane, *Wellington Foundry*, ch. X; Lincolnshire Archives, Lincoln, Amalgamated Power Engineering archive, Fosters of Lincoln, 'Twenty Eighth Annual Report, for the 10 months ending 31st December 1916'; Lane, *Britannia Iron Works*, p. 81.
8. Lane, *Steam Plough Works*, ch. 14.
9. Wigan Archives Service, Harrison McGregor archive, Directors' Minute Books, 8.11.15 – 23.5.17.
10. Rural History Centre, University of Reading, Ransomes Sims & Jefferies archive, TR RAN AD 3/11, 'Employment of Women at Orwell Works'.
11. Whitehead, *Garretts*, pp. 188, 196.
12. Lane, *Britannia Works*, p. 81.
13. Mitchell, *Statistics*, p. 27; Dewey, P., 'The British Agricultural Machinery Industry, 1914–1939: Boom, Crisis, and Response', *Agricultural History*, vol. 69, no. 2 (1995), p. 301.
14. Lane, *Steam Plough Works*, 287; Dewey, 'Agricultural Machinery Industry', p.300.
15. Whitehead, *Garretts*, p. 193.
16. Lincolnshire Archives, Lincoln, Marshall archive, Marshall 5; Foster archive, Annual Reports and Accounts, APE 1, 1913–18.
17. Rural History Centre, University of Reading, Nalder & Nalder archive, Annual Directors' reports, 1913–18.
18. Northumberland Record Office, Berwick-upon-Tweed, Elder & Sons archive, letters of Ministry of Munitions of 25.10.17; T. Hamilton (Workers' Union) 12.11.17; Box 4, Ledgers, 1904–18.
19. Wigan Archive Service, Leigh, Lancs., Harrison, McGregor archive; Annual Reports of Directors 1913–18; Directors' Minute Book no. 5.
20. War Office, *Statistics of the Military Effort of the British Empire 1914–1920* (1920), p. 363.
21. Dewey, P.E., 'Agricultural Labour Supply in England and Wales during the First World War', *Economic History Review* 2nd. ser., xxviii (1975), p. 104.
22. Dewey, P.E., *British Agriculture in the First World War* (Routledge, 1989), p. 61.
23. Dewey, *British Agriculture*, p. 66.
24. Dewey, *British Agriculture*, p. 63 and 157; Wright, P.A., *Old Farm Tractors* (David & Charles, Newton Abbot, 1972), ch. II.
25. Dewey, P.E.,'British Farming Profits and Government Policy during the First World War', *Economic History Review*, XXXVII (1984); Dewey, *British Agriculture*, p. 234, chs. 3 and 8.
26. Middleton, (Sir) T. H., *Food Production in War* (Oxford, 1923).
27. Dewey, *British Agriculture* p. 149; Wigan archives, Harrison, McGregor board minutes, 31.7.17.
28. Dewey, *British Agriculture*, pp. 150–1: Wattenberg, *Historical Statistics of the USA*, p. 351.
29. Dewey, *British Agriculture*, pp. 155–9.
30. Agricultural Engineers Association, *Report on Trade Conditions in the Agricultural Machinery and Implement Industry* (1924), pp. 1–5; Mitchell, *Statistics*, pp.726, 681.
31. Wood, J., *The Model T Ford* (Shire Publications, Princes Risborough, 1999), pp. 1–10.
32. Lane, *Britannia Iron Works*, p. 77.
33. Hornsby directors' minutes, 19.6.17, 11.9.17.
34. GEC-Alsthom, Firth Road, Lincoln, Ruston & Hornsby archive, Hornsby Directors minutes, 13 December 1917; Col. J.S. Ruston to Hornsbys, 21 December 1917; 1912–17 balance sheets for both firms; minutes of the meeting of holders of ordinary shares in Hornsbys, 19 June 1918.

35. The business of the Aveling-Barford's succesors, Barford Construction Equipment and Barford Site Dumpers, now both part of Wordsworth Holdings Plc, still occupies the same site in 2003.
36. Pointer, M., *Ruston & Hornsby, Grantham, 1918–1963* (Grantham, Bygone Grantham, 1977), pp. 10–21; Dewey, 'British Agricultural Machinery Industry', p. 310]

Notes to Chapter 10. A new world, 1919–1939

1. Mitchell, *British Historical Statistics*, p. 729.
2. GEC-Alsthom, Lincoln, Ruston & Hornsby archive, Reports of Ordinary General Meetings, 1918–21; Lincolnshire Archives, Lincoln, Foster archive, APE 1; Northumberland Record Office, Berwick-upon-Tweed, Wm Elder archive, Box 4, Ledger, 'Journal No. 1.'.
3. Dewey, *War and Progress*, pp. 74–5.
4. Rural History Centre, University of Reading, TR RAN AD 7/17.
5. GEC-Alsthom, Lincoln, Ruston & Hornsby archive, Reports of Ordinary General Meetings, 1918–26.
6. Dewey, *War and Progress*, pp. 229–30.
7. Rural History Centre, University of Reading, Ransomes Sims & Jefferies archive, Directors' Full Board Meetings, 22 Dec. 1922, 24 Jan., 21 March 1923.
8. Agricultural Engineers Association, *Report on Trade Conditions in the Agricultural Machinery & Implement Industry* (1924), pp. 3–6.
9. GEC-Alsthom, Ruston & Hornsby, Ltd archive, Thirtieth Annual Report for the year ended 31st March, 1919; Dewey, 'British Agricultural Machinery Industry', p.310.
10. Dewey, 'British Agricultural Machinery Industry', pp. 307–8; Lane, *Wellington Foundry*, pp. 138–9.
11. Anon., 'A Flying Visit to the Deanshanger Iron Works, Stony Stratford', *IMR*, 1 Feb. 1910, pp. 1231–3; Ibid., 'Roberts' Dispersal Sale', *IMR* 1 Feb. 1928; Milton Keynes Museum of Industry and Rural Life, Roberts' archive, 'E. & H. Roberts, Ltd, Statement of Affairs, 22 February 1927'; Wooton, K., and Loudon, N., 'Roberts of Deanshanger', *Old Glory* no. 31 (Sept, 1992), pp. 66–9.
12. Whitehead, *Garretts*, p. 196.
13. Barford, Edward, *Reminiscences of a Lance-Corporal of Industry* (Elm Tree Books 1972) p. 9.
14. Essex R.O., D/F 23/1/2/5, Davey Paxman archive, letter of AGE to Davey Paxman shareholders, 18 June 1920.
15. PRO, BT 31/32289/15575.
16. Whitehead, *Garretts*, p. 198.
17. Whitehead, *Garretts*, p. 198.
18. Whitehead, *Garretts*, pp. 234–5; Lane, *St. Nicholas Works*, pp. 281–3; Lines, C. J., 'The Development and Location of the Specialist Agricultural Engineering Industry, with special reference to East Anglia' (University of London M.Sc. thesis, 1961), p. 97.
19. Barford, *Lance-Corporal of Industry*, p. 15–17.
20. For the preceding two paragraphs, see Barford, *Lance-Corporal of Industry*, ch. 2; Whitehead, *Garretts*, pp. 234–5; Lines, thesis, p. 94; Bedfordshire Record Office, Bedford, J. & F. Howard archive, brief history of the firm; *Bedfordshire Times & Independent*, 19 February 1932, p.7; Kemp, *The Bentall Story*, p. 34; Evans, *Listers*, p. 166; Key, M.(ed.), *The Blackstone Collection*, No. 4, March 2002 (Michael Key, Didcot, Oxon., 2002; Anon., 'Oldest-Established Implement Manufacturers in Britain? Knapps are now in their Third Century', *IMR* 1 April 1962, pp. 489–90.
21. Dewey, P.E., 'British Farming Profits during the First World War', *Economic History Review* (1984).
22. Sturmey, S. G., 'Owner-Farming in England and Wales, 1900–1950', originally in *Manchester School of Economic and Social Studies*, XXIII (1955), reprinted in Minchinton, W. E. (ed.), *Essays in Agrarian History* (David & Charles, Newton Abbot, 1967), vol. II, p. 287.
23. Ministry of Agriculture, *A Century of Agricultural Statistics: Great Britain 1866–1966* (HMSO, 1968), p. 82; Penning-Rowsell, E. C., 'Who "betrayed" whom? Power and Politics in the 1920/21 Agricultural Crisis', *Agricultural History Review*, vol. 45, part II (1997).
24. Ministry of Agriculture, *A Century of Agricultural Statistics: Great Britain 1866–1966* (HMSO, 1968), Tables 43, 44.
25. Dewey, 'British Agricultural Machinery Industry', pp. 311–12.
26. Wilder, R.J.H., *The Wilder Story* (Ms., Rural History Centre, University of Reading).
27. Ministry of Agriculture, *A Century*, Table 64.
28. Evans, *Listers*, ch. 8; Gloucs. Record Office, Gloucester, R.A. Lister & Co. Ltd, Report of Directors to 30 September 1929; R.A. Lister & Co., *Your Works and Ours* (the firm, Dursley, 1936), p. 9.
29. Jansson, T., *A Historical review of the development of the Milking Machine with special reference to technical advances and improvements* (Alfa-Laval, Tumba, Sweden, 1973), p. 26 fol.; Whetham, E. H., 'The Mechanization of British Farming, 1910–1945',

Journal of Agricultural Economics, 21, no. 3 (1970), p. 318; Ministry of Agriculture, *A Century*, Table 30.

30. Hudson, K., *The Archaeology of the Consumer Society: The Second Industrial Revolution in Britain* (Heinemann, 1983), p. 26.
31. Jansson, *Milking Machine*, pp. 29–32; Hosier, A. J. & F. H., *Hosier's Farming System* (Crosby Lockwood, 1951), ch. 5; Taylor, D., 'Growth and Structural Change in the English Dairy Industry, *c*.1860–1930', *Agricultural History Review*, vol.35 pt. I (1987), p. 47.
32. Dewey, *British Agriculture in the First World War*, p. 63 and 157; Wright, P.A., *Old Farm Tractors* (David & Charles, Newton Abbot, 1972), ch. II.
33. Glasgow City Archives, John Wallace archive; brief company history (1957); Prospectus of Wallace (Glasgow) Ltd, TD 482/21/60; 'A visit to Cardonald Works', *The Farming News*, 13 April 1920; 'Farm Tractors: The Glasgow', *Modern Farming*, November 1920; *Tractor & Machinery*, November 1999, p. 49].
34. Moorhouse, M., *The Illustrated History of Tractors* (Quintet Publishing, 1996), p. 78; Williams, M., *Ford and Fordson Tractors* (Ipswich, Farming Press, 1985).
35. Anon., 'World Tractor Trials', *JRASE* 1930: Wright, *Old Farm Tractors*, ch. VIII; Ministry of Agriculture, *Agricultural Machinery Testing Committee*, Report No. 27, 'Vickers' Agricultural Tractor (Paraffin) (HMSO, 1931).
36. Dewey, 'British Agricultural Machinery Industry', pp. 303–4; Lane, *Steam Plough Works*, pp. 367–9, 371–3, 389; Ministry of Agriculture, Agricultural Machinery Testing Committee, *Reports on Fowler 'Four-Forty' and 'Three-Thirty' Diesel Crawler Tractors* (HMSO, 1935 and 1936).
37. Lane, *Britannia Works*, pp. 115–23; Day, R., 'Clayton & Shuttleworth Combine Harvester', *Tractor & Machinery* (Nov., 1996), pp. 23–5].
38. Wright, *Old Farm Tractors*, pp. 18, 58, and Plate 19; Painting, N., *Alldays & Onions: A Brief History* (Landmark Publishing, Aldbourne, Derbyshire, 2002), pp. 125–7; Kemp, *Bentall Story*, p. 30; Key, M., *Pick of Stamford: a history of the Pick Motor Company* (Paul Watkins, Stamford, 1994), p 37 fol.; Reprinted 1923 brochure for Ruston-Hornsby Cars, pp. 4–5, by courtesy of the archivist at GEC-Alstom, Lincoln, Mr. Ray Hooley (1991); Newman, *Good Company*, pp. 100–2; Baldwin, N., 'British Wallis', *Tractor & Machinery*, November 1995.
39. *Country Life*, 17 May 1983, p. 1439; Wigan Archives, Leigh, Lancs., Harrison McGregor archive; W. Drew, 'H.M.G. of Leigh.'; *David Brown Newsletter*, March 1957; Reports of Directors of Harrison McGregor, 1911–23; Directors' Minute Books, 31 July 1917, 11 February 1918, 19 March 1930, 9 February 1938, 9 February 1939.
40. Northumberland Record Office, Berwick-upon-Tweed, Wm Elder & Sons archive, Box 4, Ledger no.1; Norfolk Record Office, Norwich, archives of Randell, Smithdale, Springall; Public Record Office, MAF 58/162/85822, 'Allocation of Steel and Iron Castings to Manufacturers and Repairers of Agricultural Machinery.'.
41. Museum of Rural Life, Farnham, Surrey; archive of J. Gibbs & Co. Ltd; Ceredigion Archives, Aberystwyth; archive of Bridge End Foundry, Cardigan, Dyfed; Wiltshire Record Office, Trowbridge, archives of J. W. Titt & Co., and T. H. White & Co.
42. Dewey, *War and Progress*, pp. 200–3.
43. Collins, E. J. T., *Power Availability and Agricultural Productivity in England and Wales, 1840–1939* (Rural History Centre, University of Reading, Discussion Paper Series No. 1 (1996), pp. 6–7; Rural History Centre, University of Reading, International Harvester archive, 'Financial Statements 1926–1945'.
44. Condie, A., *The Marshall Fowler Album* (Nuneaton, 1995), pp. 96, 104; Anderson, P., *Three Decades of Marshall Tractors* (Ipswich, Farming Press, 1997), pp. 44–55; Lane, *Steam Plough Works*, pp. 371–3.
45. By Wik, R.M., in *Henry Ford and Grass-Roots America*, quoted by Williams, *Ford & Fordson Tractors*, op. cit., p. 54.
46. Williams, *Ford & Fordson Tractors*, pp. 63–73; Condie, A., *Fordson Model 'N' 1929–45* (Nuneaton, Alan T. Condie Publications, 1991), pp. 30–1.
47. Wright, S. J. 'World Agricultural Tractor Trials, 1930', *JRASE* 91 (1930), pp. 206–8.
48. *Censuses of Production 1930 and 1935*.
49. *Annual Statement of Trade*; Political and Economic Planning, *Agricultural Machinery* (1949), Appendix VIII.
50. Leffingwell, *Ford Farm Tractors*, pp. 89–92; PEP, *Agricultural Machinery*, Appendix X.
51. Williamson, H., *The Story of a Norfolk Farm* (Faber and Faber, 1941; Clive Holloway Books, 1986), pp. 213–4, 338; Anon., 'Producing the Ferguson Tractor, *Implement and Machinery Review*, 1 November 1938, pp. 688–91; Fraser, C., *Harry Ferguson: Inventor and Pioneer* (John Murray, 1972; Ipswich, Old Pond Publishing, 1998, p. 91.
52. Fraser, *Harry Ferguson*, pp. 98–100].
53. Lines, *thesis*, pp. 117–22.

Notes to Chapter 11. War work again, 1939–1945

1. Wilt, A. F., *Food for War: Agriculture and Rearmament in Britain before the Second World War* (New York, Oxford University Press, 2001), chs. 2, 3.
2. Taylor, A. J. P., *English History 1914–45* (Oxford, OUP, 1965), p. 427.
3. PRO, MAF 58/111, 'Tractors required in time of war', memoranda by Houghton and Manktelow.
4. PRO, MAF 58/114, 'Tractors – Storage', Memorandum by 'C.T.H' (C.T. Hutchinson); Williams, *Ford & Fordson Tractors*, p. 74.
5. Murray, K.A.H., *Agriculture* (History of the Second World War. Civil Series) HMSO/ Longmans Green (1955), pp. 86–7; PRO, Ministry of Agriculture and Fisheries, MAF 58/136, 'Memorandum on Agricultural Machinery: Supplies. Estimates 1940–41', p. 10].
6. Murray, *Agriculture*, p. 87; *Williams, Ford & Fordson* Tractors, Appendix 2; Wigan Archives, Leigh, Lancs., Harrison, McGregor archive; Board minutes, 17 December 1941; brief typescript company history, 'Harrison, McGregor & Guest Ltd, Leigh, Lancashire, England. A Subsidiary of David Brown Tractors Ltd', p. 2.
7. Murray, *Agriculture*, p. 89.
8. Ministry of Agriculture, *A Century of Agricultural Statistics: Great Britain 1866–1966* (HMSO, 1968), Tables 43, 44.
9. See Note at the end of this chapter.
10. Williams, *Ford & Fordson Tractors*, Appendix 2 and p. 81; Saunders, H. St G., *Ford at War* (Harrison and Sons, n.d., *c.*1946), pp. 53–7. I am indebted to Dr Gill Clarke, of the Centre for Biography and Education at the University of Southampton, for the latter reference.
11. Saunders, *Ford at War*, pp. 54, 57; Williams, *Ford & Fordson Tractors*, p.78; Dewey, *War and Progress*, p. 306; PRO, MAF 58/96, Memo of MAF for Exchange Requirements Committee, 9.12.40, 'Supplies of Agricultural Machinery and Implements in the United Kingdom.'.
12. Ministry of Agriculture, *A Century*, Table 30; Martin, J., *The Development of Modern Agriculture: British Farming since 1931* (Macmillan, 2000), p. 42].
13. PRO, MAF 58/162, 'Farm Machinery and Equipment in the First Ten Years after the War', p. 8.
14. Murray, *Agriculture*, p. 27.
15. PRO, MAF 58/81.
16. Parker to Sir William Tritton, 9 Feb. 1942, PRO, MAF 58/140.
17. PRO, MAF 58/140, AGM of AEA, 26.2.42; Murray, *Agriculture*, pp. 173, 199; PRO, MAF 58/162, 'Farm Machinery and Equipment in the First Ten Years after the War', section 3.
18. PRO, MAF 58/162, Appendix I, *Particulars of Agricultural Engineering Industry: Allocations of Steel and Iron castings to Manufacturers and Repairers of Agricultural Machinery.*
19. Lane, *Steam Plough Works*, pp.390–6.
20. Fraser, C., *Harry Ferguson: Inventor and Pioneer* (John Murray, 1972: Old Pond Publishing, Ipswich, 1998), pp. 103–8; Williams, *Ford & Fordson Tractors*, pp. 90–8.
21. Fraser, *Harry Ferguson*, pp. 120–2.
22. PRO, MAF 58/116; Copy of secretary's minute on A.M.574; initialled by R.D.S. (the Minister) on 20 December 1939.
23. Donnelly, D., *David Brown's: The Story of a Family Business 1860–1960* (Collins, 1960), p. 69.
24. 'Report on the David Brown Tractor', by S. J. Wright, 29.1.40, PRO, MAF 58/112; PRO, MAF 58/112, Minute by C. I. C. Bosanquet, 10.10.41.
25. Williams, *Great Tractors*, p. 85.
26. PRO, MAF 58/121. Note by Parker of 9.10.41; Burton to Wormald (MAF) 15.3.43; Anderson, *Marshall Tractors*, pp.59–60.
27. Condie, *Marshall Fowler Album*, p. 26; Lane, *Britannia Iron Works*, pp. 126–7.
28. PRO, MAF 58/79. Exports: Policy. 1939–42; memo. of 6.2.41: 'Exports of agricultural machinery.'.
29. Central Statistical Office, *Annual Abstract of Statistics No. 84* (1935–1946), Table 223.
30. Grace and Phillips, *Ransomes*, p. 10; Bell, B., *Ransomes, Sims & Jefferies: Agricultural Engineers. A history of their products* (Old Pond publishing, Ipswich, 2001), p. 104; Rural History Centre, Reading University, Ransomes archive, TR RAN AD 2/5; AD 7/13–14.
31. Bamfords, *Bamfords 1871–1971*, p. 14; Rolt, *Waterloo Ironworks*, pp. 185–8; Kemp, *Bentall Story*, pp. 34–5.
32. Rural History Centre, Reading University, TR RAN AD 2/4; Weaver, C. and M., *Ransomes: a Bicentennial Celebration* (Ransomes, Sims & Jefferies, Ipswich, 1989), pp. 72–7.
33. Gibbard, S., *Roadless: The Story of Roadless Traction from Tracks to Tractors* (Ipswich, Farming Press, 1996), pp. 63–5.
34. Sherwen, *Bomford Story*, p. 53.
35. MAF 58/97: UK Agricultural Production Committee. 1943 Farm Machinery Requirements (12 November 1942).
36. Murray, Agriculture, pp. 128–9; PRO, MAF

58/97: UK Agricultural Production Committee. 1943 Farm Machinery Requirements, 12 November 1942; MAF 58/162, Agricultural Engineering Industry – post-war arrangements.

37. Murray, *Agriculture*, pp. 161, 190; PRO, MAF 58/96: Orders for tractors from the USA under Lease-Lend for the half year ending 31 December 1941: MAF 58/97: UK Agricultural Production Committee. 1943 Farm Machinery Requirements (12 November 1942). Murray gives the total tractor order as 6,000, whereas the orders cited here, from MAF 58/96, add to a total order of 9,900.
38. Murray, *Agriculture*, Appendix Table VIII, p. 378; PRO, MAF 58/98, '1944 Requirements of Farm Machinery.' (6 March 1943); Central Statistical Office, *Annual Abstract of Statistics No. 84, 1935–1946* (HMSO, 1948), Table 186.
39. PRO, MAF 58/196. Commonwealth Economic Committee. Agricultural Machinery Enquiry 1950, Appendix C – Retail prices of agricultural machinery, at current specifications, with 'moving anterior weights' Mitchell, *Statistics*, pp. 730, 739.
40. *The Times*, 30 December 1939, Company Meetings. R.A. Lister and Company. Mr. Percy Lister's Address.
41. Rural History Centre, University of Reading, Ransomes, Sims & Jefferies archive, TR RAN AC 7/11–15.
42. Rural History Centre, University of Reading, Nalder & Nalder archive, item 343, AGM statements 1937–45.
43. Northumberland Record Office, Berwick-upon-Tweed, Elder & Sons archive, Box 4, financial statements 1925–70.
44. Wiltshire Record Office, Trowbridge, R. & J. Reeves archive, Gotch (director) to Lloyds bank 1.9.1938; 951/175, annual balance sheets and profit and loss accounts 1892–1970.
45. PRO, MAF 58/162, 'Farm Machinery and Equipment in the First Ten Years after the War', para. 10. Unless otherwise indicated, the data in the rest of this section are drawn from this report.
46. This is exactly how the firm of Reeves of Bratton described its own origins – Wiltshire Record Office., R. & J. Reeves archive, brief history of the firm (1955), and letter of Gotch to Lloyds bank, 1.9.1938; Public Record Office, MAF 58/162, 'Farm Machinery and Equipment in the First Ten Years after the War', and the 'General report on the Manufacturing of Agricultural Implements', by Mr F. A. Lemon.

Notes to Chapter 12. A very brief supremacy, 1945–1973

1. Booth, Alan, *The British Economy in the Twentieth Century* (Palgrave, 2001), Table 3.2.
2. Martin, J., *The Development of Modern Agriculture: British Farming since 1931* (Macmillan, Basingstoke, 2000), pp. 69–72, 80, 91–2, 99; Morgan, B., *Agricultural Engineering: Keyguide to Information Sources* (Mansell, 1985), p. 16.
3. Brassley, P., 'Output and technical change in twentieth-century British agriculture', *Agricultural History Review* 48, part I (2000), p. 61, Table 1.
4. M. Overton, *Agricultural Revolution in England, the transformation of the agrarian economy, 1500–1850* (Cambridge, 1966), p. 85.
5. Brassley, 'Output and technical change', Table A4, p. 84; Martin, *Modern Agriculture*, p. 131, Table 5.12.
6. Ministry of Agriculture, *A Century of Agricultural Statistics: Great Britain 1866–1966* (HMSO, 1968), Table 26.
7. Cashmore, W. H., 'Recent Developments in and the Future of Agricultural Engineering', *Transactions of the Manchester Society of Engineers* (1946–47), presented on 2 November 1946, pp. 83–93.
8. PRO, MAF 58/218, 219: 'Ford Motor Co.: proposed financial agreement in regard to capital expenditure for the construction of No. 3 Fordson tractor.'; Wilkie, J., *An Illustrated History of Tractors* (Ian Allan, Shepperton, 1999), pp. 65–7; Williams, M., *Ford & Fordson Tractors*, Appendix 2.
9. Fraser, C., *Harry Ferguson, Inventor and Pioneer* (John Murray, 1972: Old Pond Publishing, Ipswich, 1998), p. 169; Modern Records Centre, University of Warwick, MSS.226/ST/3/F/8/1/1, Letters of Interest in the Ferguson Tractor Project, MSS.226/ST/3/F/1/28, Ferguson to Black, 2 October 1945; Gibbard, S., *The Ferguson Tractor Story* (Old Pond Publishing, 2000), p. 66–7, 74–5, Appendix 2].
10. Williams, *Ford & Fordson Tractors*, pp. 103–5; Political and Economic Planning, *Agricultural Machinery*, p. 83; Fraser, *Harry Ferguson*, ch. 23; Gibbard, *Ferguson*, p. 110, Appendix 2.
11. Donnelly, D., *David Brown's: the story of a family business* (Collins, 1960), pp. 70–1; Wilkie, *Illustrated History*, pp … 68–9; Carroll, J., *Tractors of the World* (Hermes House, 2000), pp. 36–7; Rural History Centre, University of Reading, David Brown archive, TR DBT P2/A6, *'Inside Story: A summary of the design and manufacturing achievements of David Brown Tractors*

Ltd' (Huddersfield, David Brown Tractors, n.d., *c.*1972), pp. 5–7; Kudrle, R.T., *Agricultural Tractors: A World Industry Study* (Ballinger Publishing Co., Cambridge, Mass., 1975, pp. 88–90.

12. Anderson, P., *Three Decades of Marshall Tractors* (Ipswich, Farming Press, 1997), pp. 60–1,66, 74–5, 81–2, 139, 158–66, ch. 11.
13. Kudrle, *Agricultural Tractors*, pp. 88–90.
14. Williams, M., *Massey-Ferguson Tractors* (Ipswich, Farming Press, 1987), p. 50; Carroll, J., and Stuart, G., *Tractors* (PRC Publishing, 2000), p. 208; McCormick website: www.mccormick-intl.com.
15. Kudrle, *Agricultural Tractors*, p. 123; Carroll, *Tractors*, p. 13; Moorhouse, *Illustrated History*, p. 161.
16. Spokes, M. J., *Minneapolis-Moline Album* (Nuneaton, Allan T. Condie Publications, 1987), p. 4; Political and Economic Planning, *Agricultural Machinery: a report on the industry* (PEP, 1949), pp. 12, 17.
17. Gibbard, S., *Roadless: The story of Roadless Traction from Tracks to Tractors* (Ipswich, Farming Press, 1996), p. 107.
18. Carroll, *Tractors*, pp. 128–9; Gibbard, S., 'Yeoman of England: Part Three', *Tractor & Machinery*, June 2000, p. 22; Kudrle, *Agricultural Tractors*, pp. 123–4.
19. Bell, B., *50 Years of Farm Machinery: from starting handle to microchip* (Ipswich, Farming Press, 1993).
20. Bell, *Farm Machinery*, p.8, chapter 1; Carroll, *Tractors*, pp. 58–61; Williams, *Ford & Fordson Tractors*, pp. 107–119.
21. Political and Economic Planning, *Agricultural Machinery* (1949), p. 83; PEP, 'The Agricultural Machinery Industry', *Planning*, XIV, No. 275, 5 Dec. 1947, p. 174.
22. 1948 *Census of Production*, 4/1/5, Table 6.
23. Report on the Census of Production 1968, (50), *Agricultural Machinery (except tractors): Wheeled tractor manufacturing*, 81/7–8, Table 5.
24. Rural History Centre, University of Reading, Ransomes Sims & Jefferies archive, TR 9 RAN, Reports and Accounts 1967–85.
25. Gibbard, *Ferguson*, pp. 92, 128–9; Carroll and Stuart, *Tractors*, p. 188.
26. Kemp, *Bentall Story*, 35; *Acrow; The success story of achievement through team spirit* (the firm; n.d., *c.*1975), pp. 52–3.
27. *IMR*, 1 February 1947, 'Marshalls Acquire Fowlers', p. 976; *IMR* 1 January 1950,' Ransomes' New Foundry', pp. 1226–7; PEP, *Agricultural Machinery*, pp. 14–17; Wigan archives, Leigh, Lancs., Harrison McGregor & Guest archive, brief typescript history of HMG, pp. 1–2.
28. *The British Tractor and Farm Machinery Industry 1960* (Norman Karsh Publications, 1960); idem., (1970); PEP, op. cit., pp. 6–9, 11.
29. Business Monitor, PA 380, *Wheeled Tractor Manufacturing*, (1973) Table 4; Business Monitor, *Report on the Census of Production 1973*, PA 331, Table 4.
30. PRO MAF 58/237, Gordon-Smith Enquiry. *Report on the Agricultural Machinery Manufacturing Industry*. April 1948. Section II.
31. PRO MAF 58/237, Gordon-Smith Enquiry. *Appendices relating to the Report on the Agricultural Machinery Manufacturing Industry*, April 1948.
32. PRO MAF 58/237, *Gordon-Smith Enquiry: reports on leading firms*.
33. Wigan Archives Service, Harrison McGregor archive, brief typescript company history, pp. 2–3; Warren, D., *Denings of Chard: agricultural engineers 1828–1965* (Taunton, 1989), p. 42 fol.; Wedlake, R. P., 'Written Contribution', in Cashmore, 'Recent Developments', pp. 104–8].
34. Wiltshire Record Office, Trowbridge, Parmiter archive, typescript company history, 'P.J. Parmiter & Sons Ltd'; Rural History Centre, University of Reading, Bomford & Evershed Ltd, 843–885; ICC Business Ratios, *Business Ratio Report: Agricultural Equipment Manufacturers* (ICC Business Ratios, 5th edn, 1982), Table 16.
35. Rolt, *Waterloo Ironworks*, pp. 218–24.
36. Bell, *50 Years*, pp. 27, 32.
37. Ministry of Agriculture, *A Century of Agricultural Statistics*, Table 30.
38. Gibbard, *Ferguson*, pp. 134–5.
39. Acrow, *40 Years On*, p. 53.
40. Bell, *50 Years*, p. 177; Bell, B.,'Catchpole Engineering Company', *Tractor & Machinery*, July 2000, pp. 32–3.
41. Bamfords Ltd, *Bamfords Limited 1871–1971* (the firm, Uttoxeter, 1971).
42. Ministry of Agriculture, *A Century*, Tables 31, 32; Jansson, *The Development of the Milking Machine: A Historical Review* (Tumba, Sweden, 1973; Dowkes, C., 'The Mechanisation of Milking, part 3: Bulk Milk Collection', *Farm & Horticultural Equipment Collector*, Nov/Dec. 1994, p. 12.
43. Commonwealth Economic Committee, Thirty-sixth Report. *A Survey of the Trade in Agricultural Machinery* (HMSO, 1952), Tables 2, 6, 8, 17, 18.
44. Lancaster, G. A., 'Is our industry too complacent?', *Agricultural Machinery Journal*, April 1969, pp. 35–7.
45. Heath et al., *A Study of the Evolution of Concentration*, Table 9.2.

Notes to Chapter 13. Coping with the competition, 1973–2000

1. Booth, Alan, *The British Economy in the Twentieth Century* (Basingstoke, Palgrave, 2001), Tables 1.3, 2.2, 4.1.
2. Department for Environment, Food and Rural Affairs, *Agriculture in the United Kingdom* (2001), Chart 2.1, Table 2.1.
3. Ministry of Agriculture, *A Century of Agricultural Statistics: Great Britain 1866–1966* (1968), pp. 21–3; Department for the Environment, Food and Rural Affairs (DEFRA), *Agriculture in the United Kingdom 2000* (2001), Table 3.3. Minor holdings in England and Wales are excluded in 2000.
4. See the comments by Lancaster (1969) in the preceding chapter.
5. Royal Society, *Constraints on research and development in some smaller manufacturing companies: a case study of the agricultural engineering industry* (1982), paras 2.1–2.7, 4.2.
6. ICC Business Ratio Report, *Agricultural Equipment Manufacturers* (1989), p. 3; Farnworth, J., *The Massey Legacy*, vol. 1 (Ipswich, Farming Press, 1997), pp. 50–3.
7. Business Monitor, PAS 3211, 3212, 1979–83.
8. AGCO Corporation website: press notice of 25 June 2002: 'AGCO rationalises European production facilities. Plans to close Coventry, UK plant.'
9. Agricultural Engineers Association, *Farm Equipment Statistics, Autumn 1997*, (AEA, Peterborough, 1997), Tables 4, 10, 14; United Nations, Yearbook of Industrial Statistics (Geneva, 2000); Agricultural Engineers Association website – *Industry Facts* (2000).
10. AEA Data Book 1976–1986, Table 32: Keynote Publications, *Agricultural Machinery* (11th edn, 1997), table 14; p. 19.
11. Martin, *Modern Agriculture* (2000), p. 107.
12. Agricultural Engineers Association, *Farm Equipment Statistics* Autumn 1997, (AEA, Peterborough, 1997), Table 16.
13. Bell, *50 Years …*, p. 22.
14. Carroll, *Tractors of the World*, entries for Ford and New Holland; Carroll and Stuart, *Tractors*, pp. 386–93. The majority of CNH shares are owned by Fiat.
15. Lane, *Steam Plough Works*, p. 398; Lane, *Britannia Iron Works*, p. 180 fol.; Anderson, *Marshall Tractors*, ch. 12.
16. Gibbard, *Roadless*, pp. 162–6.
17. Bell, *50 Years …*; J. C. Bamford Excavators, *JCB: The First Fifty Years* (Special Event Books/J.C. Bamford Excavators, Horsham 1995), p. 108 fol.
18. The above paragraphs are based on Bell, *50 Years.*
19. The case of Massey-Ferguson is detailed in Cook, Peter, *Massey at the Brink: the story of Canada's greatest multinational and its struggle to survive* (Toronto, Collins, 1981).
20. Carroll and Stuart, *Tractors*, pp. 385–6; Company websites: mccormick-intl.com; casecorp.com.
21. In 2004, AGCO took over Valtra, the Finland tractor maker.
22. Carroll, *Tractors of the World*, pp. 60–1; Leffingwell, R., *Ford Farm Tractors* (MBI Publishing, Osceola, Wisconsin, 1998), p. 191.
23. Department of Industry, *The Agricultural Engineering Industry* (1978), paras 2.3, 4.8.
24. Keynote Market Report, *Agricultural Machinery* (11th edn, 1997), Table 6.
25. Keynote Market Report, *Agricultural Machinery*, Tables 15 and 16.
26. ICC Business Ratios, *Agricultural Equipment Manufacturers* (1980, 1982), Commentary.
27. ICC Business Ratios, *Agricultural Equipment Manufacturers* (1980), Commentary .
28. Business Monitor, *Census of Production PA 1003*, 1995.
29. ICC Business Ratio Report. *Agricultural Equipment Manufacturers* (5th edn, 1982) section 2, Tables 1,3.
30. Keynote Market Report, *Agricultural Machinery* (1997), Table 25, section 10.
31. Parmiter website, Dec. 2003; *Farmers' Weekly*, 21 March 2003, p. 69.
32. Bell, *Ransomes of Ipswich*, pp. 18–19; Anon., 'Ransomes' Catch-22', *The Independent*, 19 August 1993, p. 12.
33. The sources for the previous five paragraphs are the author's database of some 600 firms, and current company websites, with additional information from Mr J. H. W. Wilder's history of his firm [in the Rural History Centre, University of Reading], and the Milestones Museum, Basingstoke (for Taskers and Wallis & Stevens). While reliance on websites has its disadvantages and dangers, they are the modern equivalent of a nineteenth-century entry in the Post Office telephone directory, or a trade directory such as Kelly's, before the rise of the internet. As such, they are an indispensable source for contemporary business history.
34. Board of Trade, *Census of Production, 1907, Final Report*, Cd. 6320, (1912), pp. 190, 209; Office for National Statistics, *Product Sales and Trade, PRA*

29310, Agricultural Tractors, 2000; *PRA 29320, Other Agricultural and Forestry Machinery, 2000.*
35. Teanby, M., *The Roar of Dust and Diesel: A story of International Harvester, Doncaster* (Japonica Press, Driffield, 2004).

Notes to Chapter 14. Retrospect

1. Wiener, M. J., *English Culture and the Decline of the Industrial Spirit, 1850–1980* (Cambridge, CUP, 1981).
2. Thompson, F. M. L., *Gentrification and the Enterprise Culture: Britain 1870–1980* (Oxford, OUP, 2000), p. 97.
3. Chandler, A., *Strategy and Structure: Chapters in the History of the Industrial Enterprise* (1962); *The Visible Hand: the Managerial Revolution in American Business* (1977); *Scale and Scope: the Dynamics of Industrial Capitalism* (1990).
4. Booth, Alan, *The British Economy in the Twentieth Century* (Palgrave, 2001), pp. 209–15.
5. Grigg, *English Agriculture* (1989), pp. 5–6, 150; Dewey, 'Farm Labour', *AHEW*, vol. VII (2000), Appendix; DEFRA, *Agriculture in the United Kingdom* (2001), Table 3.5.

Bibliography

National Archives (formerly the Public Record Office)
(BT = Board of Trade; MAF = Ministry of Agriculture and Fisheries)

PRO BT 31/32289/155725 Prospectus for Cumulative Preference Shares of Agricultural & General Engineers Ltd, 30 Oct. 1919.

PRO, MAF 58/79. Exports: Policy. 1939–42; memo. of 6.2.41: 'Exports of agricultural machinery'.

PRO, MAF 58/96, Memo of MAF for Exchange Requirements Committee, 9.12.40, 'Supplies of Agricultural Machinery and Implements in the United Kingdom'.

PRO, MAF 58/96: Orders for tractors from the USA under Lease-Lend for the half year ending 31 December 1941.

PRO, MAF 58/97: UK Agricultural Production Committee. 1943 Farm Machinery Requirements (12 November 1942).

PRO, MAF 58/98, '1944 Requirements of Farm Machinery' (6 March 1943).

PRO, Ministry of Agriculture and Fisheries, MAF 58/136, 'Memorandum on Agricultural Machinery: Supplies. Estimates 1940–41'.

PRO, MAF 58/111, 'Tractors required in time of war', memoranda by Houghton and Manktelow.

PRO, MAF 58/112, 'Report on the David Brown Tractor', by S.J. Wright, 29.1.40.

PRO, MAF 58/112, Minute by C.I.C. Bosanquet, 10.10.41.

PRO, MAF 58/114, 'Tractors – Storage', Memorandum by 'C.T.H' (C.T. Hutchinson).

PRO, MAF 58/162, 'Farm Machinery and Equipment in the First Ten Years after the War'.

PRO, MAF 58/162, Agricultural Engineering Industry – postwar arrangements.

PRO, MAF 58/162, Appendix I, *Particulars of Agricultural Engineering Industry: Allocations of Steel and Iron castings to Manufacturers and Repairers of Agricultural Machinery.*

PRO, MAF 58/196. Commonwealth Economic Committee. Agricultural Machinery Enquiry 1950.

PRO, MAF 58/218, 219: 'Ford Motor Co.: proposed financial agreement in regard to capital expenditure for the construction of No. 3 Fordson tractor'.

PRO MAF 58/237, *Gordon-Smith Enquiry. Appendices relating to the Report on the Agricultural Machinery Manufacturing Industry*, April 1948.

PRO MAF 58/237, *Gordon-Smith Enquiry. Report on the Agricultural Machinery Manufacturing Industry.* April 1948. Section II.

PRO MAF 58/237, *Gordon-Smith Enquiry: reports on leading firms.*

Archives in county and local record offices (name of firm in brackets)

Ayrshire Record Office (Alexander Jack)

Bedford and Luton Archives, Bedford (Howard)

Cambridge University Library (Boby, Hunt, Lack, Vickers)

Ceredigion Archives, Aberystwyth (Bridge End Foundry)

Cornwall Record Office, Truro (Brenton)

Devon Record Office, Barnstaple (Gliddon & Squire)

Essex Record Office, Chelmsford (Bentall, Davey Paxman, Hunt)

GEC – Alsthom, Firth Road, Lincoln (Ruston and Hornsby)

Glasgow Record Office (Wallace)

Gloucs. Record Office, Gloucester (Lister)

Hampshire Museum Service, Winchester (Tasker)

Hampshire Record Office, Winchester (Armfield)

Leicestershire Record Office, Leicester (Vipan & Headley)

Lincolnshire Record Office, Lincoln (Foster, Marshall, Ruston & Hornsby)

Lincolnshire Museum of Lincolnshire Life, Lincoln (Robey)

Milton Keynes Museum of Industry and Rural Life (Roberts)
Museum of Rural Life, Farnham, Surrey (Gibbs)
Norfolk Record Office, Norwich (Barnard, Holmes, Randell, Smithdale, Soame, Springall)
Northumberland Record Office, Berwick-upon-Tweed (Elder)
Oxfordshire Record Office, Oxford (Samuelson)
Rural History Centre, University of Reading (Barrett Exall & Andrewes, Bomford & Evershed, David Brown, Clayton & Shuttleworth, Fowler, Humphries, Hunt, International Harvester, Ivel, Marshall, Nalder & Nalder, Ransomes Sims & Jefferies, Reeves, Samuelson, Wallis & Steevens)
Strathclyde Regional Archives, Glasgow (Wallace)
Suffolk Record Office, Ipswich (Garrett, Smyth)
University of London (London School of Economics) Pollitt Papers.
Warwick Record Office (Troth & Hillson)
Warwick University, Modern Records Centre (Standard Motor Co.)
Wigan Archive Service, Leigh, Lancs. (Harrison, McGregor)
Wiltshire Record Office, Trowbridge (Parmiter, Titt, White)
Yorkshire, East Riding Record Office, Beverley (Crosskill)

Secondary sources

Place of publication is London unless otherwise stated.

Afton, B., and Turner, M., 'Agricultural Output', in Collins, E.J.T. (ed.), *The Agrarian History of England and Wales, vol. VII 1850–1914* (Cambridge, Cambridge University Press, 2000), chapter 38.
AGCO Corporation website: press notice of 25 June 2002: 'AGCO rationalises European production facilities. Plans to close Coventry, UK plant'.
Agricultural Engineers Association, *Report on Trade Conditions in the Agricultural Machinery and Implement Industry* (1924).
Agricultural Engineers Association, *AEA Data Book 1976–1986*
Agricultural Engineers Association, AEA *Farm Equipment Statistics* (1997).
Agriculture, Ministry of, Agricultural Machinery Testing Committee, *Report No. 27, 'Vickers' Agricultural Tractor (Paraffin)* (HMSO, 1931).
Agriculture, Ministry of, Agricultural Machinery Testing Committee, *Reports on Fowler 'Four-Forty' and 'Three-Thirty' Diesel Crawler Tractors* (HMSO, 1935 and 1936).
Aldcroft, D.H., and Dyos, H.J., *British Transport: an economic survey from the seventeenth century to the twentieth* (Harmondsworth, Pelican, 1974).
Alford, L.C.G., *The Development of Industrial Lincoln*: a lecture given to the Lincoln Historical Association and the Lincoln Training College (1950), typescript.
Anderson, P., *Three Decades of Marshall Tractors* (Ipswich, Farming Press, 1997).
Annual Statements of Trade, 1853–1940.
Anon., 'Extension of Blackstone & Co.'s Rutland Iron Works, Stamford', *Implement and Machinery Review* 1 Dec., 1904.
Anon., 'The Works of Messrs. Clayton & Shuttleworth', *Engineering*, (1920), reprinted in *Road Locomotive Society Journal*, vol. 44/4, Nov. 1991, pp. 112–17.
Anon., 'A Flying Visit to the Deanshanger Iron Works, Stony Stratford', *Implement and Machinery Review*, 1 Feb. 1910.
Anon., 'A Visit to a Lincoln Engineering Works', *The Engineer*, 23 May 1913.
Anon., 'Producing the Ferguson Tractor, *Implement and Machinery Review*, 1 November 1938.
Anon., 'Ransomes' New Foundry', *Implement and Machinery Review* 1 January 1950.
Anon., 'Marshalls Acquire Fowlers', *Implement and Machinery Review*, 1 February 1947,.
Anon., 'Ransome's Catch–22', *The Independent*, 19.8.93, p.12.
Anon., 'Reports of the Committees appointed to investigate the present state of steam cultivation', *Journal of the Royal Agricultural Society of England* 2nd ser., vol. 3 (1867).
Anon., 'Roberts' Dispersal Sale', *Implement and Machinery Review* 1 Feb. 1928.
Anon., Thomas Nelson and Sons, *An Encyclopaedia of Agriculture* (1914), vol. II.
Anon., 'Through Five Generations! Cooch & Sons celebrate Their 150th Anniversary', *Farm Implement and Machinery Review*, 1 October 1950.
Anon., 'World Tractor Trials', *Journal of the Royal Agricultural Society of England* 1930.
Anon., *Acrow; The success story of achievement through team spirit* (the firm; n.d., c.1975).
Anon., *The British Tractor and Farm Machinery Industry 1960* (Norman Karsh Publications, 1960).
Anon: 'Oldest-established Implement Manufacturer in Britain? Knapp's are now in their Third Century.', *Implement and Machinery Review*, 1 April 1962.
Bailey, M., *One Hundred and Six Copper Plates of Mechanical Machines and Implements of Husbandry,*

approved and adopted by The Society for the Encouragement of Arts, Manufactures and Commerce and contained in their Repository in the Adelphi Buildings in the Strand?..carefully corrected and revised by Alexander Mabyn Bailey, Registrar to the Society. Vol. I (London, Benjamin White, 1782).

Baldwin, N., 'A-Z of Tractors; Crossley', *Tractor & Machinery* May 2001.

Baldwin, N., 'British Wallis', *Tractor & Machinery*, November 1995.

Bamfords Ltd, *Bamfords Limited 1871–1971* (the firm, Uttoxeter, 1971).

Barford, Edward, *Reminiscences of a Lance-Corporal of Industry* (Elm Tree Books 1972).

Beaumont, W.W., and Hippisley, R.J.B., 'Trials of Agricultural Motors at Manor Farm, Bygrave, Baldock, Hertfordshire', *Journal of the Royal Agricultural Society of England* v. 71 (1910).

Beckett, J., Turner, M., & Afton, B., *Agricultural Rent in England 1700–1914*.

Beckett, J.V., *The Agricultural Revolution* (Blackwell, Oxford, 1990).

Bell, B., *Fifty Years of Farm Machinery; from starting handle to microchip* (Ipswich, Farming Press, 1993).

Bell, B., *Fifty Years of Farm Tractors* (Ipswich, Farming Press, 1999/2000).

Bell, B., *Ransomes, Sims & Jefferies, agricultural engineers: a history of their products* (Ipswich, Old Pond Publishers, 2001).

Black, E., 'Plenty's taking museum pieces to steam fair', *Newbury Weekly News*, 3 May 1973.

Blake, S., 'An Historical Geography of the British Agricultural Engineering Industry, 1780 to 1914', unpublished Ph.D. thesis, University of Cambridge (1974).

Booker, J., *Essex and the Industrial Revolution* (Chelmsford, Essex Record Office, 1974).

Booth, Alan, *The British Economy in the Twentieth Century* (Palgrave, 2001).

Brassley, P., 'Output and technical change in twentieth-century British agriculture', *Agricultural History Review* 48, part I (2000).

Brigden, R., 'Agricultural Iron Foundries of the Nineteenth Century', *Foundry Trade Journal*, December 1983.

Brigden, R., 'Equipment and Motive Power', chapter 7C of Collins. E.J.T. (ed.), *The Agrarian History of England and Wales*, vol. VII (Cambridge, Cambridge University Press, 2000).

Brown, G. P., 'Crosskills of Beverley', *The East Yorkshire Local History Society Bulletin*, Autumn 1982.

Brown, Jonathan, *Farm Machinery 1750–1945* (Batsford, 1989).

Carroll, J., and Stuart, G., *Tractors* (PRC Publishing, 2000).

Carroll, John, *Tractors of the World* (Anness Publishing, 2000).

Cashmore, W.H., 'Recent Developments in and the Future of Agricultural Engineering', *Transactions of the Manchester Society of Engineers* (1946–7).

Cawood, Charles L., *Vintage Tractors* (Princes Risborough, Shire Publications, 1980)

Central Statistical Office, *Annual Abstract of Statistics No. 84, 1935–1946* (HMSO, 1948).

Chandler, A., *Strategy and Structure: Chapters in the History of the Industrial Enterprise* (Cambridge, Mass., 1962); *The Visible Hand: the Managerial Revolution in American Business* (Cambridge, Mass., 1977); *Scale and Scope: the Dynamics of Industrial Capitalism* (Cambridge, Mass., 1990).

Chapman, J., 'The extent and nature of parliamentary enclosure', *Agricultural History Review*, vol. xxxv (1987).

Clapham, (Sir) J., *An Economic History of Modern Britain*, vol. II (Cambridge, C.U.P., 1952).

Clark, Ronald H., *The Steam Engine Builders of Norfolk* (Yeovil, Haynes Publishing Group, 1988).

Collins, E.J.T., 'Harvest technology and labour supply in Britain, 1790–1870', *Economic History Review* 2nd ser., 22 (1969).

Collins, E.J.T., 'The Age of Machinery', in Mingay, G.E., (ed.), *The Victorian Countryside* (Routledge & Kegan Paul, 1981).

Collins, E.J.T., 'The High-Farming period: the Golden Age, 1850–75', in Collins, E.J.T. (ed.), *The Agrarian History of England and Wales*, vol. VII (1850–1914), chapter 2A.

Collins, E.J.T., *Power Availability and Agricultural Productivity in England and Wales, 1840–1939* in vab Bavel, Bas J.P., and Thoen, Erik (eds.), Land Productivity and agro-systems in the North Sea area, Middle Ages–20th century (Brepols, Turnhout, 2000).

Commonwealth Economic Committee, Thirty-sixth Report. *A Survey of the Trade in Agricultural Machinery* (HMSO, 1952).

Condie, A., *The Marshall Fowler Album* (Nuneaton, 1995).

Cook, Peter, *Massey at the Brink: the story of Canada's greatest multinational and its struggle to survive* (Toronto, Collins, 1981).

Corley, T.A.B., 'Barrett, Exall & Andrewes, 1818–1887', typescript Ms., University of Reading, Rural History Centre.

Corley, T.A.B., 'Barrett, Exall & Andrewes' Iron Works at Reading: The Partnership Era, 1818–64', *Berkshire Archaeological Journal* v. 67 (1973–4).

Course, Edwin, 'The Rolt Memorial Lecture, 1994: Engineering Works in Rural Areas', *Industrial Archaeology Review* XVIII, No. 2, Spring 1996.

Crosskill's Implement Newspaper, part I, July 1848. East Riding Record Office, Beverley, Crosskill archive,

DDX/347/I, abridged catalogue of East Yorkshire and Crosskills Cart and Wagon Co.Day, R., 'Clayton & Shuttleworth Combine Harvester', *Tractor & Machinery* (Nov., 1996).
Department for Environment, Food and Rural Affairs, *Agriculture in the United Kingdom* (2001).
Dewey, P., 'Economic Mobilisation', in Turner, J., (ed.), *Britain and the First World War* (Unwin Hyman, 1988).
Dewey, P., 'The British Agricultural Machinery Industry, 1914–1939: Boom, Crisis, and Response', *Agricultural History*, vol. 69, no. 2 (1995).
Dewey, P.E., 'Agricultural Labour Supply in England and Wales during the First World War', *Economic History Review* 2nd ser., xxviii (1975).
Dewey, P.E., 'Agricultural Labour', in Collins (ed.), *Agrarian History of England and Wales* VII (1850–1914).
Dewey, P.E., 'British Farming Profits during the First World War', *Economic History Review* (1984).
Dewey, P.E., *British Agriculture in the First World War* (Routledge, 1989).
Dewey, P.E.,'British Farming Profits and Government Policy during the First World War', *Economic History Review*, XXXVII (1984)
Dewey, Peter, *War and Progress: Britain 1914–1945* (Longman, 1997).
Donnelly, D., *David Brown's: The Story of a Family Business 1860–1960* (Collins, 1960).
Dowkes, C., 'The Mechanisation of Milking, part 3: Bulk Milk Collection', *Farm & Horticultural Equipment Collector*, Nov/Dec. 1994.
'Druid', the (H.H. Dixon), Saddle and sirloin (Rogerson & Tuxford, 1870)
Dugdale, J. B., 'Miscellaneous Implements Exhibited at Park Royal', *Journal of the Royal Agricultural Society of England* 64 (1903).
Edginton, D., *Old Stationary Engines* (Shire Publications, Princes Risborough, 1996).
Ernle, Lord, *English Farming Past and Present* (6th edn, Frank Cass, 1961).
Evans, D.E., *Lister's: The First Hundred Years* (Gloucester, Alan Sutton, 1979).
Farnworth, J., *The Massey Legacy*, vol. 1 (Ipswich, Farming Press, 1997).
Fraser, C., *Harry Ferguson: Inventor and Pioneer* (John Murray, 1972; Ipswich, Old Pond Publishing, 1998).
Freeman, M., and Aldcroft, D., *The Atlas of British Railway History* (Beckenham, Croom Helm, 1985).
Fussell, G.E., *The Farmer's Tools* (Andrew Melrose, 1952; Bloomsbury Books, 1981).
Gibbard, S., 'Yeoman of England: Part Three', *Tractor & Machinery*, June 2000.
Gibbard, S., *Roadless: The Story of Roadless Traction from Tracks to Tractors* (Ipswich, Farming Press, 1996).
Gibbard, S., *County: A Pictorial Review* (Ipswich. Farming Press, 1997).
Gibbard, S., *The Ferguson Tractor Story* (Old Pond Publishing, 2000).
Goddard, N., *Harvests of Change: The Royal Agricultural Society of England 1838–1988* (Quiller Press, 1988).
Grace, D., 'The Agricultural Engineering Industry', in *AHEW* Vol. VII, chapter 16A.
Grace, D.R., and Phillips, D.C., *Ransomes of Ipswich. A History of the Firm and Guide to its Records* (University of Reading, Institute of Agricultural History, 1975).
Haining, J. and Tyler, C., *Ploughing by Steam* (Model and Allied Publications, Hemel Hempstead, 1970).
Harris, J.R., *The British Iron Industry 1700–1850* (Macmillan, 1988).
Haynes, Walter, *History of Ruston & Hornsby Limited* (Lincoln and Grantham; the firm, 1954; 3 vols., typescript, unpublished)
Hill, (Sir) F., *Victorian Lincoln* (Cambridge, Cambridge University Press, 1974).
Hobsbawm, E.G., and Rude, G., *Captain Swing* (1969).
Hosier, A.J. & F.H., *Hosier's Farming System* (Crosby Lockwood, 1951).
Hudson, K., The Archaeology of the Consumer Society: The Second Industrial Revolution in Britain (Heinemann, 1983).
Hueckel, G., 'English Farming Profits during the Napoleonic Wars, 1793–1815', *Explorations in Entrepreneurial History*, vol. 13 (1976).
Hutchinson, Wm. T., *Cyrus Hall McCormick* (New York, 1968), vol. II
ICC Business Ratios, *Agricultural Equipment Manufacturers* (1980).
ICC Business Ratio Report. *Agricultural Equipment Manufacturers* (1982).
ICC Business Ratio Report, *Agricultural Equipment Manufacturers* (1989).
Industry, Department of, *The Agricultural Engineering Industry* (1978).
J.C. Bamford Excavators, *JCB: The First Fifty Years* (Special Event Books/J.C. Bamford Excavators, Horsham 1995).
Jackson, R., 'A brief history of E.H. Bentall & Company', *Farm & Horticultural Equipment Collector*, July/August 2000, p. 9.
Jannson, T., A *Historical review of the Development of the Milking Machine with special reference to technical advances and improvements* (Alfa-Laval, Tumba, Sweden, 1973).
Jeremy, D., *Dictionary of Business Biography*, entries by Phillips, D.C. (Burrell, Fowler, Howard, Ransome, Ruston, Shuttleworth), and Holderness, B.A. & Phillips, D.C. (Clayton).
Kemp, P.K., *The Bentall Story: commemorating 150 years service to Agriculture* (the firm, n.d.).

Key, M. (ed.), *The Blackstone Collection*: 'Carter Brothers Special', no. 8, March 2003.

Key, M., *Pick of Stamford: a history of the Pick Motor Company* (Paul Watkins, Stamford, 1994).

Keynote Market Report, *Agricultural Machinery* (11th edn, 1997).

Kudrle, R.T., *Agricultural Tractors: A World Industry Study* (Ballinger Publishing Co., Cambridge, Mass., 1975).

Lancaster, G.A., 'Is our industry too complacent?', *Agricultural Machinery Journal*, April 1969, pp. 35–7.

Lane, Michael, *The Story of the Steam Plough Works* [Fowlers of Leeds] (Northgate Publishing, 1980).

Lane, Michael R., *The Story of the Britannia Iron Works: William Marshall Sons & Co., Gainsborough* (Quiller Press, 1993).

Lane, Michael R., *The Story of the St. Nicholas Works: A History of Charles Burrell & Sons Ltd. 1803–1928* (Unicorn Press, Grimston, Yorks., 1994).

Lane, Michael R., *The Story of the Wellington Foundry, Lincoln: A History of William Foster & Co. Ltd.* (Unicorn Press, 1997).

Langley, J.L., 'Memories of Late Victorian Banbury', *Cake and Cockhorse* v. 2, 4 (1964).

Leffingwell, R., *Ford Farm Tractors* (MBI Publishing, Osceola, WI, 1998).

Lines, C.J., 'The Development and Location of the Specialist Agricultural Engineering Industry, with special reference to East Anglia' (University of London M.Sc. thesis, 1961).

Lister, R.A. & Co., *Your Works and Ours* (the firm, Dursley, 1936).

Maclean, I.W., 'Anglo-American Engineering Competition, 1870–1914: Some Third-Market Evidence', *Economic History Review* (1976).

Martin, J., *The Development of Modern Agriculture: British Farming since 1931* (Macmillan, 2000)

McCormick Co. website: www.mccormick-intl.com.

Measom, G., 'The Katesgrove Iron Works of Messrs. Barrett, Exall, and Andrewes?.', in *The official illustrated guide to the south-Eastern Railway* (1858, repr. 1970).

Middleton, (Sir) T.H., *Food Production in War* (Oxford, 1923).

Mingay, G.E., *Land and Society in England, 1750–1980* (Longman, 1994).

Ministry of Agriculture, *A Century of Agricultural Statistics: Great Britain 1866 –1966* (HMSO, 1968).

Mitchell, B.R., *British Historical Statistics* (Cambridge, 1988).

Moffitt, J., *The Ivel Story* (Japonica Press, 2003).

Moorhouse, Robert, *The Illustrated History of Tractors* (Quintet Publishing, 1996) Williams, Michael, *Great Tractors* (Blandford Press, Poole, 1982).

Morgan, B., *Agricultural Engineering: Keyguide to Information Sources* (Mansell, 1985).

Munitions, Ministry of, *History of the Ministry of Munitions*, vol. XII, pt. VI.

Munting, R., 'Ransomes in Russia: An English Agricultural Engineering Company's Trade with Russia to 1917', *Economic History Review* XXXI (1978)

Munting, R., 'Agricultural Engineering and European Exports before 1914', *Business History*, xxvii (1985).

Murray, K.A.H., *Agriculture* [History of the Second World War. Civil Series] (HMSO/Longmans Green, 1955).

Newman, B., *One Hundred Years of Good Company* (Lincoln, Ruston & Hornsby, 1957).

North, L., *Wishing you a year of Plenty and Progress, 1790–1958* (Newbury, the firm, 1958).

Orwin, C.S., and Whetham, E., *History of British Agriculture 1846–1914* (Longmans, 1964).

Overton, M., *Agricultural Revolution in England*, (Cambridge, Cambridge University Press, 1996).

Painting, N., *Alldays & Onions: A Brief History* (Landmark Publishing, Aldbourne, Derbyshire, 2002).

Palmer, C.J., *History of the Orwell Works 1771–1928* (unpublished typescript, Reading University, Rural History Centre).

Palmer, F., 'The Blacksmiths Ledgers of Bucklebury', University of Reading Research paper No. 2 (1970).

Penning-Rowsell, E.C., 'Who "betrayed" whom? Power and Politics in the 1920/21 Agricultural Crisis', *Agricultural History Review*, vol. 45, part II (1997).

Perkin, Harold, *The Age of the Railway* (1970).

Phillips, A., 'Early Colchester Foundries', *Essex Archaeology and History*, 14 (1982).

Phillips, A.D.M., *The Underdraining of Farm Land in England during the nineteenth century* (Cambridge, C.U.P., 1989).

Phillips, W.G., *The Agricultural Implement Industry in Canada* (Toronto, University of Toronto Press, 1956).

Pidgeon, D., 'The development of Agricultural Machinery', *Journal of the Royal Agricultural Society of England*, series iii, I, (1890).

Pidgeon, D., 'The Evolution of Agricultural Implements – I', *Journal of the Royal Agricultural Society of England* 3rd ser., vol. 3 (1892).

Pointer, Michael, *Ruston & Hornsby, Grantham, 1918–1963* (Grantham, Bygone Grantham, 1977)

Pointer, Michael, *Hornsbys of Grantham 1815–1918* (Grantham; Bygone Grantham, 1978).

Political and Economic Planning, 'The Agricultural Machinery Industry', *Planning*, XIV, No. 275, 5 December 1947.

Political and Economic Planning, *Agricultural Machinery: a report on the industry* (PEP, 1949).

Radcliffe, V., 'Entrepreneurial Success: A Local Case Study: W N Nicholson & Sons, Agricultural Engineers of Newark on Trent', *The Nottinghamshire Historian*, No. 54, Spring/Summer 1995.
Rayner, D.A., *Thomas Aveling 1824–1882: His Life and Work* (n.d.).
Reeves, M., *Sheep Bell & Ploughshare: the story of two village families* (Moonraker Press, Bradford-on-Avon, 1978).
Riden, P., 'The Output of the British Iron Industry before 1870', *Economic History Review*, vol. xxx (1977).
Riden, P., *Victoria County History: Northampton, vol. V, Cleeley Hundred* (Institute of Historical Research, 2002).
Roepke, H.G., *Movements of the British Iron and Steel Industry – 1720 to 1951* (Urbana, Illinois, 1956).
Rolt, L.T.C., *Waterloo Ironworks: A history of Taskers of Andover, 1809–1968* (Newton Abbot, David & Charles, 1969).
Rothwell, R., *Technical Change and Competitiveness in Agricultural Engineering: the performance of the UK industry* (Science Policy Research Unit, University of Sussex, Brighton, 1979: SPRU Occasional Paper Series no. 9).
Royal Society, *Constraints on research and development in some smaller manufacturing companies: a case study of the agricultural engineering industry* (1982).
Saunders, H. St. G., *Ford at War* (Harrison and Sons, n.d., c.1946).
Shearman, L., *Portable Steam Engines* (Shire Publications, Princes Risborough, 1986).
Simmons, J, *The Railways of Britain: an historical introduction* (1961).
Skehel, M., *Tales from the Showyard: Two Hundred Years of Agricultural Shows in Lincolnshire* (Lincoln, Lincolnshire Agricultural Society, 1999).
Somerville (Lord), *Facts and Observations relating to Sheep, Wool, &c.* (1809).
Spokes, M.J., *Minneapolis-Moline Album* (Nuneaton, Allan T. Condie Publications, 1987).
Sturmey, S.G., 'Owner-Farming in England and Wales, 1900–1950', originally in *Manchester School of Economic and Social Studies*, XXIII (1955), reprinted in Minchinton, W.E. (ed.), *Essays in Agrarian History* (David & Charles, Newton Abbot, 1967), vol. II.
Taylor, A.J.P., *English History 1914–45* (Oxford, Oxford University Press, 1965).
Taylor, Audrey M., *Gilletts: bankers at Banbury and Oxford: a study in local economic history* (Oxford, Clarendon Press, 1964).
Taylor, D., 'Growth and Structural Change in the English Dairy Industry, c.1860 –1930', *Agricultural History Review*, vol.35 pt. I (1987).
Thompson, F.M.L., 'An Anatomy of English Agriculture, 1870–1914', in Holderness, B.A. and Turner, M., *Land, Labour and Agriculture, 1700–1920: Essays for Gordon Mingay* (Hambledon Press, 1991).
Thompson, F.M.L., 'The Land Market in the Nineteenth Century', *Oxford Economic Papers* IX (1957), reprinted in Minchinton, W.E., (ed.), *Essays in Agrarian History* (David & Charles, Newton Abbot, 1968), Vol. II.
Thompson, F.M.L., *Gentrification and the Enterprise Culture: Britain 1870–1980* (Oxford, O.U.P., 2000).
Thompson, R.J., 'An Enquiry into the Rent of Agricultural Land', *Journal of the Royal Agricultural Society of England*, LXX (1907), Appendix, reprinted in Minchinton, W.E. (ed.), *Essays in Agrarian History*, Vol. II (1968), Vol. II.
Tyler, C., *Digging by Steam: a History of Steam Cultivation by means of the application of Steam Power to the Fork, Mattock and similar implements* (Watford, Model and Allied Publications, 1977).
Victoria History of the Counties of England.
Voelcker, A., 'On the Composition of Cream and Skim-Milk obtained by De Laval's Centrifugal Cream-Separator', *Journal of the Royal Agricultural Society of England* 2nd ser., v. 16, (1880).
Wade Martins, S., and Williamson, Tom, *Roots of Change: Farming and the Landscape in East Anglia, c.1700–1870* (Exeter, British Agricultural History Society, 1999).
Wailes, R., 'The early history of Akroyd Stuart's Oil Engine', *Transactions of the Newcomen Society*, v. 48 (1976–7).
Walton, J.R., 'Aspects of agrarian change in Oxfordshire, 1750–1880', unpublished D.Phil. thesis (Oxford, 1976).
War Office, *Statistics of the Military Effort of the British Empire 1914–1920* (1920).
Warren, D., *Denings o: agricultural engineers 1828–1965* (Taunton, 1989).
Wattenberg, B., *The Statistical History of the United States* (New York, Basic Books, 1977).
Weaver, C. & M., *Ransomes 1789–1989. A Bicentennial Celebration* (Ipswich, Ransomes, Sims & Jefferies PLC, 1989).
Wedlake, R.P., 'Written Contribution', in Cashmore, 'Recent Developments'.
Whetham, E.H., 'The Mechanization of British Farming, 1910–1945', *Journal of Agricultural Economics*, 21, no. 3 (1970).
Whitehead, R.A., *Garretts of Leiston* (Percival Marshall, 1964).
Whitehead, R.A., *Steam in the Village* (David & Charles, Newton Abbot, 1977).
Whiting, R.C., *The View from Cowley; the impact of industrialization upon Oxford, 1918–1939* (Oxford, Clarendon Press, 1983).
Wilder, R.J.H., *The Wilder Story* (Ms., Rural History Centre, University of Reading).

Wilkes, P., *An Illustrated History of Traction Engines* (Spur Books, Bourne End, Bucks., 1974).

Wilkie, J., *An Illustrated History of Tractors* (Ian Allan, 1999).

Williams, M., *Ford and Fordson Tractors* (Ipswich, Farming Press, 1985).

Williams, M., *Massey-Ferguson Tractors* (Ipswich, Farming Press, 1987).

Williamson, H., *The Story of a Norfolk Farm* (Faber and Faber, 1941; Clive Holloway Books, 1986).

Williamson, Tom, *The Transformation of Rural England: Farming and the Landscape 1700–1870* (Exeter, University of Exeter Press, 2002).

Willoughby de Eresby (Lord), *Ploughing by Steam* (J. Ridgway, 1850).

Wilt, A.F., *Food for War: Agriculture and Rearmament in Britain before the Second World War* (New York, Oxford University Press, 2001).

Wood, J., *The Model T Ford* (Shire Publications, Princes Risborough, 1999).

Wooton, K., and Loudon, N., 'Roberts of Deanshanger', *Old Glory* no. 31 (Sept, 1992).

Wright, P.A., *Old Farm Tractors* (David & Charles, Newton Abbot, 1972).

Wright, S.J. 'World Agricultural Tractor Trials, 1930', *Journal of the Royal Agricultural Society of England* 91 (1930).

APPENDIX

Production figures, 1907–2000

The first firm estimate of the value of the output of the agricultural machinery industry was that of the first Census of Production, in 1907. Other censuses followed in 1924, 1930 and 1935. There are various estimates for the Second World War period. From 1946 there were more frequent censuses of production, published by the Board of Trade, usually triennially, and with estimates for non-census years. From 1973, the census results are found in the Business Monitor series published by the Central Statistical Office. In 1995 this work was taken over by the Office of National Statistics. Currently the information is collected under the aegis of the PRODCOM enquiries of the EU.

The figures given here for the value of current output have been taken directly from the above sources. Production in real values has been estimated at 1907 prices. The prices chosen are: for 1907–73, the Board of Trade index of prices of all manufactured products (in Mitchell, *British Historical Statistics*, p. 734); for 1974–2000, the producer price index of materials and fuel purchased by manufacturing industry, rebased on 1980=100 (from Office of National Statistics, *Economic Trends*, 2002 edn, table 2.1).

Index

Index entries in *italic* type refer to illustrations